D0936907

05/2022
STRAND PRICE
$ 5.00

how we got to be
HUMAN

william h.
LIBAW

how we got to be
HUMAN

SUBJECTIVE MINDS WITH OBJECTIVE BODIES

Prometheus Books

59 John Glenn Drive
Amherst, New York 14228-2197

Published 2000 by Prometheus Books

How We Got to Be Human: Subjective Minds with Objective Bodies. Copyright ©
2000 by William H. Libaw. All rights reserved. No part of this publication may be
reproduced, stored in a retrieval system, or transmitted in any form or by any means,
digital, electronic, mechanical, photocopying, recording, or otherwise conveyed via
the Internet or a Web site, without prior written permission of the publisher, except
in the case of brief quotations embodied in critical articles and reviews.

Inquiries should be addressed to
Prometheus Books
59 John Glenn Drive
Amherst, New York 14228-2197
VOICE: 716-691-0133, ext. 207
FAX: 716-564-2711
WWW.PROMETHEUSBOOKS.COM

04 03 02 01 00   5 4 3 2 1

Library of Congress Cataloging-in-Publication Data

Libaw, William H.
    How we got to be human : subjective minds with objective bodies / William
H. Libaw.
        p. cm.
    Includes bibliographical references and index.
    ISBN 1-57392-813-5 (alk. paper)
    1. Human beings—Animal nature. 2. Psychology, Comparative.
3. Subjectivity. I. Title.

GN280.7 .L5 2000
156'.3—dc21   00-024131
                                                                      CIP

Printed in the United States of America on acid-free paper

CONTENTS

Contents

PREFACE

THIS IS A BOOK THAT does more than rehash what scientists believe about our evolution. It is about what science commonly dodges or even denies: subjective life and its development, life as it is experienced by lower animals, apes, and humans—mixing what we learn from several sciences and other disciplines with some new ideas.

You'll find the following: How the earliest animals with consciousness would have had what computers show no signs of getting: namely, Darwinian adaptiveness. Also provided are demonstrations that suggest apes and some birds use gestures, implying conscious concepts in the gourds of the gesturing creatures. About complete spoken language, it will be maintained that it came first to humankind from the mouths of children. Further, it was a valuable cultural gift, one that came generations after a preceding initially nonadaptive genetic gift of potential complete language. Normally private parts are used to illustrate this thesis: Although human females are free of animal estrus, human males have retained animal rut and amplified it with conceptual prurience to regularly, reliably, and relentlessly eroticize females. For chewing in the mind is an argument showing that the downright peculiar art of less-than-recent decades is a direct response to the swift growth of science in the last century. Such "empty" art provides images that show our fear that spreading science has emptied life of much of its meaning— meaning that had been based on subjective knowledge of supernatural agents. Lastly, throughout the text are exhibitions of what is sometimes mentioned (and quickly dropped) in science-humanities books: We have in our heads more of a Confederacy of Mind than a United States thereof.

HOW WE GOT TO BE HUMAN

So how, then, did we get from small-brained simple-minded four-legged creatures, through larger-brained occasionally thoughtful semi-upright apes, to large-brained mentally complex erect humans? Certainly, we did it largely by winning hands—and feet and much, much more—at the stern lottery of life known as evolution. In many ways more interesting to most of us, part of that evolution was an expansion of the ability to make and use conscious mental concepts (which are evidenced by gesture and language). This expansion allowed us to move from the confining flat here-and-now of most animal mentation into an ultimate human world, a world that makes us dizzy as we try to fathom the depths of its almost limitless dimensions. The following is a look at some steps in that mental expansion of self-centered, social, and sex-driven creatures.

Consider what often happens when two male wolves fight. A stalemate results when the one that's losing gets on its back as its foe stands above it. This has been wrongly described as an abject gesture by the loser that offers his life, with the winner acknowledging with the noble gesture of walking away. A simpler explanation says neither animal makes use of gestures; gestures imply (subjective) concepts in its head. The losing animal rolls to his back only to enable him to use his feet to help in a defensive effort; the winning one sees a likely stalemate. Soon both creatures abandon the fight, but they often renew it before long, suggesting no concepts in either mind about the import of winning and losing.

Consider now two chimpanzees in a somewhat similar situation after the contest has stopped. The chimp who was losing tries to get the winner's embrace: "Let's be pals again." The chimp winner denies such comfort to the loser, denies it unless the loser lowers himself to all four limbs. Then the winner stands tall on two legs and puffs up his hair with the thrill of it all. After that, the loser gets his hug. Here, both chimps use gestures to define and acknowledge their future relationship. They have corresponding concepts in their mentation, concepts that are beyond the capabilities of wolves, concepts of social power, of the significance of winning and losing fights.

Next, consider male or female apes, who, when pursued by aggressive male or female others, stop and bend over to present their bottoms in what looks like the behavior of sexual receptivity that is normally used by in-heat females. For decades, old-time comedian Henny Youngman disarmed his audiences with his one-liner: "Take my wife. Please!" The truculent ape pursuer is, in fact, disarmed by what seems almost a parody of Youngman's joke: "Take my rump. Please!" The pursuer and the pursued have transferred the sexual behavior used by females in heat. They have transformed it into a gesture of deference in the realm of social power. Both creatures understand that behind the behind-presenting gesture is the loser's variation on the social-power concept: "Me small, not much at all. But you, you big, you powerful!"

Consider finally the verbal ideologizing of some modern men with presumed expertise about the nature of women's sexuality. Some use social science's odd jargon to say that a woman has "year-round receptivity," in contrast with a female lower mammal, who is sexually receptive only during the season when she wants a male to rub her estrus-heated private parts with his. However, to say to nonscientists that a woman has year-round receptivity is to imply that she wants sexual activity all the time from whoever will provide it, suggesting she is a "slut." Others insinuate that, as "sex is a hole" (Jean Paul Sartre's words), what a woman wants may not really matter much, implying she is more like a machine with a "slot." Why do such weird ideas about women circulate? Such notions make the rounds partly because men are uncommonly prurient. Peculiar ideas also come into play because, on occasion, most humans overwork the grand and glorious ability we have to make use of concepts. By virtue and vice of language, we blow up concepts into ideological notions.

Before ancient animals first started to use concepts, they began with concept-free experiences such as those of pain and pleasure, and then developed nonconscious neural-network patterns that associated those experiences with, say, predator and prey. Some time after that, conscious concepts corresponding to those patterns came to life, thereby allowing the beginnings of what we feel as option and choice. For us, conceptual activity includes most of what is meaningful in life, much of what rattles around in our heads all of the time. Although the subjective world of any other creature (its consciousness, mentation, mind) cannot be directly detected by any of us, this book plays detective to deduce from tracks—tracks in the form of gestures, deceptive behaviors, utterances, protolanguage, and complete language—some of the concepts that populate the mentalities that accompany objective bodies that include brains.

ACKNOWLEDGMENTS

IT IS A PLEASURE TO acknowledge the help of many people with the creation of this book. I am indebted to the writers of the books and articles cited in the bibliography for their contributions to my education. The following were kind enough to read and comment on drafts of parts of the manuscript: Jay Atlas, Michael Cart, Michael Commons, Merlin Donald, Martha Fisher, Abby Franklin, Marsha Gaffery, James Kimberly, Martha Kirkpatrick, Arvid Knudsen, Oliver Yates Libaw, Rodger Lowe, Lois Lyons, Michelle Merrill, Pavittar Safir, Maxine Sheets-Johnstone, Tim Spaulding, Mike Warshaw, and JoAnne Beil Waugh. Stalwart souls who did as much for most or all of the manuscript are Cheryl Armon, Virginia Armon, William Paul Cone, Carroll Dana, Steven L. Mitchell, Ellen E. M. Roberts, Jonathan Spaulding, and Karl Westberg. Ideas, suggestions, sources, and words from Norma Yates Libaw are an integral part of this book. Neither for them nor for things far more important can I now thank her; she no longer lives.

MINDS, LIKE EVERYTHING ELSE, WERE MADE IN SLOW STEPS

> *O the mind, mind has mountains; cliffs of fall*
> *Frightful, sheer, no-man-fathomed. Hold them cheap*
> *May who ne'er hung there.*
> Gerard Manley Hopkins, "No Worst"

THESE DOUR WORDS BY GERARD Manley Hopkins give us a brief example, not of the nature of the human mind, but of what the human mind can create.

How did we come to have such a "no-man-fathomed" thing as a mind, a thing of complexity we could hardly imagine if each of us did not experience it?

CONSCIOUSNESS WAS THE START OF MENTALITY, WHICH GREW AS ANIMALS EVOLVED UNTIL IT BECAME, IN US, MIND

This book will track the expansion of mental capability that led to the human mind. The tracks can be seen, perhaps only dimly, by postulating the (subjective) mental development suggested by the expanded activities of increasingly complex creatures.

Controversy, however, is implicit even in the few words above. Look, for example, at *The Mind's New Science*, by Howard Gardner.[1] A peek into its index will show no listing for the word "consciousness." The word "science" in the title, and the book's subtitle, "A History of the Cognitive Revolution," makes this omission somewhat understandable. Both science in general and cognitive psychology in particular have considerable unease with consciousness. The reasons for this are good, but not good enough.

On the other side of the argument are the generations of philosophers and psychologists influenced by Descartes for whom "consciousness, not intelligence or rationality . . . was the defining criterion of the mental . . . [which included] seeing, hearing, feeling, pain, and pleasure."[2] Likewise, John Searle writes, "The study of the mind is the study of consciousness, in much the same sense that biology is the study of life."[3] With this, some may disagree. Not me.

A wee beast that may not be conscious, the digger wasp, Sphex, will do nicely to start our search for mentality. Though it has a system of neuron networks, the digger wasp may not experience anything, may not have any mentality at all. We'll go on from Sphex to what may be the first creatures with any consciousness, the small-brained, feeble-minded creatures called fish, creatures that have no more than a few basic and nonconceptual kinds of experiences, such as light and dark, and pleasure and pain. Then, with giant steps up some branches of the evolutionary tree, it's on to certain lower mammals and birds, to living apes, to our ancient predecessor upright apes and erectus-like humans, to early humans of our own kind, and finally to current humans, our very own selves.

In this look at how mentality has expanded until it becomes what we call, in ourselves, mind, there is only one sensible place or process with which to begin: Consciousness, that ancient riddle, mystery, and enigma that fascinates us to this day. To begin, that is, with consciousness in the basic sense that provides immediate experience to lower animals of parts of the external world and of some internal things, such as pain.

Some scientists try to step gingerly around or even deny the existence of consciousness for animals; some even do the same for ourselves. For membership in the last group, Daniel Dennett is a good nominee. In *Consciousness Explained*, he says, "[Regarding] 'intrinsic properties of conscious experiences,' 'the qualitative content of mental states,' and, of course, qualia, . . . I am denying that there are any such properties."[4] These terms all relate to the actual experiencing of things like sensations, feelings, recollections, concepts, and behavior. So his words suggest that Dennett does more than evade or avoid the reality of the subjective "world." For those on that side of the controversy, the private subjective nature of consciousness seems to bar objective scientific study. Allowing at least some legitimacy to the study of what must be admitted to be a bizarre subject is David Chalmers. In *The Conscious Mind*, he grants that consciousness "is properly a scientific subject matter . . . ," but he goes on to say, "it is not open to investigations by the usual scientific methods."[5]

Consciousness, which enables for each of us the experience of living one's life, is more significant than the presumed purity of a narrow view of science. Here's how Daisie and Michael Radner put it in *Animal Consciousness*: "Conventional wisdom is that conscious experiences, especially those of nonverbal animals, aren't amenable to scientific study because statements about them cannot be verified or falsified. [However, there are] two notions of falsifiability. The first . . . is the privacy argument, the second the predictability argument."[6] In considering the subject, scientists should move from the confining privacy argument to the broader predictability argument. It says consciousness must leave its mark on the activities of animals that have it. Not at all coincidentally, this same conflict between what amounts to the subjective realm and the objective world will be examined when we look at those aspects of ourselves in chapter 16, at the end of our exploration.

Animals, whether nonconscious or conscious, have two sides to their interaction with the world, an input side with sense organs to provide the basis for figuring out what's happening outside, and an output side with muscles to engage that world with activity. Our main subject of interest, mentation, lies invisible between the two, unknowable in any exact sense except subjectively for each of us creatures. Our own senses and muscles never show us, other than indirectly, mental processes or a mind outside of our own. So I will have to make inferences—enlightened ones worthy of a tonier name, hypotheses—about mentation and mind, based on the input and output of increasingly complex animals.

CONCEPTS FORM A FRAMEWORK OF MENTAL LIFE: IN EVOLVING CREATURES, THEY GREW FROM NONEXISTENT OR SIMPLE TO COMPLEX AND ABSTRACT

The nature of a suitable framework for mental life is as controversial as the suitability of consciousness as the basis for such a study. Robert Wright tells us how evolutionary psychologists see the mind. "Every organ inside you is testament to [natural selection's adaptive] art—your heart, your lungs, your stomach . . . all are species-typical. . . . The working thesis of evolutionary psychology is that the various 'mental organs' constituting the human mind —such as an organ inclining people to love their offspring—are species-typical."[7] About which I must ask: Is there sufficient reason to think many such specific "mental organs" even exist, let alone that they are each the result of specifically human natural selection, let twice alone that they all are (or originally were) adaptive? Apart from my preference for *concepts* as a framework of mentation, why not—given psychology's tumultuous history—why not more caution about things like species-typical mental love organs on the part of evolutionary psychology?

Another candidate for model of mentation is given us by Michael Gazzaniga, in *Nature's Mind*. He likes a *selection* model rather than an *instruction* model for cognitive activity. In the strong form of his argument, "an organism comes delivered to this world with all the world's complexity already built in. In the face of an environmental challenge, . . . what the outsider sees as learning is actually the organism searching through its library of circuits and accompanying strategies for the ones that will best allow it to respond to the challenge."[8] Although this is at least conceivable for lower animals, for me it is beyond belief that many (never mind all) of the useful concepts that I employ to cope with the world—including the very idea I here enunciate—are built into me.

Another approach to animal mentation is made by Nicholas Humphrey, in *Consciousness Regained*. He connects consciousness with the social dimension of life for more complex animals. He suggests that not only chimpanzees, but wolves and elephants, "which all go in for complex social interactions, are probably . . . conscious; frogs and snails and codfish are probably not."[9] What he means here is less than clear. I doubt that he thinks frogs don't experience pain and so are not conscious. Likewise, I doubt that he thinks chimps, wolves, and elephants have humanlike *reflective* consciousness. So perhaps he means that the more complex animals, unlike the simpler ones, make use of *concepts*, or at least *social* concepts, in their mentation.

My interest in concepts (ideas, notions, especially those that are generalized or are about a class of objects) as the central basis for mentation is challenged on yet another front by Susanne Langer in *Mind*. Like many psychologists, ethologists, and some other philosophers, she does not regard any creatures other than humans as using *concepts* in their mentation. She says "the many claims for concept formation and symbolic thinking in animals all rest on special definitions of 'concept' and 'symbol' . . . entirely inadequate to the mental processes of Homo sapiens; they do not cover man's uses of speech and logical thought. . . ."[10] Langer similarly regards the word "mind" as properly used only in referring to humans. However, she uses a related but lesser word, "mentation," for the related activity of animals. It is unfortunate that she didn't find and use a lesser word than "concept" for the similar—though less complex—mental activity of animals.

Although she does not find and use a lesser word for concept, Maxine Sheets-Johnstone[11] (in *The Roots of Thinking*) is a philosopher who does not shy from the belief that nonhuman, languageless animals can and do use concepts. Her suggestion, for example, that present-day apes cannot grasp the concept of death is suggestive only of the limitations of apes' use of concepts.

Our input senses and output muscles function because they are hung on a framework of bones. Likewise, a framework is needed for the structure of mentation leading to mind that this book presents. Concepts (which are experienced and thus mental) will form that frame. Why is that? For one thing, we humans do, on occasion, think. Whatever else we need to think, concepts are the main thing we think with. For another thing, there is continuity in evolution. If we at times think with complex concepts, concepts that are represented with language, then some of our predecessors used simpler concepts with gestures to portray them. Concepts start with simple direct ideas and move toward increasingly generalized abstractions. The latter are by no means as dry as the words might suggest. For apes they include ideas of sex, power, and safety; for humans, the limitless list includes thoughts of life, love, and death.

The sophisticates among young children have fun with the others by asking, "Why did the chicken cross the road?" The smarty answer, "To get to the other side," presumes no fancy mental footwork, but it does raise the possibility of a simple protoconcept in the chicken's mentation. Most lower mammals and most birds live in a mental world that is limited to the here-and-now. That follows from their ability to use, at most, only such direct protoconcepts to make their way in life, although they may well get by with nothing at all like a concept. Nothing like a concept, that is, except for the nonconscious, and thus unexperienced, pattern-making-and-recognizing activity of neural networks as such, which provide the basis for concepts. In

HOW WE GOT TO BE HUMAN

Matter and Consciousness, Paul Churchland discusses, not real neural networks, but artificial ones that can learn to recognize and discriminate between patterns of information.[12] These networks are computer-programmed simulations of much of the structure of networks of real neurons in animal or human brains. Necessary though neural networks are to mentation, we have no real knowledge of how they lead to mentation.

Modern apes, in noteworthy contrast to four-legged mammals, have the start of an expanded mental world of concepts—proto or otherwise—that enables them to do things beyond the capabilities of lower animals, things that include making simple tools and communicating about sex and power with *gestures*. In *Gesture and the Nature of Language*, David Armstrong, William Stokoe, and Sherman Wilcox suggest not merely that gesture was a precursor to language, but that the basic elements of syntax are intrinsic to gesture. However, they discuss "gesture" in a broader sense than that of *deliberately significant motion*, which is the (symbolic) usage that I will employ. For them, there are both symbolic gestures and nonsymbolic gestures. For example, they mention "the [nonsymbolic] gestural task of a tree-dwelling monkey reaching for a vine."[13] That task is not deliberately significant. The monkey is not trying to convey anything that it is aware of to another creature. Note, however, that for that task, the monkey's brain must have developed abstracting patterns of neural network activation that enable the action. Note also that those patterns do not come to consciousness. If they did (as they can for humans), the patterns would show themselves as (symbolic) gestures that are optional and represent concepts. Tarzan had the option of instructing a monkey (with or without the words "Grasp a vine like this, lunkhead") as he performs the act. If the monkey fails to grasp the vine and falls, its face may show fear. But the movement of its facial muscles, although *significant*, are not deliberate. So I will not use the word "gesture" to refer to the likes of such motions.

There is evidence that apes, when they interact, use gestures as intended communicative motions that are expressive of concepts. Gestures of this kind are needed to represent concepts such as acknowledging defeat to the winner with the gesture of lowering himself to all four limbs, while the winner gestures his superiority by standing tall on two legs with his hair puffed high. For monkeys and apes, utterances sometimes serve the same purpose of conveying a somewhat abstract concept to others, such as the possible presence of a predator.

At some point in our development as complex creatures—in all likelihood in our predecessor species, Homo erectus—increasingly complex mental concepts allowed the further expansion of life by sharing food, living in two-parent families, making and using hand-axes, and expanding utterance into the beginnings of language. Such early utterance was sufficiently complex and directed to warrant calling it protolanguage, wherein simple combinations of concepts could be put together in a way that—unlike the complete language we presently use—was often ambiguous and simple-minded, thus of limited value to its users.

With the arrival of our own species, Homo sapiens, mental life continued its expansion—both in and out of this world—with things like more complex sociality and sexuality, a past and a future, and magic and religion. Short-on-syntax protolanguage, at some point in our past—probably many millennia after our own species came into existence—was succeeded by the verbal version of the stuff the writer and reader of these words use to represent abstract concepts and supply meaning: complete language.

THREE FUNDAMENTAL INSTINCTS DETERMINE DIRECTIONS OF MENTAL SPACE: EGOCENTRICITY, COMMUNALITY, AND SEXUALITY

Nature's creatures cannot live by concepts alone. The simpler creatures, in particular, probably cannot live by concepts at all. They simply lack the mental apparatus to make and use concepts to make sense of the world and survive. Lower animals live and survive to produce progeny who will live another day or so, because they have—in addition to some ability to learn—instincts within them. These innate capabilities represent the originally fortuitous attitudes and activities—the results of accidental inscriptions by nature in the code for living called DNA—that proved useful in the ongoing harsh lottery of life.

Many of these heritable predispositions or knacks predate consciousness—which was itself one of them—in the evolution of life. Creatures like spiders or ants, who I suspect are nonconscious, are guided in their lives largely by inherited instincts, innate capabilities to do things like spinning solidified slime into webs or swapping stomach contents with the next fellow.

Apart from concepts in the mentation of animals and humans, how is mentality organized? In *Mapping the Mind*, editors Lawrence Hirschfeld and

HOW WE GOT TO BE HUMAN

Susan Gelman tell us about the beginnings of knowledge and agreement about certain innate mental structures.[14] However, these structures are smaller, more numerous, and less general than the ones that I will propose.

In *Reinventing Darwin*, Niles Eldredge gives us an inside look at a central debate among evolutionary scientists.[15] Opposite his side of the table are what he calls "ultra-Darwinian" evolutionary biologists, population geneticists, and, I suspect, evolutionary psychologists. They view competition for reproductive success as the real level of activity, and are suspicious of emergent phenomena, seeing them as of little use. On Eldredge's side are most paleontologists and naturalists. (They could use an overall name. I offer, at the spectrum's other end, "infra-Darwinians.") They see complex systems, such as species and ecosystems, as emergent and real. They view explanations based on reproductive success alone as overly reductive.

I introduce Eldredge's ideas here, not to mention this fascinating debate, but for his suggestion that basic to evolution are economic systems, social systems, and reproductive systems. These three objective structures are a match with three large subjective organizations that I view as basic to mentation. The three large corresponding structures that I will employ here are egocentricity, communality, and sexuality. (Later, I will add a subjective mindset and an objective frame of mind, as they, too, occupy large rooms in the house in our heads.)

The most fundamental thing common to all instincts is that they are helpful to the survival of the individual creatures that have them. All creatures, conscious or not, have such adaptive predispositions. When the creatures in question are conscious, there is a single word that sums up the most basic nature they have acquired via instinct; that word is "egocentric." The word can hardly apply to nonconscious creatures. That's because they are subjectively empty inside, they have no mentality, no beginnings of mind somehow related to the nervous system. They are zombies, nature-made robots. With nothing like an ego or self, they can hardly be egocentric or self-centered. To be an egocentric creature is to use your experiences, such as your desires and fears, to obey nature's first law to her simple-minded earliest conscious creatures: Look out for number one.

Another law from Mother Nature—inscribed with others in the strands of spiraling DNA—for us and for all the species of conscious creatures that were our own predecessors, is innate communality. Unlike some species, most of them predators, we live, as did our ancestor species, in groups of our own kind. We see communal creatures when we look at schools of (presumably conscious) fish, herds of cattle, communities of apes, and societies of humans. Members of the brainier communal species of lower animals not only live in groups but also organize themselves in a hierarchy within each of those groups, a hierarchy that in its essence defines who can win a fight with whom. With apes, this simple "pecking order" starts to expand and articulate—with mediating gestures and grunts—the concept of social power, power with respect to others of one's own kind.

A third basic instinctive drive that we have is that of sexuality. Like egocentricity and communality, sexuality starts out in lower conscious creatures as simple and direct. For animals in our line, sexuality is activated initially in the female with a seasonal desire, called "heat" or estrus, for satisfying sexual activity. For such animals, males are interested in sexual activity only when the females are in heat. That male interest—also innate—can best be called "rut." It is like female heat in that it is experienced as having lustful desires. When they sense the heat's presence in females, the male animals engage each other in fierce contests for access to them.

As creatures get more complex, subjective sexuality gets more complicated. It can now be called "prurience," which means they can have not only lustful desires, but also lustful ideas (concepts). This can be seen with apes. It must have been so for our upright-ape and erectine predecessors. Along the line to us, our upright-ape ancestors abandoned the estrous aspect of the genetic roots of sexuality in their females, while retaining the genetic rutlike roots of sex in their males. We all have evidence that the complexity of prurient sexuality is most elaborate of all for our own human species.

Most (but hardly all) of the mental aspects of living are played out in a space with directions defined by these three instinctive dimensions: egocentricity, communality, and sexuality. A note referencing Sigmund Freud's work might have been appropriate here. After all, isn't there a rough resemblance between these three instincts for animals and the Freudian id/ego, superego, and id/libido for humans? Likewise for the nonconscious here and the Freudian unconscious? Is it possible that I suppressed (in my gourd) the connection? And that I'm resistant to searching for it in the work product of his?

About suppression and resistance I can't speak, but I do find myself possessed by an urge to couch a little confession. At the beginning of *The Origins of Virtue*, Matt Ridley writes, "This is a book about human nature, and . . . the surprisingly social nature of the human animal."[16] Then he confesses that "this is an impossibly immodest task." As I, in turn, am attempting to write about not merely the social nature but also the egocentric nature and the sexual nature of humans and of our predecessors, my intentions may seem not so much immodest as megalomaniacal. All that I can say is that for egocentric creatures with both multiple mindsets and access to complex concepts, it's never possible to assert with any real finality: Megalomania, you and me is quits!

Freudian notions aside, the size of the mental playing field enlarges as creatures evolve larger brains with greater ability to use increasingly complex and abstract concepts. It is this expansion via concepts of the playing field of experienced life, with original directions determined by instinct, that is our subject matter ahead.

NOTES

1. Howard Gardner, *The Mind's New Science* (New York: Basic Books, Inc., 1985).

2. Anthony Kenny, *The Metaphysics of Mind* (New York: Oxford University Press, 1992), p. 8.

3. John Searle, *The Rediscovery of the Mind* (Cambridge, Mass.: The MIT Press, 1994), p. 227.

4. Daniel C. Dennett, *Consciousness Explained* (New York: Little, Brown and Company, 1991), p. 372.

5. David J. Chalmers, *The Conscious Mind* (New York: Oxford University Press, 1996), p. xiv.

6. Daisie Radner and Michael Radner, *Animal Consciousness* (Amherst, N.Y.: Prometheus Books, 1989), p. 191.

7. Robert Wright, *The Moral Animal* (New York: Pantheon Books, 1994), p. 26.

8. Michael S. Gazzaniga, *Nature's Mind* (New York: Basic Books, 1992), pp. 4–5.

9. Nicholas Humphrey, *Consciousness Regained* (New York: Oxford University Press, 1984), p. 35.

10. Susanne K. Langer, *Mind: An Essay on Human Feeling*, vol. 2 (Baltimore, Md.: The Johns Hopkins University Press, 1972), p. 138.

11. Maxine Sheets-Johnstone, *The Roots of Thinking* (Philadelphia: Temple University Press, 1990), p. 203.

12. Paul M. Churchland, *Matter and Consciousness*, rev. ed. (Cambridge, Mass.: The MIT Press, 1988), p. 156.

13. David F. Armstrong, William C. Stokoe, and Sherman E. Wilcox, *Gesture and the Nature of Language* (New York: Cambridge University Press, 1995), p. 47.

14. Lawrence A. Hirschfeld and Susan A. Gelman, eds., *Mapping the Mind:*

Domain Specificity in Cognition and Culture (New York: Cambridge University Press, 1994).

15. Niles Eldredge, *Reinventing Darwin: The Great Debate at the High Table of Evolutionary Theory* (New York: John Wiley and Sons, 1995), p. 203.

16. Matt Ridley, *The Origins of Virtue: Human Instincts and the Evolution of Cooperation* (New York: Viking, 1997), pp. 5-7.

PART ONE

ANIMALS LOWER THAN APES AND MONKEYS

INTRODUCTION

IN THESE FIRST CHAPTERS, WE will look at the early work of Mother Nature that lessened the dependance of her creatures on their inheritance, that is, their genetic heritage of capabilities. To accomplish this loosening of her binding apron strings, nature provided animals with new tools they would need to do a better job of living on their own.

Making nature the subject in the sentences above is not meant to suggest that either natural selection or nature is a doer, a real agent who is motivated to act in our world. No doubt Richard Dawkins likewise did not intend to imply that genes are active motivated agents, as suggested both in the title of and within *The Selfish Gene*.[1]

Figurative language, however, is so powerful that sometimes, like it or not, we start to respond to such expressions as if they were literally true. Listen to Dawkins himself as he says (in a subsequent book, *The Blind Watchmaker*) he speculated in the earlier book, "that we may be on the threshold of a new kind of genetic takeover. . . . The new milieu of cultural tradition opens up new possibilities for self-replicating entities. The new replicators are not DNA. . . . They are patterns of information that can thrive only in brains or the . . . products of brains."[2] That sounds truly exciting, doesn't it? What, you may wonder, are these powerful new entities? Dawkins tells us that he named these new replicators "memes," and that he did so to distinguish them from genes. Although Dawkins's books include many other notions that are distinguished, here we have some perfectly adequate names for these patterns of information that spread, names like "ideas" and "concepts." However,

29

to suggest that ideas and concepts self-replicate by spreading from mind to mind would be to invite a response of "Wow, 'zat so? So what else is new?" What is new is that "meme" borrows authority by sounding a little like "gene." Also, a kind of conceptual contagion is spreading Dawkins's new word, *meme*. Daniel Dennett, in *Darwin's Dangerous Idea*, shows his fondness for this not-really-dangerous idea by using it frequently.[3] Ideas that have new names generate excitement; this one is diving into dozens of books.

In one such book, *The Meme Machine* by Susan Blackmore, Dawkins writes in his foreward, "Since 1976, when the word was coined, increasing numbers of people have adopted the name 'meme' for the postulated gene analogue."[4] We may have a fun word in "meme," but we also have a poor analogue for "gene." After all, a superior new gene helps its possessors to have more progeny than does the competition. But if a meme is (as it is defined) an element of a culture that is passed on by nongenetic imitation, there is no reason to regard it as adaptive, even in a nonbiological sense. A meme may be a mere flash-flood fad that for a while joins—but does not replace—other concepts or behaviors. I think the word "meme" makes our mental water muddy by obscuring and trivializing the word "gene."

To get back to "the selfish gene," Ridley says that behind the phrase lies the idea that "individuals do not consistently do things for the good of their group, or their families, or even themselves. They consistently do things that benefit their genes, because they are all inevitably descended from those that did the same. None of your ancestors died celibate."[5] This is an inadequate argument. After all, it is probable that some of Ridley's ancestors' descendants (and likewise some of mine) denied this consistence and inevitability. How? By dying celibate.

Mother Nature did not really do anything. Rather, it was ancient creatures, who possessed commonly adaptive mutations, that changed the genetic inheritance of their descendants. We humans and animals are the only doers, the only agents—creatures with intentions to have their way with the world. However, making doers of indifferent forces and passive objects comes easy and "natural" not just to the likes of Dawkins, Dennett, Ridley, and me, but to all humans. I digress here because it illustrates so nicely our human ability to project our active nature into passive objects and processes. This ability will be explored when we look later at projection, which is central to magic and religion.

The main new tool from nature was consciousness. Before elaborating, I must pause here because not everyone believes that consciousness is really a tool. As Mary Midgley remarks in *From Brains to Consciousness?* "many educated people today do find it hard to admit that the subjective viewpoint has an important influence on the world."[6] For example, Steven Pinker, in *How the Mind Works*, is of two minds about that idea. He sounds like a believer when he says that emotions (which are certainly conscious) have been "prematurely written off as nonadaptive baggage."[7] Yet he shows his disbelief (about consciousness being a tool) by writing, "consciousness and choice inhere in a special dimension or coloring that is somehow pasted onto neural events without meshing with their causal machinery." Something that does not mesh with causal machinery can't be a tool of any kind, neither an incidental one nor an adaptive one. I am with the first of Pinker's minds in regarding consciousness as useful, as a tool.

This new tool from nature was consciousness in the basic sense of the word—immediate experience of parts of the external world and of some things internal to us creatures, such as pain—that will be examined (along with reflective consciousness) in chapter 1. Before mutation brought consciousness to life, creatures relied mostly on innate abilities, commonly known as instinct. Additionally, they had a limited ability to learn from their participation in the world in which they lived. With consciousness, and with the larger brains that allowed a greater ability to learn, animals were more on their own, freer to move farther from Mother Nature's skirts.

In the first of the following chapters, we will look at basic consciousness itself and try to see what cannot be seen in other books about either evolution or consciousness: How it could have been helpful to the first creatures that had it. Gerald Edelman (in his *Bright Air, Brilliant Fire: On the Matter of the Mind*) offers a diagrammatic model of this basic kind of consciousness (which he calls "primary consciousness"). The connections between certain blocks of this model are labeled "Reentrant Loop Connecting Value-Category Memory to Current Perceptual Categorization."[8] These same connections, however, are also labeled "Primary Consciousness." No explanation is offered that connects the first label (which is about the brain) with the second label (which is about the mind). I can't fault Edelman for this, since no one has an explanation of how consciousness derives from brain activity. However, he does not suggest how well or poorly the model would work if only the first label were true, that is, how well brains would work without the minds that are related to them. That lesser task, generally unencumbered by scientists' explanations, is one that I will take on below.

I will also look at reflective consciousness, which is available to us and may have been available to our erectine predecessors, but was beyond the upright apes who preceded them. In the second chapter, I will look at lower animals: not only four-legged mammals, but also birds, even though we can't claim the latter as ancestors. All such creatures profit from an awareness of parts of their lives and from more learning—learning at the College of Hard or Helpful Knocks, that is. We will also see some birds, but not beasts, that begin to use some of the *concepts* like those that make human life, and to a lesser extent ape life, so complex.

In *Animal Minds*, Donald Griffin points out that only recently have concepts and other mentalistic terms started to become acceptable to psychologists and others who study animals.[9] That's true enough, but even more recently cognitive scientists have been co-opting mentalistic terms while still ignoring their basis in consciousness. *Nonconscious* abstractions, generalizations, and categorizations, let me remind you, are constructed by networks of real or simulated neurons. They are the basis for nonconscious processes in lower animals, such as classifying moving objects and reacting to them as predators or prey.

If animals did engage in conceptual activity, how would we know it? The only way is if there were (indirect) evidence, such as deceptive behavior and bodily or vocal gesture. Such evidence for what amounts to private (subjective) mental activity is usually less than clear and often arguable. Most experts don't endorse the notion that creatures other than humans use concepts mediated by gestures. They simply ignore this subject. For example, Ann Zeller's article "Communication by Sight and Smell" is about primate communication.[10] Yet, even here, the difficulties of addressing conceptual activity are apparent. Zeller discusses olfactory communication (by urinating) and visual communication (by gesture). The complex issues of whether or not either such communication is conscious, and also conceptual, are not addressed. Despite all that, I will show evidence that apes and some birds do use concepts and gestures. That is to say, these country cousins of ours sometimes use ideas to *think*.

Lastly, in chapter 3, we will take a quick look at the most domesticated of animals—other than ourselves—the odd couple that are "our" cats and dogs. One oddity about this couple is that they illustrate what it means to be a social animal: cats are not, dogs are.

Before we explore consciousness and the lower creatures that—most of us could agree—experience some parts of their lives, I want to show you one small beastie that may well be on the other side of the abyss that separates the aware, separates them from the mental zombies that are the nonconscious creatures. Let's examine the digger wasp by looking at some highlights from what Marian Dawkins tells us about it in *Through Our Eyes Only? The Search for Animal Consciousness.*[11]

This kind of creature, otherwise known as a species called Sphex, was studied by entomologist Jean Henri Fabre about a century ago. When the female digger wasp is ready to part with her single egg, she digs a hole in sand that will be the first home of her soon-to-be scion. Then, using her instinct to insure the initial survival of the larva that with luck will hatch from her egg and someday become a wasp, she leaves the burrow in order to provision it with food. Nature provides the wasp not only with the instinct to get food for her progeny, but also with the means to keep the food fresh. The Sphex finds a suitable small insect, stings it with a venom that paralyzes without killing, and takes it to the site of the burrow, which she had closed when she left after laying her egg. The Sphex then leaves the torpid-but-alive prey near the burrow's entrance, reopens the hole and goes back in, where she presumably reinspects to see that everything is still shipshape. Her regular-as-clockwork behavior after this inspection is to drag the preservative-filled food inside, drop it, exit again, and close the hole once more. She normally proceeds to bring several food items into the burrow in precisely this manner. This part of her behavioral repertoire is performed "religiously." (I grossly misuse that term solely to highlight the activity that Fabre examined.) Then, when her biological duty is at last done, without a thought for her tyke, she seals the burrow for the last time. Getting out is her offspring's problem.

What Fabre did was to intervene in the middle of her provisioning pattern, while the digger wasp was inside inspecting her burrow, by moving the future meal an inch or so from where it had been. When the wasp came out, her first act was to position the paralyzed prey back to where she had left it at the entrance. Her next act was not to bring the food inside, as we might expect, but to repeat her role as Sphex, Inspector General. Satisfied once more that all was in order, she again came out to bring the meal within. But while she was inside, the diabolical scientist again intervened to move the prey. This prey-moving by the human outside while the wasp was inspecting inside happened again and again. Fabre was persistent, he moved the nourishing morsel more than forty times. The Sphex was "resolute," and she repeated her prey-repositioning and parlor-reinspection each time. At ths point, Fabre may have felt his obligation to science was satisfied, and he quit.

HOW WE GOT TO BE HUMAN

What is to be said about all this? If nothing else, we have seen the very model of a less-than-modern major general pattern of rigid instinctive behavior. Does this mean the digger wasp is incapable of learning? That may not be so; it could be that she has learned some and can learn more, but perhaps not to alter a pattern of behavior that undoubtedly works well enough, by and large, for her and her kind. Even there, forty trials is not a large number for learning new behavior. Nor did Fabre sling any other outrageous arrows her way to dissuade her from her habit. Neither did he try to determine whether or not the wasp was conscious. At least he didn't try to apply what might be, say, a painful experience to see if she would depart from her fixed behavioral routine, thereby suggesting (to me) that she might be conscious. In chapter 1, we will see why an animal's unusual reaction to an unexpected and uncommon negative input may imply that it is conscious.

As for the number of trials, I'm sure there are countless experiments where both carrots and sticks were used to get animals to learn, say, the best path out of a maze, countless such experiments where far more than forty trials were needed to change behavior reliably. Learning is not easy, especially when some uncivil servant of science would like you to learn something that is outlandishly weird and far out from your normal life.

But is the life of a female digger wasp such that consciousness could help her to live better? So far, like the stone Sphinx, the Sphex is silent regarding that question. There may be some lower threshold size and organization for tiny brains within conscious creatures. Below that threshold, experiencing life may not add enough in the way of possible alternative behaviors—that are within the means of those brains—to make it adaptively worthwhile.

I should not leave the nonsocial digger wasp without a few words about its thousands of related social ant species. In *Journey to the Ants*, Bert Holdobler and E. O. Wilson write that "the great strength of ants is their ability to create tight bonds and complex social arrangements with tiny brains. They have done so by queuing to a limited array of very specific stimuli. A few dozen such signals lead the single ant through her daily social rounds. . . . [However, this] concatenation of simple cues is also a source of major weakness. Ants are easily fooled. Other organisms can break their code and exploit the social bond merely by duplicationg one or several key signals."[12] Although these authors write about ants being "subjectively aware," all this suggests to me that, like the digger wasp, they may be too feeble-brained to afford or take advantage of the subjective experience of even a feeble mind.

In an article entitled "Towards Welfare Biology: Evolutionary Economics of Animal Consciousness and Suffering,"Yew-Kwang Ng suggests that all conscious species are plastic and that all plastic species are conscious.[13] This idea provides a tempting measure for consciousness; certainly it is worthy of more exploration. Offhand, however, it seems possible that many nonconscious kinds of creature are capable of some minimal learning, which implies some plasticity that lets them shape their activities to some extent. (The author also quotes a few words from an attack by linguist Noam Chomsky on behaviorist B. F. Skinner, words that I can't resist repeating: "A cynical onlooker might be tempted to say . . . that psychology, having first bargained away its [spirit] and then gone out of its mind, seems now, as it faces an untimely end, to have lost all consciousness.")

NOTES

1. Richard Dawkins, *The Selfish Gene* (New York: Oxford University Press, 1979).

2. Richard Dawkins, *The Blind Watchmaker* (New York: W. H. Norton and Company, 1987), pp. 157–58.

3. Daniel C. Dennett, *Darwin's Dangerous Idea* (New York: Simon & Schuster, 1996).

4. Susan Blackmore, *The Meme Machine* (New York: Oxford University Press, 1999), p. viii.

5. Matt Ridley, *The Origins of Virtue* (New York: Viking, 1997), p. 17.

6. Mary Midgley, "One World, but a Big One," in *From Brains to Consciousness?* ed. Steven Rose (Princeton, N.J.: Princeton University Press, 1998), p. 263.

7. Steven Pinker, *How the Mind Works* (New York: W. W. Norton, 1997), pp. 370, 562.

8. Gerald M. Edelman, *Bright Air, Brilliant Fire: On the Matter of the Mind* (New York: BasicBooks, 1992), p. 120.

9. Donald R. Griffin, *Animal Minds* (Chicago: The University of Chicago Press, 1992).

10. Ann C. Zeller, "Communication by Sight and Smell," in *Primate Societies* (Chicago: University of Chicago Press, 1987).

11. Marian Stamp Dawkins, *Through Our Eyes Only? The Search for Animal Consciousness* (New York: W. H. Freeman and Company, 1993).

12. Bert Holldobler and Edward O. Wilson, *Journey to the Ants* (Cambridge, Mass.: Harvard University Press, 1994), pp. 71, 123.

13. Yew-Kwang Ng, "Towards Welfare Biology: Evolutionary Economics of Animal Consciousness and Suffering," *Biology and Philosophy* 10 (1995): 255–85.

1

CONSCIOUSNESS
THE GREAT RIDDLE-MYSTERY-ENIGMA

IN A RADIO BROADCAST ON October 1, 1939, Winston Churchill described Russia as "a riddle wrapped in a mystery inside an enigma." Steven Pinker uses the same expression for consciousness as a single problem.[1] Here, I will separate the consciousness problems we have into three distinct parts: a riddle, a mystery, and an enigma.

A riddle may be profound or simple, but even the profound ones usually have answers. The riddle of consciousness is: What's it good for?

If you believe, as I do, that we are entirely natural creatures with biological evolutionary origins for our basic capabilities, then we might agree that there ought to be an answer to that riddle. Consciousness should be useful. We should have more than mere hope that there is potentially demonstrable Darwinian adaptiveness for it. Donald Griffin's 1981 book, *The Question of Animal Awareness*, was one of the first attempts by a life scientist to depart from the numbing assumption that science should not look behind animal behavior to help understand animals.[2] Unfortunately, at that time he was not able to present anything like a theory of adaptive value that might sustain itself against the fact that computers can be programmed to perform virtually any algorithmic procedure that can be suggested as the basis of the adaptive value of consciousness.

If consciousness is a natural phenomenon that some creatures have and find valuable, then it should somehow help them make a living for themselves so they can make a life for their potential progeny. We do have a bit of negative evidence for that. One of the few things we know about basic consciousness for us is that, when we're quite ill or very tired, it just disappears. We "lose" it. This suggests that, like every other human activity, it requires energy, energy that we can't afford when we're way below par. Perhaps large, complex, and "rich" species like us can sometimes afford a luxury, but that's not likely to have been the case for the small, simple, and "poor" species in which it first came to life. If it was a luxury when it first appeared, the overloaded creatures carrying it would have dropped out of the race for survival.

Yet few have looked for and none have found such a function for consciousness. Joseph Mortenson, in *Whale Songs and Wasp Maps: The Mystery of Animal Thinking*, has a lot to say about the activities of a wide variety of animals.[3] He does not even mention, let alone examine, the possible Darwinian survival value of consciousness for those creatures. Marian Dawkins, in *Through Our Eyes Only? The Search for Animal Consciousness*, is negative about what consciousness is good for: "Every single, so-called 'function' of consciousness that has been proposed so far could just as well be carried out— so it would seem—by an unconscious organism or even by a machine programmed to behave in complex ways."[4] She also writes, "Consciousness remains an intractable and even embarrassing problem for biologists. . . . They sometimes deny it is a scientific problem" (p. 7). So no use or adaptive value for consciousness (or awareness, which I will use as its synonym) has as yet been found. It is often more a risk to be avoided than a riddle to be solved.

Given the absence of past success, perhaps trepidation should make me stutter my bold claim. I will offer an answer to the Darwinian riddle of consciousness. Here is a clue to that answer: For complex creatures, it may be too difficult at this time to tease out the thread of its adaptive value. So even though consciousness is the core of our human life, I will not seek to learn its survival value for us sapiens. For the small-brained creatures that first had it, however, the basic ability to experience life may have been valuable, *in itself*, in a way unavailable to any computer we have today. So we will look at some presumed ancient conscious pea-brains for its adaptiveness. I think such creatures responded better to *unexpected* adversity than any programmed computer or robot can.

I call the second problem of consciousness a mystery. Mysteries are sometimes as deep as or deeper than riddles. Also, they may or may not have solutions. The mystery of consciousness is: What creatures, other than sapiens, live a life that is conscious? This mystery is occasionally addressed by life scientists, but none are so incautious as to suggest which creatures are conscious and which are not.

Consciousness: The Great Riddle-Mystery-Enigma

I will offer no more than a half-answer to the mystery of awareness by drawing a rather arbitrary line, and saying: Above this line creatures are conscious, below it the uncertainty remains. With some misgivings, I will also suggest a means of exploring the murky depths below my facile line. The uneasiness arises because the only weapon I can conceive of that might separate the grain of conscious animals from the chaff of nonconscious zombielike creatures is, of all things, the whip of pain. The two groups will separate themselves by the kind of behavior they engage in. But how many would fancy the role of scientist as sadist?

An enigma may be even tougher than a riddle or a mystery, for it may well be inexplicable, a riddle with no answer, a mystery with no solution. The enigma of consciousness is how mere matter, whatever the natural laws that guide such indifferent stuff, stuff without evidence of the least whiff of feeling, sensation, or conscious thought, can give rise to something as unlike itself as creatures experiencing parts of their lives.

There have not been many attacks by scientists on the enigma of consciousness. Here's a glance at one that connects it with another enigma—one regarding fundamental physical particles—but solves neither. In *Evolving the Mind*, A. G. Cairns-Smith describes a number of quantum "theories" of consciousness.[5] All such attempts to relate consciousness to quantum physics suggest that two great scientific enigmas—consciousness, and the seeming collapse of a fundamental particle's condition from pseudoexisting as mere probabilities of being at various places, collapse to actually existing at such-and-such a place—are related. Since none of these notions provide evidence of a relationship between the two, I see no value in confronting both problems at one time.

Many philosophers relish the meaty problems associated with enigmas. They chew them endlessly, their descriptions of them are often smoother and occasionally more seasoned than mashed potatoes, but they are not given to supplying answers for enigmas. John Searle's article, "The Mystery of Consciousness," provides a smooth and tangy review of recent scientific and philosophical books about consciousness, none of which considers solutions to the riddle and the mystery of consciousness, let alone its enigma.[6]

As for me, I'll say it flat out: I don't do enigmas.

CONSCIOUSNESS: THE NATURAL BASIS FOR BOTH MIND AND SPIRIT

Prior to addressing the riddle, mystery, and enigma of consciousness, a little needs saying about the meaning of consciousness. That is not easy with this strange subject. Think for a moment of the difficulties we'd have explaining the concept to intelligent, naturally evolved nonconscious androids that came to earth from elsewhere and somehow learned enough of human language to allow discourse, but had no understanding of the words "consciousness" and "awareness." Decades ago, when the religion-rooted film *The Song of Bernadette* was made, large billboards announcing it proclaimed, "For those who understand, no explanation is necessary. For those who don't, none is possible." In this respect, consciousness is far more difficult to explain than religion. The explanations that follow are intended for conscious humans, rather than for nonconscious androids from elsewhere or for homegrown not-really-human zombies.

Donald Griffin offers two definitions that are really for *human* consciousness: first, "the recognition by the thinking subject of its own acts or affections," and second, "the state or faculty of being mentally conscious or aware of anything."[7] More elementary than either of these two definitions is the *basic* consciousness lower animals share with us: that which provides experience of parts of the external world (such as noise) and of only some internal things (such as pain).

Griffin's first kind, our awareness of our own doing, thinking, and feeling, I will call "reflective consciousness." I will also call on Mithen, so that I may disagree with his suggestion in *The Prehistory of the Mind* that those engaged in crafts may not be *aware* of the skills and technical knowledge they use. "They often have difficulty explaining what they do unless they can provide a demonstration. . . . This emphasizes the importance of verbal teaching of technical skills. . . . When knowledge is acquired by verbal instruction it . . . becomes available for [reflective] consciousness."[8] However, people with technical or artistic skills normally can and do have reflective experience of their skills, but that experience is without words.

Reflective consciousness refers to mental processes *about* mental processes. That is something we humans can do, but animals cannot do. Our human brains are evidently of suitable structure and size to entertain some neural activity that generates experience as reflective consciousness, that is, thinking about our experiences. This conscious thinking about some aspect of our conscious selves is certainly an expansion from what simpler creatures can do. It is an expansion of the *things and processes* of which we can be conscious. There is no reason to believe it includes or requires any change in the nature of *basic* consciousness itself. Likewise, consciousness in the social sense of understanding the mentation of others need be considered no more than an expansion of reflective consciousness about ourselves.

Griffin's second kind of consciousness, which includes both the basic and the reflective kind, I will call "overall consciousness." It enables the process of experiencing anything, feeling or consciously thinking, knowingly deciding or doing anything. This covers the spectrum from, say, sensing the blueness of the sky, through responding to the bright blue sky by putting on sunglasses, to cogitating about the ability to experience blue sky.

We sometimes casually use consciousness as merely another broad word for mind. To do so seriously is to make the word too fuzzy. Although we shouldn't muddy mental waters by conceiving consciousness as identical with mind, it is the basis for mind. Mind is everything that is or was mental for humans, all our present experiences and our use of past experiences. Mental processes, although they are presumed to be always accompanied by neural processes, are to be distinguished from those neural processes, which are not, as such, experienced. It is the development of mentation and mind in this sense that is an overall theme of this book.

As elsewhere with this vexing subject, there are those who disagree with me about mind being based on consciousness. In *Brain Size and the Evolution of Mind*, Harry Jerison says, "My definition of mind as 'knowing' reality might suggest that consciousness is necessary for mind. This is not true. There is excellent evidence of knowledge without awareness, which can be understood only in terms of unconscious processes."[9] It is certainly true that we often use the word "knowledge" with no precision. We say things like, "There's plenty of untapped knowledge in this old brain." I find those usages, and his, to be too figurative and fuzzy for serious discourse. I find more useful a sharp divide around the subjective world, with no "knowledge," "mind," or "mentation" without consciousness.

Consciousness is also the basis for a decidedly peculiar aspect of the experience of human life, by which I mean the "spirit" within each of us. Pioneer anthropologist E. B. Tylor held a similar belief a century ago. Ancient thinkers inferred "that every man had two things belonging to him, . . . a life and a phantom [a mental representation] as being its image, . . . both separable from the body."[10] Tylor believed that primitive people then regarded the "spirit" as the combination of the "life" and the "phantom" conceptual image of life.

Many times in the early days of our long past, a reflective human mind may have noted the seemingly total difference between itself and the physical structure within which it is embodied, or the difference between a remembered friend and that friend's dead body. That perceived difference may have gradually come to suggest not only a "thing" called life, but a thing called a *spirit* within. When we first began to take heed of ourselves as subjective mental entities, perhaps one or two hundred thousand years ago, this wondering about the seemingly nonphysical nature of the wonderer and others may have been a puzzle. That initial hard puzzle, making sense of feeling "my body is not me but rather a place or thing that I inhabit," may have subsequently been softened over the millennia by conceiving these inner selves—that reflective consciousness shows us—to be unearthly spirits living within our earthbound abodes of flesh. After another long while, regarding ourselves as spirits that were not made of the stuff of this earth might have suggested that we, as such subjective spirits, could survive the death and destruction of our bodies. I don't know about you, but for me, the concept of a future life after death is—to say the absolute least—a wonderful idea. I just can't find any reason to accept it.

There is indirect evidence that some early humans, tens of thousands of years ago, believed that their subjective selves survived death: the presence of fossilized valuables near buried bones suggests as much. Despite an overabundance of opinion, there is no evidence regarding the truth of the proposition that the spirit survives bodily death. What is evident is that the idea of unearthly spirit, once we get past the jolly-good notion of surviving death, just bristles with prickly problems. If these spirits, and the minds that they are a part of, are not "refleshed" after death, how could they sense, feel, think, or move with no replacement for the dead bodies with their fleshly senses, brains, and muscles? After all, our before-death spirits and minds don't function without bodies, do they? If spirits and minds are refleshed—reincarnated, that is—is the new body young or old? Does it age and die, is it reinvested with the information and patterns that once filled the old brain?

Another solution to the beyond-death body and spirit problem is noted by philosopher Paul Kurtz: "For some believing Christians the only sense that can be given to immortality is to say with St. Paul that at some point in the future, the body, including the soul, will be physically resurrected. This stand at least avoids the issue of whether the soul is separate from the body, and the believer merely needs to say that some divine being will in the future ensure the survival of the whole human being."[11] Even with this, there is a problem. One who lives with both that belief and with an enfeebled body or mind still has a conceptual problem. Will her beyond-death body-and-soul be the somewhat disabled one of the time of her demise, or will it be an earlier more vigorous one that lacks her recent experiences?

There are some who believe that the spirits of the newly dead are not refleshed at all. Rather they are rejoined with an all-encompassing sea of spirit that invisibly surrounds the living. For me, this is thickening muddy waters with transfiguring topsoil, thereby reducing them to a soothing (but opaque) mud bath.

Believers in supernaturally powerful agents or other seemingly miraculous means of surviving death might say our minds are too feeble to fully understand such matters. I won't quarrel with someone choosing such an answer, but that kind of response allows anything we can conceive of, and more, to exist. So it does rather limit reasonable discourse on the subject of life after death.

As for the contrasting idea (to which I cannot do other than subscribe) that spirit, as an aspect of consciousness of self, is part of our earthly nature, such an idea suggests that nothing spiritual (or mental) survives death. I must note that a rigorously objective writer, such as David Hufford, in *The Terror That Comes in the Night*, shows us that it is possible to make scientific studies of what seem supernatural experiences, such as the near-death experience, without taking any position regarding the reality of a supernatural realm.[12]

No one has ever developed a satisfying explanation for the enigma of consciousness itself, other than the vague idea that awareness not only interacts with neural activity, but somehow emerges from it—like a pseudospectral butterfly from a still-living caterpillar. Despite this, some scientists make claims that their studies suggest there is some progress in understanding consciousness to be near, if not upon us. For example, Gerald Edelman developed a theory of "re-entrant signaling."[13] Interesting though this theory is, it does not live up to his book's subtitle (*A Biological Theory of Consciousness*). It describes only the activity of the "caterpillar" neural networks, and not at all that of the "pseudospectral butterfly" of consciousness that emerges from and interacts with those networks. My reading of such works suggests an increased understanding of brain function, but not of consciousness.

Furthermore, it's far from clear that we'll ever make sense of being conscious of things and activities. As Antonio Damasio says, "Perhaps the complexity of the human mind is such that the solution to the problem [the reach of science as concerns the mind] can never be known because of our inherent limitations. But much as I have sympathy for those who cannot imagine how we might unravel the mystery (they have been dubbed 'mysterians') . . . I do believe, more often than not, that we will come to know."[14] Like Damasio, I am of two minds on this subject; unlike him, it is seldom that I think we will solve the problem. (Although I am mostly a "mysterian," I don't care for that designation, as it retains an aura of magic. The only alternative I can think of is to coin the word "enigmatician," even though it, in turn, has a scientistic ring to it.)

Will we someday have evidence that my hue of "red" is perceived by me just as yours is by you? Will we find an understanding of how my experience of "redness" rises from the arrival on my retina of certain high-frequency light waves? How to make machines that are like us—that is the even tougher problem addressed by Marvin Minsky, in *The Society of Mind*. First he acknowledges that "we are still far from being able to create machines that do all the things people do."[15] Then he goes on to say, "But this only means that we need better theories about how thinking works." Note how the word "only" suggests better theories might be just around the corner.

What I think is that we should give ourselves a brake and slow down our burgeoning belief that the great god science can solve everything. The enigma of consciousness may be beyond my ken and every other human's ken. After all, there's little reason to think that human minds, which may have evolved to improve the search for—or, rather, the searcher's share of—meat and potatoes, will be capable of making sense of such an arcane enigma.

For all that, there are things that can be useful in approaching the riddle and the mystery of being conscious. Let's look at some of them. For humans, to be conscious of something is to be engaged in one of certain activities. We engage in activities such as feeling gut pain from too much lunch today, seeing the grim grayness of the sky from which no dessert pie descends, recalling what breakfast was yesterday, dreaming tonight of a divine dinner tomorrow, or—may the thought perish—being aware of oneself as too much a mere ingestor of food. There is little reason to think that consciousness is a thing or a place in the brain where certain kinds of knowledge can be temporarily stored, as in "I searched my consciousness for that memory." Rather than a thing or a place, it is more likely a process, one that can sometimes accompany certain types of brain activity.

When a person is doing several things at once, the most important of the several doings is accompanied by consciousness. Thus, when driving a familiar carefree and car-free road with the radio on, awareness may be of the music and not of the road; the driving is often done nonconsciously. But when the road gets cluttered with heavy traffic, the driver changes his treatment of his mind's activities, becomes aware of the road, and loses his awareness of the music. As both the sound of the music and the sight of the road are activating the driver's brain, it is awkward to think of consciousness as a place occupied first by music and then by road conditions. It's more straightforward to think that certain kinds of brain activities, like driving and hearing music, can be either conscious or nonconscious, and that the activity that is most important at the moment is the one that becomes conscious.

This suggests an intimate connection between awareness and attention. Perhaps the two are virtually identical for conscious animals, but more likely they are not, as even some almost microscopic creatures, which I suspect are nonconscious, seem to pay attention to their activities, in the sense of responding to their environment. For us, both consciousness and attention seem necessary when we are learning something, such as how to ride a bicycle or play a piano. However, once learned, such an activity is best performed by supplying neither attention nor awareness to the details of the doing, as any performer can tell us.

Few would dispute that being conscious of things is central to the lives of each of us. Indeed, we have no adequate word to describe just how central it is, for without consciousness each of our bodies would be quite empty, like a tree-stump or like a stone, devoid of inhabiting mind and spirit.

IT'S UNLIKELY THAT CONSCIOUSNESS IS AN EPIPHENOMENON: IF IT WERE, WE'D BE VOYEURS WITH ONE-WAY WINDOWS TO OUR ZOMBIE BODIES

There is one last task—a fun one, with which we can pleasure ourselves—before looking at the riddle of just how it is that consciousness is adaptive:

HOW WE GOT TO BE HUMAN

To address what seem like the bizarre beliefs of some that it may not be of any such value. Their beliefs suggest that consciousness may be an epiphenomenon—a mere side effect that has no influence, none at all, on our bodies (which include our brains), or on our activities in the big world outside.

Do you doubt that some people believe such a thing? Listen again to Pinker's discussion in *How the Mind Works*. Starting with that title, he sometimes uses the words "mind" and "brain" almost interchangeably. In some ways that is more than fair; much of the mind's work is dependent on the brain's work. Yet he says, "The mind is what the brain does . . . processes information."[16] But if it is the brain that does the information-processing, what does the mind itself do? Listen again to Pinker's mind struggling with, not *how* the mind works but with *whether* the mind does any work. First he tells us the mind "makes us see, think, feel, choose, act. . . ." Although it sounds like all these kinds of experiences have some effect on the doing brain, that may not be what Pinker means. His later words suggest something totally different: "Consciousness and choice [are] somehow pasted onto neural events without meshing with their causal machinery." Got that? The mind just sits there, doubtless experiencing some of the brain's work, but not at all influencing the brain's activity. Here, Pinker regards consciousness as a useless icing on the layered cake of the doing brain.

This seven-syllabled concept—epiphenomenalism—is (as Johnny Carson used to say about many things) really weird. It makes us into two distinct beings. One is subjective mind and spirit that has all and only the experience of living, with no part in activity. The other is an objective body (including brain) that has all—and only—the activity of living, with none of the experiencing.

If epiphenomenalism is correct, then conscious subjective life can be said to suggest a voyeur within each of us that looks through a one-way window. The voyeur, through the eyes of a nonconscious objective creature on the other side of that window, sees the objective outside world. For humans (and more narrowly for animals), it experiences some of the activities of the nonconscious creature. This voyeur's experiences are more than merely visual; they include everything subsumed under the concept of overall consciousness. The nonconscious physical creature on the other side of the one-way window includes a brain—but not a mind or spirit. It is nothing more than a zombie that has no experience whatsoever of its life. The voyeur may get kicks and jollies; it feels the unexperienced activities that only seem to be the pains and passions of the objective zombie that houses the voyeur. The window is only one-way, so that physical zombie body, its brain included, can neither experience anything nor be influenced by the observing voyeur's mental and spiritual experiences. The voyeur has all the experiences, but that's all it has; it lacks in particular any way to influence the zombie, let alone the world. The zombie does all the interacting with itself and with the world, but does it nonconsciously, like a tree or a Venus's-flytrap, and nothing comes to it through the other-way window from the voyeur.

Can it be proved that we are each not that combination of a voyeur and a zombie? I won't even think of trying to do such a negative task. Nonetheless, I'll raise one question. It's a large question, one that seems like a language problem, but it's also a concept problem, and therein lies the rub. How could a zombie—even one with a passive voyeur looking over its shoulder— living in an objective world populated only with other zombies, how could it possibly have the word "consciousness" in its vocabulary?

Clearly, it would have that word in its objective world—look, if you think you are a subjective voyeur looking through objective zombie eyes, you've just seen it printed on objective paper in the preceding sentence! Let us grant (but only for the moment) that a zombie might employ language in its zombie-world. (The *meaning* and *import* behind language and gesture require consciousness.) But *consciousness* is neither part of nor explainable in its robotic world. That word is indeed estranged from its empty-of-aware-ness realm. So why on earth or anywhere else would a zombie employ that strange word?

It makes no sense—no sense unless the window is not really one-way—unless what is wrongly thought to be a zombie has been penetrated by what is wrongly thought to be a voyeur. As military doctors in sex lectures to new male recruits say, "Any penetration, no matter how slight, can have major consequences." The consequence in this case is that—whatever peculiar creatures we animals and humans may indeed be—we are not the forced marriages of voyeurs and zombies. We are bodies and minds, with spirits that are parts of those minds.

Before we move on to other matters, let me acknowledge someone who dislikes the separation of minds from bodies in discourse. In *The Roots of Thinking*, Maxine Sheets-Johnstone tries to restore each of us humans to an essential unity rather than the frequent falling asunder into separate mind and body. To support that unity, she writes, "Thoughts and feelings are indeed *manifestly* present in bodily comportments and behaviors. 'The mental' is not hidden but is palpably observable in the flesh."[17] This is a view with which I cannot agree. The mental is only indirectly (and with difficulty and uncertainty) observable, by means of gesture and symbol. Furthermore, we must also deal with behaviors that may be those of nonconscious creatures such as digger wasps and other forms of low life. So our interest in asserting the unity of individual living animals and humans should not overwhelm our interest in consciousness, its nature, and its bearers.

THE RIDDLE OF CONSCIOUSNESS: HOW IT IS ADAPTIVE

How does awareness contribute to the survival of a creature that has it? Now that we have computers and robots that can do many of the things we ourselves can do (but are empty of anything like inhabiting consciousness as source of mind and spirit), the question becomes more vexing. We can point to many things we can do that are also things that we can at least imagine robots might be constructed to do, like communicating and behaving with what seems like intelligent purpose. Nonetheless, whatever awareness is, surely it is not mere hot air, not hot air that comes from voyeurs possessed by zombies nor hot air that intervenes to make far objects seem to shake and shimmy like hootchy-cootchy dancers.

Nor should anxiety about the need to be narrowly scientific deter us. "The reasons usually given for excluding consciousness from the domain of scientific inquiry—that it is ill-defined, untestable, and superfluous—are themselves open to attack."[18] Yet presently we cannot trace the mystery of consciousness back through our evolutionary history by using some objective criteria to find the first conscious creature. If consciousness is adaptive, it makes no sense to assume that rocks or plants or even micro-organisms can be conscious of something. Of what use would it be to them? We can also agree with no one less that Darwin that dogs and cats and other mammals are conscious.[19] They are experiencers of things like pain; why else are we concerned about cruelty to animals?

I'm obliged to note that even here, there are doubters. In his article "Do Animals Feel Pain?" Harrison examines the three common arguments for pain in animals: (1) certain animal behaviors suggest mental states, (2) similarity in structure and nervous system between humans and animals, and (3) evolutionary theory suggests no sharp discontinuity between humans and animals. After pointing out some weaknesses in those arguments, Harrison states his case for restricting pain to the human species. He simply asserts that pain is a mental state, "and mental states require minds," where "minds" are clearly humanlike mentalities that can reflect on their own activity.[20] This amounts to declaring that pain is excluded from animal mentation.

Although we may agree that mammals have some experiences, what about scorpions or ants or worms, or what about the digger wasp? Are they nothing more than living machines with no awareness? We cannot answer with any authority, for they are too unlike ourselves for us to have more than a guess.

If we extrapolate the possession of awareness backward from the mammals, an arbitrary but reasonable place to stop is one of the first kinds of creatures that had both a backbone and a nervous system—fish. So let's assume that, like lower mammals, fish experience some aspects of their world and that they have some sensations and feelings. However, it is uncertain that they do even a little real thinking, such as employing something like (conscious) concepts for the categories of predators and that of fellow fish of their own kind. I will further assume that creatures below fish on what we commonly think of as the ladder of evolution may not be conscious of any things or events. For them, there may be no subject at home to respond when we knock on the doors of their objective bodies.

I am hardly insisting on accurate or firm divides here. Thus, a "ladder" is a poor metaphor, as evolutionary biologist Stephen Jay Gould reminds us in *Full House*: "Evolution is a copiously branching bush with innumerable present outcomes, not a highway or a ladder with one summit."[21] Also, as Mortenson points out, creatures other than vertebrates, such as the octopus, may also be conscious.[22] He tells us that if a diver is too rough with an octopus, even without hurting the creature physically, sometimes the animal seems to go into a state of shock and then dies. It's certainly hard to understand why a creature that did not experience some of its life would do that. On the other hand, perhaps Nicholas Humphrey is right when he says, "Invertebrates, without a sensory cortex in the brain, are unlikely to be conscious."[23] About consciousness in low-level animals, the only clear thing is that our knowledge of the situation is a muddy mishmash.

One piece of what may be scientific information that we do have about consciousness is the result of experiments that tried to measure the difference between the time for a person's nervous system to act and the time for the person to have the awareness that accompanies that activity. Benjamin Libet, in "Subjective Referral of the Timing for a Conscious Sensory Experience," tells us that for an incoming stimulus through the senses to the brain, roughly an additional half-second is said to elapse before awareness occurs.[24] In a subsequent article, Libit reports that for activity outgoing from the brain, the experiments suggest that about a half-second after the initiation of a certain activity in the brain, a person gets the awareness that seemed to accompany the initiation of the activity. I will ignore the broad implication here that—because brain activity always seems to occur before awareness—we are subjective voyeurs who learn, after a half-second delay, what our zombie bodies are doing. Instead, let's consider the narrow implication of the evidence that whenever we are first conscious of something, the neural activity related to that awareness started about a half second earlier.

This has large ramifications for past and present humans and animals with lives that include an element of rapid activity. Whatever the adaptive value of consciousness is, these experiments suggest it is of no help to the fighter who has a fist or a weapon likely to be coming at him within a half-second. Nor is it of use to the batter with a ninety-mile-an-hour baseball coming his way, or to the hunter with boar tusks or saber teeth that are within a half second of finding his flesh. In those kinds of situations we may function for the moment without the benefits of consciousness, whatever they might be. So let's rule out the possibility that being conscious of things is adaptive for the life of rapid activity as such.

What else is there to life besides rapid action and reaction? Well, somewhere on a limitless list there is less-than-rapid activity and there is also decision-making about activity. The latter in particular is the place to look.

Being conscious of things has adaptive value when one is deciding what to do. Decision-making involves values, a creature making a decision bases its choice on the value to the creature of each alternative. When the stage gunman told comedian Jack Benny (or rather, told Benny's stingy persona) to choose between "your money or your life," Benny finally responded, "I'm thinking! I'm thinking!" Here the performer was showing us his supposed values. So evaluating, choosing, and making decisions is the area where consciousness makes the creatures that have it more apt to prosper.

John Searle offers a related hypothesis: "One of the evolutionary advantages conferred on us by consciousness is the much greater flexibility, sensitivity, and creativity we derive from being conscious."[25] Although this may be so, Searle's search for evolutionary advantages of consciousness might be better served by looking for them closer to where they originated, in an animal much less complex than a human. For all that, he is one of the few philosophers and scientists who finds that the adaptive value for consciousness is worth looking for.

Listen now to the presumed voice of proponents of artificial intelligence and designers of computers and robots. They should say that decision-making is one of the things that programmed electronic machinery is really good at. Computers spend a large part of their time making what seems like decisions. Even an extremely simple machine, like the combined heating-and-cooling system in some of our homes, makes these "decisions." Such a machine senses the temperature and compares it with the temperature setting. The system "values" temperatures that are near the setting positively, while those below or above the zone near the setting are "valued" negatively. It then engages in activity appropriate to those "values." That is, it turns off the heater or air-conditioner, or it turns the appropriate device on.

Note that I use what are called "fright quotes" to do more than hint that machines don't *really* decide and value things. There are those, however, who not only deny you fright-quotes but assert that machines have something like mental states, or they actually have them. Look at philosopher Daniel Dennett's "Precis of The Intentional Stance."[26] His *intentional stance* allows him to attribute mental states and events to, say, a heating-cooling system. This lets him view the activities of some machines as indistinguishable from the behavior of some conscious creatures.

In *The Conscious Mind*, Chalmers goes not only further, but "all the way." He asks (with a verbal reference to Thomas Nagel's 1974 paper about the nature of the experience of other creatures, "What Is It Like to Be a Bat?"), "What is it like to be a thermostat?"[27] His remarks suggest he really thinks that such a device may have simple experiences, and thus be conscious. No one can prove him wrong. Nor, for that matter, can anyone prove that the water-level sensor-controller in a toilet (a device that functions similarly) does not have experiences. (Although I don't, perhaps you think both machines experience a nonverbal "More, please" and "That's enough, thanks.") I can't step where Chalmers does. It leads to every organ in my body having its own experiences, likewise to the cells within those organs, and on and on down to quantum particles.

Nonetheless, creatures could, and perhaps do, likewise make "mechanical" decisions. To be conscious of the basis of such choice seems unnecessary. What does it add?

HOW WE GOT TO BE HUMAN

That sounds like a tough question, but perhaps it is not as tough as it seems. To understand my answer, we look for a moment at life and its evolution. Change the example from a home temperature controller or toilet water-level regulator to a hypothetical nonconscious small-brained lowly animal.

Let's call this early foot-soldier in life's struggles with the surrounding world, this creature who might have been a fish; let's call it a "grunt." This *nonconscious* grunt has, among others, two things it "values" positively: Getting enough to eat in the way of, let's say, bugs, and avoiding exposure to cold, wet weather. Assume the grunt has these two nonconscious "values," and assume it also has a means of weighing, "valuing," one against the other. They have been built into its genetic structure after countless generations of life in a climate that has mild temperature and only occasional cold rain. These "values" are adaptive. (I use quotes around this word and others, not for their attention-getting merit, but to represent the words' nonconscious not-really equivalents.) They serve the grunt and its potential descendants well. Its weighing of the two is such that occasionally it spends low-on-nourishment ("hungry") time in its burrow to avoid the bad weather, but mostly it goes out and gets plenty of bugs to eat. The grunt prospers, as do its descendants for many generations.

Assume next that the climate changes, and does so suddenly compared with the time needed for evolution to select an appropriate random genetic change in the weighing of the two grunt "values," a change that would make the creature better suited to the changed weather. Genetic changes that would result in a thicker fur are outside the bounds of this thought experiment. Assume the new weather is frequent cold rain, and ignore its effect on the bug population. How fares the current generation of nonconscious grunts?

The animals, of course, never feel hungry nor do they ever feel cold, but they change their activity to strike a new balance—based on old weighings of "values" suitable for the old climate—between their pursuit of bugs-for-the-belly and the burrow-beneath-the-wet-cold. The new balance requires them to spend much more time in their burrows, without food, avoiding the rain. Only occasionally do they go out to brave the elements for a few bites of bugs. Thus before long they are thin, little more than skin and bones. They have fewer progeny, their numbers diminish, and they may soon face extinction.

Picture now another small grunt, its brain largely (but not completely) identical to that of the grunt above; however, consciousness accompanies some of its brain's activities. What is it like to be a *conscious* grunt? Unlike scientists and philosophers who ask a similar question about creatures who live in the natural world, I (having "created" this grunt) can answer with confidence: It can experience hunger and cold, and weigh the two.

Before the climate changes, its behavior is much like the activity of its nonconscious counterpart, but associated with its tiny brain somehow reside felt values. Pleasant, to say the least, is a bug-full belly, as is mild dry weather. Unpleasant indeed is an empty gut or a cold wet pelt. What happens to this grunt's descendants after the climate changes for the worse?

I suggest that, not all, but most grunts that experience some of their lives will accept more of the unpleasantness of a cold wet skin to avoid some of the sharper pangs of hunger. Their built-in nonconscious weighing of "values" is at least partially overridden by a new weighing based on these new-found experiences. So they suffer somewhat because they are often wetter than they like when they are out pursuing bugs. They suffer somewhat because often their bellies are less full than they like when they are drying out in their burrows. They are leaner than their ancestors who lived the good life in the good weather, but they are not mere skin and bones. They feel more of the slings and arrows of outrageous fortune than did their also-conscious predecessors, but they survive in good number, and they do not become extinct. I am suggesting this is so for most conscious grunts; I am not suggesting that it is so for all of them, nor even that it is always so for most grunts. In other words, I am suggesting that consciousness had adaptive value, by and large, for the earliest creatures that were conscious.

At this point, the strong advocate of artificial intelligence might rise to point out artfully to the court that the weighing of "values" for machines or nonconscious animals can be changed by programmers or by natural selection. I was hoping he would raise that point. It lets me dramatically say: Your honor, I believe I can soon rest my case.

Of course, programmers and natural selection can change nonconscious "values." But the public defender of such means of change misses the real point, which is that the remote delayed hand of natural-selection's "programming" is nowhere near the equal of the immediate feelings of a conscious creature. Animals living in a suddenly changed world cannot wait the countless generations of their descendants' lives that might be needed before a useful mutation shows up. It is my belief that when random genetic mutation brought conscious animals into being, a new and powerful means of adaptation to rapid change may have been initiated. Now, Mother Nature's gifts of instinct and some slow learning no longer dictated all of the decisions an individual creature made. Life was launched on a path of increasing freedom, and of increasing responsibility and choice on the part of individual creatures.

It is not unlikely that diehard members of the artificial intelligentsia will take their case to an appeals court. There they could say computer algorithms could be written into artificial animals' "brains" that would cause them periodically to take stock of, and perhaps alter, their "valuative" procedures so that these creatures could bend them to accommodate sudden environmental change. Before that court is convened, I suggest they write such algorithms for a small "brain," and test them to see if they show a bit of "common sense." We don't want them displaying behavior like that of complex useless car alarms, do we?

To summarize, it is my view that (nonreflective) basic consciousness is adaptive, or at least, it was adaptive for the relatively simple creatures who first adopted it. It supplements inheritance and aids learning by letting an animal make decisions based on its experience of its situation in the world. Such creatures more often can respond appropriately to adversity and opportunity. Awareness provides the spark of felt life that ignites the long-burning ego, and thus enables a creature's egocentricity.

THE MYSTERY OF CONSCIOUSNESS: WHICH CREATURES HAVE IT

But-but you may well say, if basic awareness is adaptive, we should be able to do some kind of scientific testing to determine whether or not members of a given species are, in fact, conscious. Which brings us to my seemingly sadistic potential solution to the mystery of which creatures have come to be conscious and which have not.

Shakespeare has Hamlet say:

> To be, or not to be—that is the question:
> Whether 'tis nobler in the mind to suffer
> The slings and arrows of outrageous fortune
> Or to take arms against a sea of troubles,
> And by opposing end them.

Let's switch from the bard's poetry about the questioning mind of a noble prince to the prose question of which animals in troubled seas or on distressed land are conscious. I think the answer may result from examining whether or not creatures take up *new arms* in response to the hooks and traps of *sudden new* outrageous misfortune, with the latter form of misery to be provided by less-than-empathetic scientists.

We saw above that experiencing life might arm an animal with feelings that would help it to survive a sudden turn for the worse in the weather of its world. The most sudden—and if it is conscious the worst—turn in the "weather" for, say, a fish or a muskrat, is an approaching predator with sharp-toothed jaws. However, how animals respond to predators is not pertinent to our question. This is so because natural predators are a common problem for animals. Predators have been a problem for countless generations of prey species. This means that Darwinian selection of useful mutations may have provided such animals, conscious or not, with fairly good (although surely imperfect) instinctive solutions to their problems. Likewise, slow learning provides imperfect solutions for animals—conscious or not—fortunate enough to survive their first encounters with predators.

What might separate android sheep from conscious goats is infrequent, unpredictable, and unexpected negative stimuli—hooks for fish, traps for muskrats, and other suitable forms of sudden uncommon outrageous misfortune for other creatures. In response, the nonconscious android sheep will have negative "valuations," the conscious goats will experience pain. Let's assume short-on-empathy scientists are willing to risk administering pain to promote the advance of science. What is to be learned?

Nonconscious small-brained animals will have what could be called a normal small behavioral repertoire. This repertoire would be based on both the genetic inheritance of the animals and on what little they have learned slowly from observing others and from their own activities.

Such a repertoire, although useful for the creatures in their normal world, would be unlikely to include anything like an instinctive panic-button response to the hooks and traps and other uncommon unfortunate outrages they could have thrust upon them by experimenters. I see no reason for their biological heritage to include such a panic response to completely uncommon "pain" stimuli. We're hardly talking about anything like millions of microbes that can, over many generations, respond to deadly antibiotics by mostly dying out to be replaced by the many descendants of a few mutated creatures among the original number that were indifferent to the once-deadly medicines.

What "should" you do, if you are a nonconscious fish with a new-fangled hook in your mouth? What if you are a ghoul muskrat in a world that now includes a recently developed low-tech trap that has closed on your leg like shark jaws? The chances are, no matter what you do, you will soon die. If you are such a zombie fish and you have a one-in-a-billion panic-button genetic mutation that makes you swim strongly against a "painful" hook, as some real fish indeed do, is that apt to improve your chances of surviving and leaving progeny? Perhaps you may get lucky and survive, but unless most fish of your kind and their descendants are subject to frequent and ongoing fish-hooks, the means by which you achieved your victory will not be passed on to most members of future generations of your species of fish. Nor, if you are an uncommonly trapped nonconscious muskrat, will your genetic panic response profit your fellow muskrats' descendants.

Other than having the equivalent of an innate panic response, is there any other kind of genetic adaptation that could help a nonconscious creature cope with a sudden uncommon pseudopainful situation? There is one: the ability to learn quickly, to learn immediately to do something like trying anything to improve the situation. If the nonconscious you were an animal with a rare quick-learning-in-troubled-times mutation, you might be able to cope with outrageous misfortune as well as a conscious creature can. But for quick learning of the kind we know of in higher mammals, a larger and more complex brain is required. Whereas such brains have evolved only in slow smallish increments. They have evolved in animals that are already equipped with consciousness—and thus in less need of such a mutation for the particular purpose of coping with the likes of fish hooks and muskrat traps.

So neither a mutation for a panic button nor one for rapid-learning is likely to have been adaptive for a zombie version of a fish or a muskrat. Such mutations are highly unlikely to be built into such creatures. Thus the above panic scenarios are simply unlikely to be true for nonconscious fish and muskrats.

On the other hand, if our poor fish and our miserable muskrat are conscious creatures, their situation is quite different. They are in pain. To me, that means they might do anything, anything at all that has any possibility of changing their situation and, perhaps, even alleviating their distress. Swimming against the hook's force might pull it, with or without some of your flesh, from your body. Gnawing off your own limb might similarly free what remains of you from the trap. It's true that struggling against the hooks and traps might not help you—consciousness as an adaptation provides no promise of victory over outrageous fortune; it might not save you. However, it might well increase the likelihood that more of your kind will survive.

My point here is more about panic behavior than it is about adaptive value. The point is that panic behavior is not likely to be part of the repertoire of a nonconscious creature. Whereas a conscious creature, in response to pain, will take up arms beyond its normal repertoire of behavior. Intense pain will make it seek desperate out-of-repertoire solutions.

Thus, pain may be a magic bullet that will show whether or not a given relatively simple creature is conscious. Even if pain were not the only or the best adaptive response, it could still have arisen in and been useful to animals, according to Robert Rosenfeld's article "Parsimony, Evolution, and Animal Pain."[28] To me, a hooked fish's abnormal leaping and tugging is out-of-repertoire and thereby shows that the creature is conscious. Likewise for the trapped muskrat that chews off its limb to free what's left of itself. A nonconscious fish or muskrat would respond to uncommon adversity by staying within its normal repertoire of behavior; staying within that repertoire until many similar adversities results in some slow learning and behavior change.

Might it be that the above network of ideas has holes? For example, could there be creatures, perhaps ants or the above-discussed wasp, the Sphex, that have awareness accompanying some activities, but that are not equipped by nature with the means to experience pain? More importantly, what if random mutation in one stroke brought about what to us seems two wildly different things: what if it brought about a certain kind of brain configuration or process with new neural capabilities of its own, and also brought about the capacity to be conscious? Then how could the latter be separated from the former? Then wouldn't my attempt to compare two grunts that are mostly identical except for consciousness amount to separating inseparable Siamese twins with but one brain and mind?

THE ENIGMA OF CONSCIOUSNESS: HOW IS IT BASED IN THE INDIFFERENT PHYSICAL WORLD?

I have already confessed a lack of worthy ideas on the enigma of consciousness, its presumed but hardly imaginable or understandable relation to the indifferent physical stuff that is fundamental to all things. I'll try now to make as short a shrift as possible of my understanding of work on this problem in the area of physical science. Before so doing, I'll attempt to inoculate you against losing your sense of "reality": John Gribbin warns that "reality" in the everyday sense "is not a good way to think about the . . . fundamental particles."[29] Also, Nick Herbert writes, "the Copenhagen interpretation holds that in a certain sense the unmeasured atom is not real; its attributes are . . . realized in the act of measurement."[30]

Roger Penrose has searched assiduously for a fundamental connection between consciousness and the basic quantum-physics enigma of the peculiar semireal existence of fundamental particles.[31] He sees a parallel between the two states of quantum particles—I'll call one "nonreal" and the other "real"—and the two potential states of certain brain activities, nonconscious and conscious. According to my reading of Penrose, it may be more than a parallel. Perhaps nonreal particles in the brain are somehow responsible for nonconscious brain activity. Likewise, real particles in the brain may be responsible for conscious brain activity. Penrose further suggests that some future theory of quantum gravity may shed more light on a connection between the quantum world and the conscious world. It remains to be seen whether such a theory will prove even less real than are quantum particles.

Some scientists have attempted a different kind of solution to the enigma of consciousness. Theirs seems to be a psychological solution. Note that there are unsophisticated people who would regard, say, a human-made satellite as enigmatic. With time and learning, the enigma will disappear for them. Suppose consciousness results from a certain kind of, say, reverberation in a group of neural networks (Edelman's idea[32]). These scientists might suggest that there seems to be an enigma only because we have been unfamiliar with the solution. As we get used to such a theory, perhaps it will seem less and less an enigma; it will be less and less necessary to view any profound problem to exist. Daniel Dennett, for example, in *Consciousness Explained*, seems to suggest something like that.[33] To burlesque him a bit, he seems to think that as we get used to the idea that we only seem to have consciousness, we will realize that the latter is not a profound problem—it is a nonexistent problem.

For all that, proclaiming the basis of consciousness to be "thus and such" has been done in the past. Humphrey accepts the idea that consciousness may be based on some kind of neural reverberation.[34] What if it is indeed demonstrated about consciousness and such reverberations that whenever you have the first you have the other and vice versa, but no explanation is developed of why this is so? Would it be "fair" to demand more? What if we were told, "Don't ask why or how it happens; we can't answer such questions"? Wouldn't there then be a further (unanswerable) question: "Is that the case because our human minds are too limited, or is it because there cannot be an answer, even for beings far brighter than us in a galaxy billions of light years away?"

Although it's hard to argue with some who write about consciousness, with others—such as Dennett in *Darwin's Dangerous Idea*—the task is easier. He writes about *intentionality*, the capacity of mentation to point beyond itself, to have content, import, meaning, to be about things. He seems to believe the "real" intentionality of conscious creatures and the "derived" intentionality of computerlike machines are the same thing. This lets him offer us two choices. First, "you should be ready to draw the . . . conclusion that you yourself never enjoy any states with *original* intentionality, since you are just a survival machine designed, originally, for the purpose of preserving your genes until they can replicate. . . . The intentionality of our selfish genes . . . [makes *them*] the Unmeant Meaners, not us!"[35] As the presumed intentionality of our presumed selfish genes results from his (and Dawkins's) projecting selfishness into them, this is hardly a satisfactory explanation. Dennett's second option is, "Acknowledge that a fancy-enough artifact . . . *can* exhibit real intentionality. . . . It, like you, has taken on a certain autonomy, has become the locus of self-control and self-determination, not by any miracle, but just by confronting problems during its own 'lifetime' and more or less solving them." This would be okay only if this robot or android is made of stuff and processes that allow it to experience part of its life and *thus* have intentionality. So, as we have no idea of how such a fancy artifact could be made, it's not really okay.

Much as we'd like to, we have no knowledge regarding the nature of the enigma of consciousness. As Chalmers puts it, "Consciousness is the biggest mystery. It may be the largest outstanding obstacle in our quest for a scientific understanding of the universe."[36] With no knowledge, we have only words that, as symbols, do not point to anything useful; we have verbosity.

Some of us also have hope. In *Leaps of Faith*, Nicholas Humphrey expresses some: "All but a few contemporary psychologists agree that there will eventually prove to be some sort of satisfactory monist theory of mind-brain relationships; a theory, that is, which does succeed in showing how mental activity and brain activity can be one and the same thing. But at present there really is very little consensus about the form, let alone the substance, of this theory-to-come."[37]

NOTES

1. Steven Pinker, *How the Mind Works* (New York: W. W. Norton & Co., 1997), p. 60.

2. Donald R. Griffin, *The Question of Animal Awareness* (New York: The Rockefeller University Press), 1981.

3. Joseph Mortenson, *Whale Songs and Wasp Maps: The Mystery of Animal Thinking* (New York: E. P. Dutton), 1987.

4. Marian Stamp Dawkins, *Through Our Eyes Only? The Search for Animal Consciousness* (New York: W. H. Freeman and Company, 1993), p. 8.

5. A. G. Cairns-Smith, *Evolving the Mind* (New York: Cambridge University Press), 1996, p. 255.

6. John R. Searle, "The Mystery of Consciousness," *The New York Review of Books*, 2-16 November 1995.

7. Donald R. Griffin, *Animal Minds* (Chicago: The University of Chicago Press), 1992, p. 10.

8. Steven Mithen, *The Prehistory of the Mind* (London: Thames and Hudson), 1996, p. 190.

9. Harry J. Jerison, *Brain Size and the Evolution of Mind* (New York: American Museum Of Natural History, 1991), pp. 4, 21.

10. Daniel L. Pals, *Seven Theories of Religion* (New York: Oxford University Press, 1996), p. 25.

11. Paul Kurtz, *The Transcendental Temptation* (Amherst, N.Y.: Prometheus Books, 1991), p. 404.

12. David J. Hufford, *The Terror That Comes in the Night* (Pennsylvania: University of Pennsylvania Press, 1982), p. 255.

13. Gerald M. Edelman, *The Remembered Present: A Biological Theory of Consciousness* (New York: Basic Books, 1989).

14. Antonio R. Damasio, *Descartes' Error: Emotion, Reason, and the Human Brain* (New York: G. P. Putnam's Sons, 1994), p. xviii.

15. Marvin Minsky, *The Society Of Mind* (New York: Simon and Schuster, 1986), p. 19.

16. Pinker, *How The Mind Works*, pp. 21, 24, 562.

17. Maxine Sheets-Johnstone, *The Roots of Thinking* (Philadelphia: Temple University Press, 1990), p. 308.

18. Daisie Radner and Michael Radner, *Animal Consciousness* (Amherst, N.Y.: Prometheus Books, 1989), p. 192.

19. Charles Darwin, *The Expression of the Emotions in Man and Animal*, 3d ed. (New York: Oxford University Press, 1998).

20. Peter Harrison, "Do Animals Feel Pain?" *Philosophy* 66 (1991): 25-40.

21. Stephen Jay Gould, *Full House* (New York: Harmony Books Division of Crown Publishers, 1996), p. 21.

22. Mortenson, *Whale Songs and Wasp Maps*, p. 43.

23. Nicholas Humphrey, *A History of the Mind* (New York: HarperCollins Publishers, 1992), p. 214.

24. Benjamin Libet, "Subjective Referral of the Timing for a Conscious Sensory Experience," *Brain* 102 (1979): 193.

25. John Searle, *The Rediscovery of the Mind* (Cambridge, Mass.: The MIT Press, 1994), p. 109.

26. Daniel C. Dennett, "Precis of The Intentional Stance," *Behavioral and Brain Sciences* 11 (1988): 495–546.

27. David J. Chalmers, *The Conscious Mind* (New York: Oxford University Press, 1996), p. 293.

28. Robert P. Rosenfeld, "Parsimony, Evolution, and Animal Pain," *Between Species* 9, no. 3 (summer 1993): 133–37.

29. John Gribbin, *In Search of Schrödinger's Cat* (New York: Bantam Books, 1984), p. 4.

30. Nick Herbert, *Quantum Reality* (Garden City, N.Y.: Anchor Press, 1985), p. xiii.

31. Roger Penrose, *Shadows of the Mind: A Search for the Missing Science of Consciousness* (New York: Oxford University Press, 1994).

32. Edelman, *The Remembered Present*.

33. Daniel C. Dennett, *Consciousness Explained* (New York: Little, Brown and Company, 1991).

34. Humphrey, *A History of the Mind*.

35. Daniel C. Dennett, *Darwin's Dangerous Idea* (New York: Simon and Schuster, 1996), p. 425.

36. Chalmers, *The Conscious Mind*, p. xi.

37. Nicholas Humphrey, *Leaps of Faith* (New York: Basic Books, 1996), p. 195.

2

FISH, FOUR-LIMBED BEASTS, AND BIRDS

THREE OF NATURE'S COMMANDMENTS, EGOCENTRICITY, COMMUNALITY, AND SEXUALITY: PERHAPS FIRST ISSUED TO FISH

A GLANCE AT OUR FAMILY tree tells us that once upon a time we were apes, and long before that we were grungy little mouselike mammals trying to make a living where small dinosaurs might not notice us. Long before that, our kind of folk were just plain fish, like lots of folks were back then. Besides a backbone, one thing all these ancestors of ours had in common with us was a central nervous system designed to help them stay alive.

In my hypothesis, all creatures that have a central nervous system can be conscious of some of the things with which that nervous system is involved. When its brain is preoccupied with getting food, a fish may feel hunger, and then, if it gets lucky, maybe brief satisfaction with a tasty bit of lower or left-over life. When its networks of neurons are attending to a passing predator, the fish may feel fear and the strong intent to move elsewhere.

Thus, the first thing to say about our species' conscious fishy ances-tors is that they were egocentric in behavior. The old Mother God, nature, may have nudged them out of the nest (by giving them consciousness), but she didn't let her head go to her heart. So nature's first command-ment to her at-first feeble-minded conscious offspring was: "Thou must look out for number one."

HOW WE GOT TO BE HUMAN

A confession on my part is called for here. I enjoy the use of figurative language (metaphor) such as referring to Mother Nature and to her commandments above. Likewise, I find it refreshing when a scientist does the same, as when Steven Pinker tells us the human brain does *computation*.[1] (At a more factual level, it is not our brains, but *we* who compute.) Such usage does more than enliven what might without it be dull discourse. More on this subject is said by Alan Wolfe: we can't "capture the complexity and interconnectedness of human society without metaphors."[2] This very power of metaphor, however, can lead us astray when it tempts us to take it too literally. Thus, the novelty and capability of computers moves us to accept Pinker's words as simply true. However, the notion that what our brains do is "compute" is no more true than is the idea (which once excited us) that we are run by "clockwork" brains. My own use of commandments from Mother Nature is intended to be too pseudobiblical to deceive anyone about it having any literal truth.

As for "looking out for number one," it is doubtless true that not only conscious animals but also nonconscious creatures (and even some plants) exhibit activity that seems self-protective. Nonetheless, such creatures and plants are not egocentric. How can they be? "Inside" they have neither egos nor selves, nor for that matter, anything else. They are only zombies, going through the motions, motions that may indeed extend their lives but motions that they do not experience.

Another commandment from nature to the creatures of some of the species that experience their lives, including those in our own line, was: "Thou shalt make the important part of thy life with others of thine own kind."

These ancestor fish of ours may have initiated the genetic change that made them and us communal. Each individual fish felt better (safer or more at ease) in the midst of a community of its own kind. Fish swimming in "schools" thereby provide the evidence of their communality. In such fish schools, as in bird flocks and cattle herds, the animals travel together, without being led by a leader and without intended communication, by staying close to neighbors and steering away from trouble.

Note that being communal is already a departure from pure egocentricity. It may conflict with the desire for food or safety. If you are a laggard communal fish and a predator stands between you and your fellows, you are in conflict about which way to swim. The implications of conflict within the mentality of an animal will be explored throughout this book. Here, I'll say that the way to make sense of conflict is to posit not one integrated mentality within, say, a fish, but several relatively unintegrated mentalities, "subminds," each with its own attitude, its own axe to grind. Another consequence of communality and its balancing act with egocentricity is the "pecking order," the hierarchy that develops in communal creatures. That, too, will be examined in this and later chapters.

One more commandment from nature completes the triplet that provides some of the basic rules for the game of experiencing life: "When opportune, seek sexual satisfaction with a suitable other."

Like "higher" creatures—those with four or two legs, with whom it is easier for us to identify—when one of a couple of fish releases eggs and the other releases sperm to cover those eggs, they both may find it gratifying. This may disappoint (or more likely relieve) you, but I won't try to guess just how fish experience the joy of sex. However, the significance of sex for conscious creatures will be examined more and more closely in these pages as mentation expands.

Lacking capabilities such as abstract concepts and symbols, our fishy forefathers had no mental tools with which to make recollections of things past, or fantasies of things future. They lived within a right-here-and-right-now world. The nervous system was driven by the current state of the glands and organs, and by what currently confronted the eyeballs and other sensors. So their cognition was limited to the current brief episode or situation.

You may be thinking here: "Not remember things? Then how could they learn? Surely they could learn something." Well, yes, they can learn, and to learn, one must remember. Merlin Donald discusses two kinds of memory, procedural and episodic.[3] The more archaic procedural memory "preserves the generalities of action, across events." Episodic memory "preserves the specifics of events." Fish memory is procedural, a remembering of how to do things, one that shows up only in the next doing. Like our own remembering of how to stand and walk upright, it is not episodic, and it is without recollection. In our case, it is without recalling either the details of balancing or the falls and frustrations that were party to the learning.

Let me hasten to add that there are no straight lines or sharp divisions in nature. So we'll see exceptions to the animals that live exclusively right here and right now. We'll take note, later in this chapter, of some of the birds and apes who are takers of upward pathways that require concepts in their mentation.

LOWER ANIMALS: ALMOST NO USE OF ABSTRACT CONCEPTS, GESTURES, SYMBOLS

It is the quality of mental life that distinguishes human life from that of lower animals. By that I mean that lower animals, by and large, have no use of (more-or-less-abstract) concepts and the gestures, deceptions, and symbols needed for concepts to leave their marks in the world. It is the lack of such capability that restricts lower animal life to the immediate here-and-now. Without the concept—in what passes for your mind—for a different place or another time, you can't deal with either. Nor, without such means, can you comprehend and cope with the complexities of those of your kind. We will see some evidence for this viewpoint below.

Before we get to that testimony, it is worth noting that some who write about the life of animals seem to have the implicit view that many animals can pretty much do whatever humans do in the way of using their minds. They do the same kinds of things as we do, but fewer of them, or they do them on a simpler scale. One recent book (by Jeffrey Masson and Susan McCarthy) suggests that lower animals may experience love, compassion, and loneliness, and they might also make and enjoy art.[4] I must suggest that, for animals lacking the ability to use concepts, the opposite is closer to the truth. We'll see later that lower animals, even apes, do not experience reflective consciousness—awareness of their own mental processes. They have no awareness that they themselves have mentality or mind (although some apes have awareness, evidenced by their responses to mirrors, that they have bodies). So we must ask: How can love and compassion exist within a creature that doesn't know that it, and thus its presumed love object, have any mental processes or mind? Come now, that can't be so. Loneliness in creatures that lack the means to think about any object that is out of sight or scent? Please think again. Creation of art from a creature that doesn't know that its experiences can be viewed as the basis for the comedy and tragedy as well as the beauty and banality of its life? Surely this is not so.

Yet it must be admitted that the mentation of nonhuman creatures is a very slippery concept for us to grasp. Although less than likely, it is conceivable that some of them, even the denizens of the briny deep seas, could use complex concepts and symbols. We know that some observers credit dolphins and whales—who are after all, mammals—with such humanlike capabilities. But what about a fish—say, a big fish and a bad one like a shark? Could it have concepts about what to eat and what not to eat?

The most chivalrous fish of the ocean,
To ladies forbearing and mild,
>Though his record be dark,
>Is the man-eating shark
Who will eat neither woman nor child.

He dines upon seamen and skippers,
And tourists his hunger assuage,
>And a fresh cabin-boy
>Will inspire him with joy
If he's past the maturity age.

A doctor, a lawyer, a preacher,
He'll gobble one any fine day,
>But the ladies, God bless 'em,
>He'll only address 'em
Politely and go on his way.

This lark in its entirety is called "The Chivalrous Shark."[5] Leaving its likelihood aside, the road to human life is paved with concepts representing good or bad perceptions and intentions. Those increasingly complex concepts allow creatures to conceive of things that don't meet the eye or any other sense. Lower animals cannot think about things like elsewhere-others or superiority and inferiority, let alone things like minds within bodies, or like love, compassion, and loneliness. All such things are abstractions unavailable to the attending animal. Even young-enough human children lack the means to pay them heed. Direct representations of the outside world that feed direct intentions can't begin to do this kind of dimension-adding to a flat life. To understand abstract entities, concepts of such things and processes are needed within the mentation, as are gestures and symbols needed to represent them to others in the outside world.

Now let's observe some things lower animals do—and also, some things they are said by observers to do—and see how they can be explained as the behavior of egocentric communal sexual creatures with concerns limited to the here-and-now, conscious creatures without abstract concepts.

Territorial animals, for example, are often said to "mark" their territory boundaries, commonly by urinating, to warn away others of their kind. This notion may have been originated by pioneer ethologist Konrad Lorenz in *King Solomon's Ring*. He wrote there of the wolf victor of a fight urinating to mark the battlefield.[6] But such message-sending is symbolizing behavior oriented to the past and the future. It presumes a mental quality that is beyond such creatures. A simpler and other-things-equal better explanation is that such an animal gets nervous as it approaches the boundary of its territory. Susanne Langer explains that this is because things get less familiar and the unknown is upsetting.[7] When an animal gets excited in this way, it often follows an ancient—but hardly hallowed—custom. It prepares for flight or fight by lightening its load, that is, it urinates or defecates.

What about other animals' reactions to these evacuation products, products that need not be markings? As anyone who has seen a dog near a tree knows, animals are especially interested in the scents provided by their own kind, so they will pause to sniff the odor of pee or poop, and need not be "reading warnings" from whoever left the scent. One bit of information we can glean from Harry Jerison's *Brain Size and the Evolution of Mind* is that (relative to brain weight) dogs and other four-legged mammals have olfactory bulbs that are huge compared to those of humans.[8] This enables them to sniff out knowledge of prior visitors to the local area, and something about what they ate. It's true that such "markings" might also be nonconscious instinctive activity that can serve an adaptive purpose of warning. However, there is no reason to believe these animals experience them as serving such purposes, and it is mental activity—the experiencing of life—that is our main subject matter.

It was Lorenz (in *King Solomon's Ring*) who first gave us the story suggesting that wolves engage in symbolizing behavior after they fight. The basic story has the losing wolf rolling onto his back and exposing his throat to the winner. The winner, seeing that behavior, walks away instead of killing his foe. In *The Roots of Power*, Maxine Sheets-Johnstone buys right into the same story: "A vanquished wolf . . . may either twist and lower its head, offering its neck to the victor who is above it, or it may lie flat in front of the victor."[9] Then she goes on to say, "That domestic dogs use the same gestures of submission is common knowledge."

I must disagree with Lorenz about wolves employing gestures. I must disagree with Sheets-Johnstone about it being common *knowledge*, rather than common *illusion*, that dogs and wolves use gestures. Implicit in the anecdote from Lorenz, and explicit in Sheets-Johnstone's account, is the idea that a wolf can make a gesture in the sense of a *deliberately significant motion*. Although the word "gesture" has a broader meaning as well, it is this meaning that I will use. "Deliberately" implies that the behavior in question has a *mental* component, and "significant" implies that the behavior in question has *conceptual* mental activity behind it.

A simpler explanation for the wolves' behavior suffices. The losing wolf rolls to his back to permit defensive fighting with his legs as well as with his jaws. This results in a stalemate, or at least it staves off total defeat, because the winning wolf has only his teeth as weapons. Supporting this simpler explanation is the evidence that soon after such an event, the animals are often at each other's throats again. If they had the intellect to use such gestures of submission and superiority, they would be able to remember and abide by them for a while at least, as apes do.

It is Donald's remarks (in *Origins of the Modern Mind*), namely, that "apes in the wild do not possess even a rudimentary system of voluntary gestures or signs,"[10] with which I now must disagree. I can find support for this disagreement in *Gesture and the Nature of Language*: "There can be little doubt that chimpanzees have well-developed abilities to communicate using signs (symbols, indices, and icons)—whether gestured or manufactured."[11] As will be elaborated on in chapters ahead, chimps sometimes make themselves *small* (by getting on all four limbs) or *large* (by getting up on their legs). Sometimes such behavior is an *iconic* gesture that suggests actual smallness or largeness. The same behavior can be an *indexical* gesture, small size suggesting little power, large size suggesting great power. That apes seem to understand the significance of such moves is evidence that their behaviors represent conceptual knowledge.

To return to our friends, the dogs, and to our enemies, the wolves, symbolizing behavior is something they cannot do. Dogs and wolves, like other creatures with a pecking order, do remember and act on their position in the hierarchy vis-à-vis another. They commonly engage in submissive behavior in the presence of a higher ranked animal. But that behavior is not optional nor is it done to convey information. It is not any kind of symbolizing activity, it is not gesture. Rather, it is behavior directly expressive of the animals' certain knowledge: they are low and weak, the "top dogs" are high and mighty, and the closer they get to those high-ranked animals the more they will feel safe and secure.

FEMALE SEXUAL HEAT, ESTRUS, AND
MALE RUT—THE COOLIDGE EFFECT

For the survival (within future individuals) of the genetic material within the cells of an individual animal, sexual activity is certainly valuable, but lower animals do not experience the knowledge that they have cells with genetic material, let alone the wish that such stuff survives. Nonetheless, some of the high points in the experiences of lower mammals must be associated with the S-word, sex. So nature's third commandment (transformed from a phrase for fish to one for lower mammals) reads: "When thine privates itch, seek relief by rubbing against another's mating part." (This commandment and the others are only figurative fancies. Nor are the instincts behind them specific and literal instructions; rather, they are genetic patterns which individuals of each species form into what seem like rules.)

Sexual opportunity knocks for such animals only when the female is "in heat," a condition otherwise known as estrus. The word "estrus," which refers to the portion of the ovulating cycle during which the female desires sexual activity, comes from the Greek word for gadfly, an insect that lays eggs under the skin of cattle. The eggs soon become irritating, squirming larvae. The word "estrus" suggests the females are itching for sexual relief; it does not imply any *conceptual* activity regarding sex.

What about the sexuality of the males? Maxine Sheets-Johnstone has some insightful things to say about the examination by scientists of the sexuality of females versus that of males (although her remarks are about primates and early hominids, they are of interest here as well): "Male genitalia are seldom remarked upon as a separate topic on par with female genitalia. In contrast with sexual swellings in females, virtually nothing is said of sexual swellings in males."[12] I interpret these and other related remarks of hers as suggesting that ethologists, primatologists, and other social scientists often neglect to see that they themselves are part of a culture that still has roots in patriarchy, and thus has prurient interest in female sexual organs, but not in those of males.

About sex (sex for lower animals at any rate), one thing is certain. As members of most human couples—especially those who find pleasure in South-American-dancing competitions—after a fracas may tell each other, "It takes two to tango!" For the male lower mammal, the sight or scent of an in-heat female is extremely attractive, attractive in almost the literal sense whereby iron filings are attracted to a magnet. Estrus is a signal that the female is receptive—its apprehension makes the male in turn go into *rut*, experience overwhelming desire, and itch for sexual relief, again without implying any conceptual activity.

As it happens, we are fortunate in having another word about sex, "prurience," a broader word than either female estrus or male rut, a word that not only refers to both sexes, but implies the possibility of conceptual activity as well as desire. To be prurient is to have lustful *desires* and/or to have lustful *ideas*. The word will be useful when we advance to apes, and even more useful when we look at humans.

Female estrus induces male rut in lower animals that are in our line. Both are of genetic rather than learned origin. Also, both are (decidedly) experienced. However, the evidence from their behavior suggests that the males usually experience desire more intensely than do the females. When we look at the sexuality of apes, we will see that there is reason to believe apes have prurient concepts in addition to prurient desires. Later, in examining our predecessors, the upright-apes and early Homo species, we will see that, after estrus was genetically abandoned, female prurience had become, for the most part, an *option* available to individual creatures. In contrast, the rut-retaining males—far from giving up prurience—learned to expand it; part of their mentation was engaged by sex much of the time.

For lower mammals, however, there are no concept-implying symbolic sexual messages sent by one and interpreted by the other. The males have to contend and fight with other males, but such struggles are not symbolic. Such males are not sending optional messages to one another; they have no choice but to struggle for what they want. Although a female's condition of being in "heat" is, in one sense, an overwhelmingly loud message to the male, she has no option regarding its transmission, nor is it her intent to send it. In the language of philosopher Susanne Langer, that message is a *signal* rather than a *symbol*. Langer also comments that symbols are instruments of conception that are not used by animals. However, I think that if she had had access to recent work by scientists on ape gesture, she might have changed her mind about symbol-using by apes, monkeys, and even some birds. Without symbols or the concepts they represent, animal sexual activity is only here-and-now, if that term includes the distance and time needed to track down what an animal's eyes or nose tells it is a "significant other."

For many lower creatures, male sexuality has come to be referred to as the Coolidge Effect. The effect is that such males are always interested in estrous females. After finding a female who is in heat, and after one or more copulations with that female, the male may lose sexual interest in her. If he finds a new estrous female, however, he resumes full interest and regains full performance. In contrast, females in heat remain interested in copulation without much regard for whether the male is "old" or "new."

The name comes from a supposed account of a visit by President Calvin Coolidge and his wife to a chicken ranch. Donald Symons recounts this near-legend of the presumed repartee between the president and his wife.[14] Mrs. Coolidge reportedly asked if the rooster she was observing copulated more than once a day. When told that it indeed did so, dozens of times, she then asked that her husband, who was behind her on the tour, be told about the rooster when he arrived. When the president showed up and was told of the rooster's frequent sexual activity, he in turn asked if it was always with the same hen. When told it was a different hen each time, he suggested that the information quickly be relayed ahead to his wife. It is to laugh: Har-har and tee-hee. (Although some of our recent presidents may seem to have demonstrated the Coolidge Effect, neither heart-felt lust for forbidden females nor Oval Office hanky-panky is a qualifier for roosters' league sexuality.) The Coolidge Effect is of importance to us here because it is another variation of the genetic-originated rut that is basic to the prurience of hominid male sexuality. We will look at this male rut and prurience in greater detail when we explore the sexuality of apes, of our upright ape and erectine predecessors, and also—I must not shrink from duty—that of humans.

There is a Darwinian adaptive explanation that has been developed for the difference between male and female sexuality in many animals. It goes as follows: The female ripens but one large nutrient-full, and thus "expensive," egg at a time. A female's heat is nature's way of gaining a high likelihood that the egg gets fertilized by some male's sperm, so that a new creature with some of her genetic material will come to life. In contrast, the male has a large stock of small, "low-investment" sperm. Nature directs him to broadcast them widely to increase the likelihood that his genetic inheritance will survive him. This sounds more than reasonable. When fleshed out as a full explanation with some supporting evidence, it may well be as scientific as we can expect social science—which must deal with complexities foreign to the physical sciences—to get. We will encounter other seemingly scientific explanations, some of them with little flesh between skin and bone, in the pages to come. In a field as far as this one from what seem to be the "solid flesh" physical sciences, many people who study animal behavior, especially behavior with mentation behind it, seek to put meat on bones by claiming, even when the claim is not clearly strong, a piece of the body of Darwinian evolutionary explanation.

Lower animals live largely in the local present. However, some of them have some part of their lives that are not so limited and confined. How do they do that? How do they get out of the local jail where others serve life sentences? There are two ways for creatures to break out of the here-and-now, to break into interesting elsewhere places, future events, and to have more ways of interacting with others of their own kind.

The first way amounts to new developments of innate biological capability, otherwise known as good old reliable instinct. One example of such enlarging instinct is migration, wherein creatures—whose ancestors traveled in daily closed paths—seasonally travel huge distances to and from greener pastures as a result of accidental mutations that proved to be an adaptive response to the separation of continents.

The other, and more interesting, way out of the here-and-now is the ability to use more abstract concepts, together with gestures or other symbolizing activity such as deceptions, to see behind and beyond things as they are right now and right here. A fascinating example of this (which will be closely examined below) is the apparent ability of some ground-nesting birds to lure predators away from their nests with what's called the broken-wing ploy, whereby they behave as though their wings are not in working order.

Merlin Donald discusses human emblematic gestures, where emblems "are stereotyped signs that can be interpreted outside of language."[15] He notes that, among different cultures, "the same representational functions are served by a variety of gestures." In this chapter, I suggest that some birds have both gestures and concepts. When a female plover uses several different behaviors that result in predators failing to find her nest or nestlings, that suggests both that those behaviors are gestural and that behind them is a concept in the bird's mentation. In *Animal Minds*, Donald Griffin discusses the complex "broken-wing" behavior of the plover.[16] However, he associates it with the seemingly complex bower-building of the bowerbird and the seemingly complex "food-dance" of the honeybee. I think the bower-bird's building is relatively rigid, although not as rigid as the endless nest reinspection of the wasp Sphex, nor as rigid as is the bee's dance. In contrast, the plover adapts her behavior to suit that of the predator she seeks to divert. That makes her behavior a gesture.

There may be something about using concepts about concepts that sends the user "over the top." I think Robert Wright, in *The Moral Animal*, sometimes does as much. For example, he discusses the differences in care by parents of children who are "natural" versus those that are adopted. "[Evolutionary psychologists] have written, 'Perhaps the most obvious prediction from a Darwinian view of parental motives is this: Substitute parents will generally tend to care less profoundly for children than natural parents. . . . Parental investment is a precious resource, and selection must favor those parental psyches that do not squander it on nonrelatives.'"[17] A simpler explanation is available, one that does not make the relentless demand (usually made by evolutionary psychology) that adaptive genetic roots, branches, and flowers exist for most things mental. Larger and larger brains were successively adaptive. Individual humans, all with large brains, are given to having ideas and concepts. Some human parents may conceive their adopted children as less "theirs" than their natural progeny. With others, that's not the case. Why make anything more of it?

BIRDS OF MANY KINDS: THEY ALL LIVE MOSTLY IN THE HERE-AND-NOW

Before examining birds of different feathers and other fancy features, a brief explanation is in order as to why they are included in a chapter that purports to examine animals that might be like some of our own animal ancestors. The simplest way of relating animals is to see how they are related on what is called the phylogenetic tree. In a loose way, these chapters are organized along this idea of descent of one kind of creature from another.

There is another way of looking at the development of increasingly complex creatures. That other way suggests that there are stages of development or complexity—apart from a single connected line up the branching tree of evolution—into which animals can be sorted. I have suggested consciousness is a first giant step in the development of mentation. As discussed in the previous chapter, we can't even point to where consciousness arose in a single evolutionary line, let alone point to its possible coming to life at points in several such lines. I have suggested that a subsequent expansion, via what might be a set of steps in different directions, allowed animals to begin to use abstract concepts to escape from lives limited to the local present. Now I plan to show how some creatures begin to use concepts—evidenced by gesturelike behavior—to direct their activities to elsewhere and to the future. Ethologists' studies of some birds are handy sources of evidence for some of this beyond-the-here-and-now living. There may not be any lower mammals that have similarly expanded their outlook, so I propose we now take a tour of birdland.

Let's start by first stepping back to a couple of here-and-now-ish birds, hens and bowerbirds. Hens can't cherish it, but they have the honor of having initiated, in our thinking, the concept of pecking order as hierarchy. As those of us unacquainted with live hens can learn from Marian Dawkins, when unacquainted hens are first placed together, the first thing they do is fight, fight hard, and often they draw blood.[18] It takes them weeks to sort out who can lick, or rather peck, whom. Thereafter, each knows where she stands and when she must stand aside from food and roosting sites or else get pecked.

The presence of a pecking order in a species implies the existence of a communal impulse, a biological force on each of the creatures to stay together, but there is no reason to believe the impulse to form a hierarchic order is itself of hereditary origins. Rather, that pecking order comes from the mentation and the learning of the individual creatures of a species with sufficient brain power. Members of another communal species, perhaps of fish that swim in schools, may not have enough brains to use their mentality to satisfy both their egocentricity and their social nature by recognizing and making pecking orders. Thus, the concept of such a social order—if it exists in chicken mentation—is an abstraction in itself from the bare facts of the physical world of fowl, which shows big and small hens, but no pecking order.

The pecking order, however, may or may not be a concept (the word "concept" is used in these chapters to denote abstractions or general notions that are conscious) for chickens. It may be only a nonconscious pattern, an unexperienced abstraction formed by the chicken's neural networks. Nevertheless, these birds live a more complex life when their world has not only physical facts, but a social fact that you can't directly point to, such as the hierarchy of the peck order.

Bowerbirds live in New Guinea and Australia. In *The Animal Mind*, James Gould and Carol Gould tell us all about them.[19] The males of this species build "stages" in mating season, on which they place large numbers of colorful small objects. The males also gather berries and with their beaks press them to "colorize" the stage further. When they can find burnt wood they use it, mixed with saliva, to similarly "decorate" the stage. A female seems to compare both bowers and male display behaviors to aid her in selecting a male who will father her offspring. Once she has found her male and had it off with him, she leaves both him and his bower. By herself she builds a nest and takes care of both eggs and progeny.

Some people like to think such bowers are works of art. Certainly, they may look as artful as the work of some human artists. But bowers are no more art than are colorful sunsets. It is people, at least some of them, who may find bowers and sunsets more moving than much of the work of modern artists. Birds don't have the ability to conceptualize their world as humans do. Nor can they use their own activities to represent what is significant about their world. So why should we dilute the word "art" by spreading it out thin enough to cover a bowerbird's droppings of damp objects? As for bowers providing more than a minor enlargement of life to the birds that build and select them, I doubt that. For bower birds, it's still the same old story; the fight for love and glory is not conceptually different with bowers. Nor do the concepts "love" and "glory" have meaning for any bird or four-legged beast.

Bowerbird females act as though their motto for selecting males is "by their works shall ye know them." In contrast, female black grouse (discussed by Marian Dawkins) act out the assumption: By their handsome white tail feathers, as they spread and fan them, shall ye know the best males.[20] Ethologists who study these birds have found, that by one criterion, the females do a better job of male selection than the bird-watchers themselves can. This criterion is the likelihood that the males will still be alive six months after selection. It presumably suggests a better combination of nature and nurture within the survivor males. By means unseen by the watchers, the females generally picked males who would survive at least a half year. The ethologists themselves could not pick likely male survivors nearly that well. For all that, there is no reason to think the black grouse lives elsewhere than in the here-and-now. The females, however, do see things, here and there, now and again, in their males that are not apparent to their human watchers.

In contrast with the birds above, male white-throated sparrows have added something new in their fight for love and glory. There is evidence (described by Marian Dawkins) that, in the breeding season, the males recognize the songs of other particular males.[21] In itself, that may not be exceptional. More clearly unusual, males will take vocal issue (by singing in turn) with the issuer of such songs only when the other singer in fact represents a threat. The males set up more or less adjacent breeding territories. Once they are so established, those possessing adjacent territories are not a threat to a particular territory holder. (This presumes that none of the latter's neighbors got stuck with a really ratty hunk of real estate, and that none has a raunchy red neck under the white throat feathers and is eyeing the well-to-do settler's land and the mate he has attracted.) So a landholder does not respond in kind when he hears his neighbors sing the songs he has learned are theirs. The only real threat under normal conditions is from a landless outsider. When such a stranger arrives nearby, evidenced by his unfamiliar song, a landholder sings his own hostility, which usually sends the carpet-bagger packing. Sometimes threat is not enough, so the owner has to chase the stranger away.

A nosy ethologist noted all this and then, as diabolical as many of that breed, made tape recordings of individual white-throated male sparrows, so to enable him to mess with the minds of those males. The ethologist learned that when a particular recorded song was played from the recording artist's territory, the male under evaluation did not respond. When the same song, however, was broadcast from a location where it did not belong, the subject male responded as vigorously as he did to interloping strangers. From this, it is clear that this sparrow has added a new note to his fight for love and glory. He has somehow learned that "you gotta know the territory" before you decide to defend your own piece of it. The ethologist had shown that there are maps in the minds that he messed with. Interlopers beware: the white-throated sparrow males mark where another sparrow's territory falls. Known for getting "inside the mind" of another species are Dorothy Cheney and Robert Seyfarth (in *How Monkeys See the World*).[22] They, too, used recording equipment. They played deceptive recording of the utterances of a given monkey and observed the responses of other monkeys.

Using the word "mentation" to cover anything that is experienced, clearly there is here an expansion in the mentation of these sparrows compared with other birds. Is the expansion one of instinct-based "desires" alone, or is there something like a new concept as well in this bird's brain's mind? It might take a devilishly clever ethologist to parse this problem. My own hunch is that no new concepts have invaded this sparrow's mentality. The evidence is that he has only two classifications for other males: benign neighbor singing on his own front porch, and outsider threatening from where he doesn't belong. It seems to me that once you have the means in your mentation for making concepts, you ought to be able to make more than a single pair for classifying your fellows. You ought to be able to distinguish a lost neighbor from a total stranger. On the other hand, lost neighbors may never be a problem when nosy ethologists are not around. Perhaps the sparrow cannot form a new concept without a good number of experiences where such a concept would help him make sense of his world.

The small bird known as a chickadee seems to have an even more impressive map (perhaps merely brain map, perhaps also mental map) than does the white-throated sparrow. From Marian Dawkins's account, we learn that when food is plentiful, these birds squirrel it away, as do the mammals we call foxes and, no surprise, squirrels.[23] The chickadees do not wait for hard times or cold weather to retrieve these resources; tomorrow will do nicely. So there seems not to be in their mentation any of the conceptual hard work—hard for me, at any rate—of planning ahead. Rather, a genetic adaptation that may not be mental is the likely basis for their storage and retrieval of food. "That's just what we chickadees have genes giving us the inclination to do."

Ethologists, this time industrious as well as devilish, created a bank of simulated tree trunks, each with a number of drilled holes as cylindrical "safe-deposit boxes" for the birds. Appropriately enough for such boxes, each of them was provided with a little door, made not of metal but of velcro, that a bird could open before it either deposited or retrieved food. The experimenters then allowed the little chickadees to enter the aviary that contained all this clever construction, where now all velcro doors had been opened. The birds carried seeds to deposit in whichever boxes they selected among the seventy-two that were available. The ethologists watched and noted the location of the fifteen that the birds happened to choose. With no more depositors in sight, the bank was then closed. The birds were chased out of the aviary, the food was removed from the chosen holes, and all velcro doors were closed. The next day, the chickadees were allowed in as the experimenters watched. The birds (who had banked on retrieving their goods) pulled open the doors of and searched inside only the particular fifteen where they had previously deposited the seeds that were no longer there. Further artful tests were performed to show that remaining odors of food or bird at the chosen sites were not the means by which the birds selected holes to search. So the evidence suggests the chickadees had maps (perhaps mental, perhaps only neural) and memory of which holes they had chosen to store their pitifully small savings. (As for the deprived little chickadees—embittered birds who called the bank to complain that the boxes were not secure at all—they were put on "hold" and had to listen to recordings extolling the safety of the bank's boxes.)

PLOVERS, HONEYGUIDES, PARROTS: PERHAPS THEY USE SOME GESTURES AND SYMBOLS

Except for the use of maps and memory, all the activities just described were those of creatures limited mostly to the here-and-now. Earlier in this chapter, I mentioned an exceptional behavior, that of a ground-nesting bird that seemed to be torn between fleeing a predator and protecting its young. Fleeing a predator is certainly narrowly egocentric, but protecting its young is not. Can we explain how an egocentric creature can care about its nestlings? We can't, not unless the meaning of its egocentricity is expanded. And the most convincing and parsimonious way to do that is to assume that the bird regards its young as a part of itself. As the young—for all of us mammals as well as for birds—come out of the mother's body, it is not hard to see that the mother can regard her young as a detached part of herself. Thus, she has concern for them almost as she does for her own being, without attaching any conceptual meaning to them as her "dear children" or the like. As Susanne Langer observes, empathetic hearing on the part of an animal mother may make her young and their calls seem like a part of herself.[24]

Apart from being an early example of how egocentricity can expand to include a measure of altruism, the "broken-wing ploy" has two other interesting tales to tell us. The first has to do with what often seems a strictly human skill, deception. The second tells us how difficult for scientists is the task of understanding animals.

Many people believe that one of the ways in which animals differ from humans is with regard to deception or truthfulness. Among them, some believe animals are too noble or too innocent for such underhanded behavior; others think creatures simpler than us don't have the brains and minds for it. That suggests animals may have in their hearts something like the end of pompous Polonius's advice, in Shakespeare's *Hamlet*, to his son:

> This above all—to thine own self be true,
> And it must follow, as the night the day,
> Thou canst not then be false to any man.

For a few creatures, the facts of the matter are different than the king's advisor suggests. Some animals, from birds to apes, act as if they knew what politician Polonius may have told his son in private: "There are times when thine own self will truly gain if the occasion suggests thou be false to some men."

HOW WE GOT TO BE HUMAN

Are these exceptional animals all deliberate deceivers? That is doubtful. Those at the less brainy and less mindful end of the range have, without understanding, simply stumbled onto behavior that occasionally profits them. At that low end are some birds who, when eating in the midst of their fellows, will occasionally issue a "predator alarm" call when no predator is in sight. The deceiver's neighbors, however, will credit the alarm and take off for elsewhere. The cheater then feasts on its ill-gotten grains and fruits. Not a nice story for young boy scouts, nor for innocence-seeking amateur bird-watchers. Does the alarm-issuing bird comprehend that it lies? As yet, we have no reason to think it does. The situation requires a double-dealing ethologist to figure out how the birds deal from the bottom of their decks.

At the high end (apart from humans) of the range of deceivers are apes. They comprehend their own occasional deception, they use deceit with understanding. The activity of apes is described in the next part of this book. Some of their behaviors provide good examples of high-end animal deception, so we'll look briefly at them here.

There is anecdotal evidence that sometimes a chimpanzee will pause in the middle of an encounter with an antagonist. Then, facing away from the opponent, he uses his hands to manipulate his own facial features, thus removing an expression that ethologists normally call a "fear grimace." Following that, he resumes his agonistic encounter. The ape's alteration of his appearance may be intended to deceive his opponent. Perhaps he is also engaging in the act and the pause that refreshes his own courage. The ape is either trying to deceive his opponent, trying to deny his fear to himself, or both. Note that, for the second of these, the struggle is between two parts of his mental apparatus.

A particular chimpanzee in the presence of food within sight of high-ranking other apes practiced deception of his superiors. Human observers saw the ape look at the food only briefly. After this momentary flirtation with temptation, the chimp studiously avoided peering at the goodies, let alone trying to retrieve them, until the powerful others left the location. Then he quickly sidled up to the food items and gobbled them.

Note that chimps engage in both of the deceits cited above by virtue or vice of what must be gestures. Restoring with your hand a fearless expression to your face is optional and deliberate—like a gesture—as is studiously looking away from food. However, chimps use other kinds of gesturelike behaviors besides those that falsify their intention, as the next chapters will show. The very fact that they employ gestures of different types strengthens the evidence that abstract concepts lie behind the behaviors.

Now let's turn again to the ground-nesting plover's broken-wing ploy. Informed opinion on this activity has gone through three stages. Originally, this behavior was thought to be a deliberate act of deceiving predators who were on the prowl for the hatchlings or even the eggs in the bird's nest.

Then the tide of such belief went out of ethological acceptance. Langer examined this problem.[25] She thought scientists indulged a clear anthropomorphism by believing that a ground-nesting bird could pretend to have a broken wing in order to decoy intruders away from her brood. The bird could not be employing a symbolizing gesture of faking a broken wing. To do that, it was thought the bird would have to understand how the predator sees its world, especially how it sees the bird itself. Such mental activity seemed far too complex for a bird brain.

So broken-wing behavior and its like came to be regarded as conflict between two instinctive nonconceptual impulses. (Recall the really rigid behavior we saw, perhaps totally mindless and thus zombielike, when we looked at the digger wasp endlessly repositioning its preserved prey—which had been repeatedly moved by a scientist—and endlessly reinspecting its burrow.) The mother bird was experiencing conflicting pressures: to save itself and to save its offspring. On the one hand, the mother wants to be near its nestlings where she can protect them. On the other, she fears the approaching predator and wants to get away from it, to fly away. These concrete instinct-based impulses in different part of her mentality compete within her and generate conflicting behavior. She spreads a wing but does not fly away. She moves away from the nest to save herself when the predator is too close. She moves toward the nest to protect her young when the dangerous creature is further away from her.

New ethological observations on the ground-nesting plover have brought back the tide of belief that the broken-wing ploy is, in fact, the result of deliberate deception. James Gould and Carol Gould describe in detail the evidence that ground-nesting plovers do pretend to have problems flying, and thereby tempt predators to follow them away from their nestlings.[26]

When the momma plover spots a far-off potential predator, she moves well away from the nested eggs or hatchlings. It is thought that she then selects from or sequences between four artful artifices. The first is to sneak silently toward the approaching brute and then run in a direction away from the nest while squeaking noisily. In a second sequence of what looks like deceptive gesturing moves, she quietly moves to an exposed area away from the nest. There she acts like she is incubating a future brood. When the would-be consumer of her young sees and moves toward her, she departs in a direction away from the real nest. Her third ruse is first to move well away from her nest, then to call loudly. When the predator has spotted her, she alternates moving toward and away from the savage beast, but the latter moves are also away from the nest. The plover's last means of deceit is the aforementioned broken-wing simulation. When she is up to staging a peak performance, she first flaps a wing as if trying to take off, following that she flops over, and then in her third act she acts wingless by running rather than flying away.

In these plover ploys, the evidence shows that the mother bird does not do these things simply in response to the interloper; rather, she takes the lead in the encounter. This makes it unlikely that the plover is just acting out conflict and not knowingly decoying her antagonist. Need she then understand the mental picture her opponent has of her? That would be going too far. It is more likely that she is a behaviorist; she has learned to understand the behavior of the predator and to take advantage of it. How did she get such knowledge? She learned not what evil lurks in the minds of the would-be predators, but what behavior they engage in to pursue their ends. That suggests she once did not have such knowledge. She may have learned to some extent from watching other mothers and to some extent on her own. Perhaps with her earlier broods she did engage in behavior resulting only from conflict between instinctive impulses. Perhaps as a winged creature in frequent overview of ground-level activity, she saw other plovers from whom she learned.

Although the plover is a brilliant bird, I doubt that she's as bright as Daniel Dennett implies.[27] He suggests the plover thinks out the solution to its problem. Considering how difficult it is for humans to think out solutions, it seems more likely that a plover learns from its own initial *accidental* acting out of the solution (luring the predator away from the nest). Likewise she may learn by watching from the sky as other plovers cope with predators.

Before leaving the plover ploys, I must give you the views of Henry Plotkin. He believes those ploys are the result of a gene mutation that was adaptive for the survival—not of the plover—but of that gene itself. In *Darwin Machines and the Nature of Knowledge*, he says that if the broken-wing display resulted in the mother plover getting "caught and killed . . . the offspring . . . may survive because the predator has been distracted. . . . The altruistic behavior is therefore explained by it being an adaptation for the good of the parent's genes."[28] Plotkin is insistent: "This explanation of 'altruistic behavior' . . . is also almost certainly correct." I must ask, what if the young orphans cannot yet get food for themselves? What if they are still encased in their eggshells? What about the many more broods with her genes that the mother will now never bring to life?

What *is* almost certain is that more work by bright bird-watching ethologists will help to resolve these issues. Cynthia Mills, in her article, "Unusual Suspects," describes the recent work of a scientist who videotaped field assistants walking amongst plovers. "The plovers flew or ran in front of intruders quietly, then seemed to wait until they were seen. . . . When an intruder stopped following, the birds . . . squawked louder and flapped harder."[29]

Another interesting example of what seems to be the use of abstract concepts is the activity of a bird called the African honeyguide. This bird also interacts with four-legged animals in a way that seems to require abstract concepts. James Gould and Carol Gould tell us how the honeyguide interacts with an animal called the honey badger.[30] This interaction is not one of conflict; rather, it is cooperative. Individuals from each of the two species seem to collude to harvest the work product of African bees. The honey badger, it will come as no surprise to hear, gets the honey. The honeyguide, more surprisingly, gets the beeswax, the larvae, and when it can, some of the insects. The leader of this strange team is the honeyguide.

Tree cavities are used by the bees to enclose their nests. The honeyguides, by themselves, are not able to break into those delicacy-filled places. For that they need the honey badgers. Such a bird first locates a nest, then finds a badger, and starts a noisy display to get its attention. If successful, it flies a short distance and again gives voice to its desire to attract the badger. This continues, for as much as a mile or so, until the nest is in sight of the badger. There the guide waits while the badger breaks in and gorges on honey. Then the bird does likewise with the larvae and the wax.

The question is: Does the bird use concepts to get its share of the looted nest? Does it understand what it is doing? The answer may seem to be "yes," but we can't be too sure. The fact that the birds feed on wax, and can digest it, is unusual; it suggests the honeyguides are specialized in living off bees. As their name suggests, the honey badgers may also be so specialized. What if the two species had coevolved to their present estate? Both would have benefited by mutations that made them sensitive to the other's behavior, all without either of them using "new-fangled" concepts to think about what they do for a living.

This is not at all like the situation of the plover and its predatory foe. The latter two could not have coevolved together. In that pairing, only the plover benefits, while the predator plays the fool. Nature raised none of her children to specialize in wearing the fool's cap, since that's hardly adaptive. (If haphazard mutation brought full-time fools to life, fewer and fewer would have survived long enough to have foolish progeny.) But the honeyguide and the honey badger may have coevolved without either being able to entertain concepts. Think of birds that annually migrate thousands of miles each way. Need we regard them as thinking something like, "The weather's changing, it's time to move to my other home faraway?" Might they not have evolved with greater and greater distances to fly as the evolution of the earth's continents moved the latter apart? If so, what had once been more or less circular local feeding paths became grand circuitous migrations, all without conscious thinking with concepts.

So we should be suspicious about the honeyguide's ability to employ abstract concepts. Perhaps we should still have some doubt about such activity in the plover's mentation. Such suspicions are warranted because the ability to use concepts seems to be a general-purpose capability. Why doesn't the plover similarly use its head in some other areas of its life? Why doesn't it offer us multiple-use evidence as do the chimps? The latter use gestures to deceive (looking elsewhere then where they are interested), to ingratiate (getting down on all fours to kow-tow), to ask for something (by extending an open hand), and to enlarge their bodily threats (shaking or breaking branches). As far as we know, the plover's unusual ability is singular, like that of the chickadees who store food in places that they can later find. Perhaps the plovers, like the chickadees, are each sort of an animal idiot savant, able to do only one kind of trick well and everything else poorly. Perhaps their tricks are mostly nonconscious rather that conceptually conscious. Alternatively, enterprising ethologists might well find more conceptual heavy metal if they can find the cunning to plow beneath the surface of the plover's and honeyguide's minds.

Jane Goodall gives us another couple of anecdotes that suggest the difficulty of understanding deceptive behavior. She discusses a chimp who hurt his hand in a fight with the alpha male. "The following week he limped very badly—but *only when in the alpha's sight*."[31] Then she tells us that one of her dogs, after hurting its paw and receiving her attention and sympathy, likewise later limped only in her presence. I suspect that the chimp had a concept in his mind, but the dog had merely made an association in his brain. However, making a case for my belief would require evidence of a more elaborate pattern of behavior by the chimp. Such data would be difficult to get, expecially as deceptive behavior is not used routinely.

For a last demonstration from our fine feathered friends, let's listen to a bird that can talk—a parrot. Is it a complex-concept-free idiot savant? Or is it a savvy employer of concepts symbolized with human words? The bird in question is an individual African gray parrot named Alex, whose activities are described by Marian Dawkins.[32] Though Alex's color is, as parrots go, plain, his utterances are astonishingly fancy. He has learned to vocalize a good number of English words and phrases.

Alex was trained by Irene Pepperberg.[33] She trained Alex by letting him overhear the talk and see the acts of a trainer (Pepperberg) and her human fellow-trainer "shill." The trainer would ask, "What is this?" about each of several food or play objects for the parrot that were held in front of the shill. The shill's answer to each question was a name, either correct or incorrect, for the object. A correct answer brought both the object and praise to the shill. After an incorrect answer, the shill was told "No!" and not given the object. Two people alternated occasionally in the roles of trainer and shill.

Alex-the-parrot soon bought into the routine and he began to shill for himself. In so doing, he learned to use the names for six toys and three food objects. He also got to play with the toys and to eat the food. By similar means, Alex learned the names of numerals "1" through "6," and he learned to apply those same names to the quantities of different kinds of objects. Additionally, the parrot learned to say things like "No!" and "I'm going away!" at times that seemed appropriate. Pepperberg has been training Alex for some two decades. As she leaves him each evening, he says, "Bye. I'm gonna go eat dinner. I'll see you tomorrow."

Alex's most masterful performance occurred after he was taken to a veterinarian's office for lung surgery. When his trainer turned to leave, the bird called out, "Come here. I love you. I'm sorry. I want to go back."

The words certainly tear at your heartstrings. Should we believe the words have meaning to the parrot, meaning that is anything at all like what their meaning is to us? The overly empathetic or the naively projective, overcome by what seems the bird's persuasive rhetoric, might say, "but he uses the words so well, so appropriately. Why shouldn't we think he means them as we would?"

To help us see why we shouldn't, let's look at one set of experiments intended to explore the use of language by chimpanzees. The language in question, appropriate for apes, made use of variously colored and shaped blocks to represent human words. Once the chimps learned to use the blocks, they did quite well at employing them to interact with the trainers and to get things they wanted. As most would agree that chimps may well be more intelligent than parrots, the following negative results can tell us something about the conceptual shortcomings of both species. Let us ask if the chimps gave to the language symbols the same meanings the experimenters did.

Unfortunately we can't address the question to the chimps; on the flip side of fortune, the same experiments were run with creatures of another species, a species whose answers to the question of the meaning of the blocks tell us a whole lot. This second species was the well-known Homo sapiens sapiens, with sub-subspecies, college student. Some of the students so tested said the blocks meant to them pretty much what they meant to the experimenters. But—and here's the rub to the underlying clearheaded about parrots' understanding human language—other students had no idea what the blocks meant. They simply learned there were some rough patterns in the use of blocks, patterns that could be used to get certain results desired by the students. If some college student experimentees (who we must believe had fairly advanced understanding of the verbal language of the experimenters) failed to deduce the meanings of the blocks, why should we think any blockhead apes, let alone any bonehead parrots, would understand what they meant?

The above is a negative argument, intended to cast doubt on the ability of chimps and parrots to use the concepts behind human language. Another negative argument points out that modern Homo sapiens use concepts *widely*—to absurdly understate and compress the case into one word— something parrots show no sign of doing. To answer the question with a more rhetorical inquiry: if parrots are capable of such conceptual gymnastics, why haven't Alex and his peers in the wild built even the rudiments of the symbolic world we humans largely live in and symbolize with the word "culture?" Why don't parrots have something in the way of a protolanguage—perhaps very simple utterances or gestures—of their own? To burlesque the point, why is there no gesturing to one another by spreading a wing or two, thereby to suggest new arrivals on a branch should leave? Why no cocking of a head, to doubt another's credentials for landing on a twig so close? Why not fan the air, to clear away another's subterfuge?

I hasten to add that Alex-the-parrot, although he may not have anything like human understanding of words, may have as much conceptual capability as plovers do. Plovers seem to use gestures or deceptive behavior to get desired behavior from a predator. Alex's imploring words at the veterinarian's office can be regarded as verbal gestures or deceptive behavior that would tempt most of us to "protect" him from the vet.

CONTESTING MALES OF BEASTS LIKE DEER AND WOLVES DON'T HAVE THE OPTION OF GESTURE

For another animal, a mammal this time, that cannot engage in gestures to symbolize its activity, we turn to the males among the well-studied red deer who live on a coastal island in Scotland. Marian Dawkins describes the noisy seasonal bouts between belligerent males.[34] Look again at the good old male fight for maximum sexual activity—which is as close as most male mammals get to glory and love. Here we see that the struggle between male red deer is often not a literal fight. Rather, it is a battle of blood-curdling bellows by each of the antagonists. It is apparent that bellowing is not actual combat. In my opinion, neither is it vocal gesturing or symbolic utterance.

In the fall, when the red deer females go into estrus, the males get more red-blooded. They depart from the more or less graceful grazing that is their main occupation during other parts of the year; they go into rut. At the start of that fall mating season, the more powerful (usually larger) males round up groups of as many as twenty or so females, the "ownership" of which they defend vigorously against other males. The smaller males at times challenge the larger ones; they also seek and find stray females with whom, given cooperation, they mate. For as many as five or six weeks, the powerful "propertied" males are so busy facing down challengers that they have little time to eat. Thus they lose weight and strength while their sexually hungrier competitors have more time to eat, regain strength, and, more important to them, to reenter the sexual marketplace. So the struggles go on throughout the mating season, with many an initially rich-in-females animal losing his high rank, losing much of his property, and, more importantly, forfeiting his opportunities to mate. Many a poor plebeian among the males rises, not to make a fortune in females—it's a bit late for that—but to make out with a number of the femmes.

The fights between males are sometimes both physical and serious in terms of wounds from the sharp antlers of opponents. A lot of the interaction, however, is not physical, but rather consists of roaring contests. In the latter, the males take turns bellowing as loudly and forcefully as they can. Ethologists tested the idea that such bellicose bellowing served as more than mere threats to opponents. The scientists thought the roaring might provide means for the males to gauge their opponents' strength against their own. This would allow two contestants to decide on a winner without actually fighting. The ethologists were able to show that, by listening to the resounding racket, they (the scientists) were able to predict winners of such contests. This suggests that the participants reached a decision on the same basis: Only when two were closely matched in roaring did they go on to a physical fight.

The frequent use of bellowing matches to replace real fights represents an enlargement, from a presumed equally pugnacious but more direct ancestor, in the brain-work of the deer. However, what is involved in a fight with roaring is no new note but rather an expanded version of the same *kind* of activity. The roaring does not represent a new and more complex conceptual mental process, one with optional utterance that may or may not be used by a male, a male who intends to communicate information.

A male having heard—say, from a smaller foe—bellowing better than his own cannot choose to switch and fight rather than to retreat. For the deer, the braying is not gestural. It is not bluster, not symbolic, not a message *about* the state of the struggle, one that may be accurate but perhaps is deceitful. Bellowing is not something that can be ignored or not done. Like the aroma that the wind wafts to the rutting males from one of the estrous females, it represents the facts of the matter rather than a comment on the subject. A male cannot regard roaring, roaring that is better than he knows his own to be, as mere "big talk" or "talking trash" from an opponent who might not really be so tough. A contesting pair regard themselves to be precisely as "big" as their braying.

For me and I hope for you, consuming the movable feast of works about animals and puzzling over the import of what such creatures do is an enjoyable armchair activity. For the ethologists who study animals like the fish, four-legged beasts, and birds we have looked at here (and for the primatologists who study apes, let alone the anthropologists who deal with humans), discerning the significance of a creature's behavior is a large and difficult problem that is central to their work. They have two main tools available to them. The first—narrowly Darwinian adaptive explanation—is scientifically valuable for its power, and politically valuable for its authority. The second, controlled projection of themselves into the beasts they observe, carries no authoritative power but rather the opposite. Its use often brings criticism from their peers. Thus, for example, anthropologist E. E. Evans-Pritchard scorned the work of many of his predecessors who theorized about primitive human beliefs, scorned such work by calling it pure speculation, examples of the "if-I-were-a-horse" mistake. However, Evans-Pritchard's disdain has since been criticized by others. For example, Daniel Pals (in *Seven Theories of Religion*) suggests that, in a sense, the projective method "is the only one we have when we wish to understand the motives and actions of other people."[35]

Perhaps the real problem is not *that* the projective method is used, but *how* it is used. The use of Darwin's legacy alone may not be adequate to the task of understanding animals. Too much of their activity cannot be explained as simply behavior that is genetically adaptive. The use of thoughtful projection—if scientists can find suitably subtle means to confirm its work—can help them to break new ground, although there is also a risk of falling into a sump-hole beneath the less-than-solid new ground.

NOTES

1. Steven Pinker, *How the Mind Works* (New York: W. W. Norton & Co., 1997).

2. Alan Wolfe, *The Human Difference: Animals, Computers, and the Necessity of Social Science* (Berkeley, Calif: University of California Press, 1993), p. 164.

3. Merlin Donald, *Origins of the Modern Mind* (Cambridge, Mass.: Harvard University Press, 1991), pp. 150–51.

4. Jeffrey Moussaieff Masson and Susan McCarthy, *When Elephants Weep: The Emotional Lives of Animals* (New York: Delacorte Press, 1995).

5. *101 Plus 5 Folk Songs For Camp*, compiled and edited by Mike Cohen (New York: Oak Publications, Inc., 1966), p. 60.

6. Susanne K. Langer, *Mind: An Essay on Human Feeling,* vol. 2 (Baltimore, Md.: Johns Hopkins University Press, 1972), p. 146.

7. Ibid., p. 148.

8. Harry J. Jerison, *Brain Size and the Evolution of Mind* (New York: American Museum of Natural History, 1991), pp. 23, 51.

9. Maxine Sheets-Johnstone, *The Roots of Power* (Chicago, Ill.: Open Court Publishing Company, 1994), p. 46.

10. Donald, *Origins of the Modern Mind*, p. 126.

11. David F. Armstrong, William C. Stokoe, and Sherman E. Wilcox, *Gesture and the Nature of Language* (New York: Cambridge University Press, 1995), p. 217.

12. Maxine Sheets-Johnstone, *The Roots of Thinking* (Philadelphia, Pa.: Temple University Press, 1990), pp. 94–95.

13. Langer, *Mind,* vol. 2, p. 290.

14. Donald Symons, *The Evolution of Human Sexuality* (New York: Oxford University Press, 1979), p. 208.

15. Donald, *Origins of the Modern Mind*, p. 221.

16. Donald R. Griffin, *Animal Minds* (Chicago, Ill.: The University of Chicago Press, 1992), p. 2.

17. Robert Wright, *The Moral Animal* (New York: Pantheon Books, 1994), p. 103.

18. Marian Stamp Dawkins, *Through Our Eyes Only? The Search for Animal Consciousness* (New York: W. H. Freeman and Company, 1993), p. 36.

19. James L. Gould and Carol Grant Gould, *The Animal Mind* (New York: Scientific American Library, 1994), p. 85.

20. Dawkins, *Through Our Eyes Only*, p. 32.

21. Ibid., p. 38.

22. Dorothy L. Cheney and Robert M. Seyfarth, *How Monkeys See the World* (Chicago: University of Chicago Press, 1990).

23. Dawkins, *Through Our Eyes Only*, p. 40.

24. Langer, *Mind,* vol. 2, p. 123.

25. Ibid., p. 107.

26. Gould and Gould, *The Animal Mind*, p. 136.

27. Daniel C. Dennett, *Kinds of Minds* (New York: Basic Books, 1996), p. 122.

28. Henry Plotkin, *Darwin Machines and the Nature of Knowledge* (Cambridge, Mass.: Harvard University Press, 1994), p. 92.

29. Cynthia Mills, "Unusual Suspects" *The Sciences* (July/August 1997): p. 34.

30. Gould and Gould, *The Animal Mind*, p. 146.

31. Jane Goodall, *The Chimpanzees of Gombe* (Cambridge, Mass.: Harvard University Press, 1986), p. 581. Emphasis in original.

32. Dawkins, *Through Our Eyes Only*, p. 119.

33. Irene M. Pepperberg, "Cognition in an African Gray Parrot," *Journal of Comparative Psychology* 104, no. 1 (March 1990): 41–52.

34. Dawkins, *Through Our Eyes Only*, p. 28.

35. Daniel L. Pals, *Seven Theories of Religion* (New York: Oxford University Press, 1996), p. 225.

3

CATS AND DOGS

AN ODD COUPLE OF
DOMESTICATED CREATURES

CATS AND DOGS ARE TRULY an odd couple of creatures. Cats live with us on a separate and equal basis. Dogs are inseparable from us, and in their own eyes, we will see, unequal to us.

Most cats live their own lives near us. When they can find work, they work for themselves, although their work may benefit us. They work as rodent removers, where the work brings its own rewards. As for what we give them, they usually do us the favor of accepting it but frequently they don't.

Dogs live as close to us as they can get. More often than not, they are unemployed. When they do work, they work for humans, mostly by protecting or guiding us, and for a decreasing number, by hunting or herding for us. Whatever we give them, for the most part they act as if they're glad to get it.

CANINE COMMUNALITY: OBEISANCE AND SHAME, BUT NOT MUCH GUILT

Why this difference between cats and dogs? The origin of their difference as pets is the communal component of their genetic inheritance. Cats don't have it; they are asocial. Doggies do have it. Like us, they like to be part of a community.

In Shakespeare's sonnet 29, the narrator says:

HOW WE GOT TO BE HUMAN

> When in disgrace with fortune and men's eyes
> I all alone beweep my outcast state,
> And trouble deaf heaven with my bootless cries,
> And look upon myself and curse my fate.

Cats never do so, but dogs sometimes act as if they think themselves unfortunate enough to be in disgrace with their owners' eyes. When dogs are scolded, say, for making a mess or making one in the wrong place, their hangdog expressions that beg for a reprieve, their beseeching belly-crawling, their desperate attempts to ingratiate themselves by hand or foot licking, all these clearly show that they feel themselves in disgrace, and want out. However, unlike Shakespeare's protagonist, once they are fortunate enough to escape their powerful-other's eyes, dogs are quite untroubled. They feel no need to howl heavenward to moan their fate; they abandon disgrace and return to their normal state.

By this behavior, dogs tell us something about the origins of the human emotions of both shame and guilt. These emotions come from our feeling that we are members of a community. They come from our belief that we have obligations to the group we belong to, obligations that include acting in accord with its values. These obligations are particularly strong toward the group's leaders, who often articulate its values.

We have come a long way since we were small four-legged mammals something like dogs, but through the many millennia, our line has always been communal. To feel yourself in disgrace with the eyes of others, especially with the eyes of the leader of the group, is to feel shame. It is something like shame that dogs experience, shame in the presence of the powerful leader that, to a dog, its "owner" is. Even when they are innocent of what we scold them for, they accept our judgment, and are ashamed.

Nor is it even now entirely different with us. Forgive me for rudely articulating something so unthinkable, but think how you would feel if somehow, at the office, an insurgent gut forced you to pass to your drawers, not just gas but a bit of something less gaseous. Of course, for one of us, the approach of a sharp-eyed, keen-nosed boss would only heighten the already existing embarrassment and, underneath, shame, whereas a dog that, say, drops an indoor load may feel at most a slight disquiet. A slight disquiet, that is, before its approaching owner's scowling eyes suggest the need for a resolution: minor micturating and the boss's acceptance of major kissy-facing occur before the reestablishment of a state of comfort, if not of grace.

Cats and Dogs: An Odd Couple of Domesticated Creatures

In contrast with shame, guilt is beyond the range of a dog's mentation. Unlike Shakespeare's protagonist, a dog abandons its outcast state when it is alone, or at least, out of sight of its master. Pooches, like most mammals below the human estate, have a mind only for the present. With regard to their masters, and everything else, its mostly out of sight, out of mind. As with everything mental, it is extraordinarily difficult to demonstrate confirming behavior. Nonetheless, guilt, which amounts to present acknowledgment to oneself of past sins, is out of the mental range of dogs, cats, and other lower mammals.

Shame, like most behaviors that are not simply instinctive, calls for some (conscious) mental activity. However, that experience of being ashamed does not require any *concepts*. Rather, it is a *feeling*. Although dogs can be shamed, they can't think about being ashamed. In contrast with dogs, apes seem able to entertain the concept, in addition to the experience, of shame. The evidence for this claim about apes is that they can often handle deference to a superior, a common component of shame, as an *optional* matter. For it to be optional, they must be able to think about whether or not to perform an act of subservience that seeks expiation. As will be seen in chapter 4, apes who have lost a fight with another ape are *sometimes* unwilling to employ a gesture, one of physically lowering themselves, to signify obeisance to the winner of the contest. When the loser finally bends his body below the victor, who then stands tall and puffs up his hair, it is because only the submissive gesture will bring the desired consequence, an accepting embrace from the fight's winner.

For humans, all but the youngest ones, shame exists at the level of things that can be thought about. At times a young child may experience an attempt by other children to induce shame. The other kids use the traditional gesture of rubbing one finger over another. The top finger, in effect, disallows the other, implying a naughty no-no. Perhaps the child cannot articulate its belief that the gesture is an improper accusation. In that case, it may choose to respond with a defiant arm-raising gesture rather than with an accepting head-lowering. So even a young child can think about shame, as well as experience it.

In themselves, feeling of shame and other things are mental states without concepts. Dogs experience nonconceptual shame without having further mental states about shame. Thus, they cannot engage in behavior that is about shame, behavior whose enactment amounts to a comment on a situation rather than a representation thereof. That gives them no choice. They are obliged to seek the forgiveness of their human shamers with the nongestural behaviors of cowering and toadying.

ASOCIAL FELINES:
ONCE EGYPTIAN DEITIES AND MEDIEVAL DEMONS,
NOW ASSOCIATED WITH WOMEN

We all know that cats are, in some fundamental sense, different from dogs. It is their asocial nature that is at the root of the difference. It's as if we humans were members of a club, one that we are mostly willing to let dogs join. We are accepting of dogs' applications for admission, that is, as long as they know their place, which is at the bottom of the ranks. As for cats, they don't even recognize the existence of the club, let alone wish to join it. We, in turn, don't know whether to be affronted, aghast, or amused at their disinterest and indifference to the opportunities for advancement they could have if they would just wise-up.

As bottom-ranked members of the social club, dogs are a comfortable part of our "natural" social world, and have been so for the, perhaps, tens of thousands of years that we have palled around together. Just how many tens of millennia has that been? In *The Hidden Lives of Dogs*, Elizabeth Thomas says two tens of millennia.[1] Evidence more recent than hers suggests dogs may be descended from wolves that were tamed more than 100,000 years ago. Steven Mithen notes that, in Scandinavian hunter-gatherer cemeteries dating a mere seven millennia ago, "we find dogs which had received burial ritual and grave goods identical to those of humans."[2] This suggests that not only were we long-standing friends in this world, but we planned to be pals with them in the next world.

In this role of mutual friends with us, cats seem less natural. At times, they seem unnatural. Millennia ago, people regarded cats not merely as less natural or unnatural, but as having out-of-this-world connections. In *The Body Language and Emotions of Cats*, Myrna Milani tells us that, in ancient Egypt, cats were thought to be supernatural beings.[3] When gods walked the earth as felines, their human hosts were honored and flattered by whatever those august and inscrutable creatures deigned to do, do with them, do near them, or even do to them.

From Desmond Morris's *Catlore*, we learn that a stunning reversal of fortune in medieval Europe changed cats from wonderful beings to terrible ones.[4] They were the devil's agents, evil made flesh. Their meat was poisonous, their teeth venomous. Their breath caused consumption and destroyed human lungs. The cat was associated with the sorcerer, the church encouraged persecution and prosecution of both. In 1484, the pope gave the Inquisition the power to burn cats and—hey, why take chances—cat lovers. Note, however, that for all that, cats were still, as they had been in Egypt, supernatural. They were strange creatures with power beyond that available to mere mortals.

We live now in times more secular than did the ancient Egyptians or the medieval Europeans. Now, it is less than credible to describe cats as deities or demons. Perforce, we need more natural ways of representing these unnatural-seeming felines. We humans still live, however, in what amount to patriarchal communities, where the often threadbare views of powerful males still line the social fabric, communities where the coin in the realm of ideas commonly clinks with masculine mettle. How, then, shall we now represent the strangeness of cats?

Cats are like females. Females (as we shall see in chapters to come) have been a frequent mystery to males ever since some prehuman hominid species abandoned female sexual heat. That made them a mystery ever since males could no longer know in advance whether or not an attractive female would be receptive to male sexual advances. So, hey, we have sort of a syllogism here, don't we? Felines are strange; females are strange: therefore, felines are like females. You want evidence? We don't seek evidence on this issue, but try this: Gossip is what women do, and those who do it are catty. Likewise, coy women are called kittenish, and angry women are said to "show their claws" with insufficient cause. We even use a vulgar slang word for a woman's genitals that is identical to another word for cat. Of course, this short list ignores the evidence that young men may call some young women "dogs," and some women are called by the name for female dog. Nonetheless, I suspect that most people regard cats as more feminine than dogs.

CAT AND DOG SEXUALITY: TOO CLOSE FOR COMFORT

We have ambled close to the subject. Let's mosey even closer and get nosey about the sexuality of dogs and cats. Unlike dogs, whose ovulation leads to heat and then to copulation, cats' ovulation is not prior to copulation; rather, it results from it. If the female does mate, her interval in heat ends after about four days and does not recur until a half or full year has elapsed. If she does not mate soon, get some cotton for your ears: Her time in heat may last as long as ten days and recur every two or three weeks for a while.

HOW WE GOT TO BE HUMAN

Females in estrus, when they get a whiff of male cat urine, often jump with joy and frolic in a frenzy. They behave in the same manner just after having had sex. They similarly seem be on a "drug high" in response to catnip, a plant of the mint family, which provides a pseudosexual aroma to most cats of either the male or the female variety. Cat lovers who give catnip to their pets unknowingly give them the feline concept-free equivalent of what, for human youth, were once called "dirty pictures" and is now known as "adult entertainment."

Cats copulate for only a few minutes, but they precede the act with a longer interval of foreplay. Dogs, on the other hand, are not given to much in the way of foreplay. For them, the sex act itself is both short and long. It's short because males need little time to reach orgasm, long because they may need a half hour or more before male detumescence allows the pair to physically decouple.

As most of us know, humans are occasionally the targets of the sexual interest of male dogs, who are not fussy about the gender of the human whose leg they hope to hump. Among cats, it is the estrous female who sometimes develops the hots for a human. In her case, it is a human male whom she pursues. When a door between them gets closed, she yowls and trembles against the door for him.

Our own response to the sexual activity between dogs or between cats is also interesting. We commonly don't like to see them performing the act. Likewise, we often try hard to keep our children from being exposed to the sight of copulating cats or dogs.

A man who was my neighbor had a strong distaste—about the sexuality of the short-haired male dog that lived with me—that went even further. This dog would sometimes roll over onto his back and wiggle back and forth on the grass. The neighbor walked over to me one day and asked, as if it were something I should have taken care of without him having to ask: "Can't you keep your dog from exposing himself like that?"

The sexual activity of cats and dogs is, for most of us, too close for comfort. The reason for this is our common habit of nonconsciously putting ourselves in others' shoes and others into our shoes. We find ourselves seeing not just a copulating animal, but rather *another* creature having sex, one that doesn't have the sense to do it in private. The distaste is rooted in the fact that we humans are almost all of the opinion that our own sexual activity should be performed only in private. (The basis for attitudes about sexual privacy will be looked at with regard to different hominid species in chapter 9.)

Cats and Dogs: An Odd Couple of Domesticated Creatures

For some humans, those who go overboard with our ability to identify with our pets, cats and dogs can seem as smart as humans. Daniel Dennett notes that the stupidity of pets is largely ignored by their fans.[5] Thus, for example, a dog can never figure out how to unwind a leash after it runs around a tree or post. He might have added that a cat can hardly figure out how to handle a leash at all.

Let me tug briefly on a related cord in the web of concepts with which we construct our human minds. This cord is that of our *occasional* distaste when we consider how interested dogs are in the butts of other dogs.

This somewhat scatological subject may be offensive, at times, to some of us, which is part of the reason it is included here. Before attempting to explain my seeming perversity, I will raise the ante by describing a clearly vulgar joke in a recent cartoon by Callahan. Picture a security guard (who is also a dog) seated behind a desk and in front of a door marked "Private." To a person (who also is a dog) wearing a coat and standing in front of the guard, a person who is clearly becoming irate, this guard-and-dog says: "I'm sure you're probably who you say you are, but regulations require me to sniff your butt."

What is of interest here is not the fact that some of us find such minor scatology interesting while others do not. What *is* of interest is that, depending on the finder's frame of mind, both categories frequently reside in the self-same person. The point is that dogs do not have the option to sniff or not to sniff. As the security guard cartoon implies, they must always sniff (unless the potential sniffee threatens them with bared teeth and a growl).

Whereas humans—who live largely in a world of concepts, concepts that include lowly butt-sniffing—have an option, an option either to ignore (or disdain) such scatology, or to involve themselves in it (and enjoy it by fig-uratively sniffing butt). We also have the option with regard to literal butt-sniffing, as most mothers of babies in diapers can tell us. In the chapters to come, we will see that animals with larger brains and mentalities than dogs, such as apes (and of course humans), can use gestures and utterances as *optional* symbols of their concepts, optional because concepts (as I use the word) are always conscious.

Our human interest in privacy and distaste for public performance, both sexual and eliminative, are learned as we grow up. Our attitudes about sex come throughout childhood. Some basic attitudes about our waste products are commonly learned at about the age of five. In *Signs of the Flesh*, Daniel Rancour-Laferriere notes that children of age four or younger seem not at all distressed by the odor of their feces.[6] The same children at five or six have a strong dislike for the stench of their stool. We transfer these attitudes by projecting them into other humans and animals, or by introjecting them from other creatures.

NEOTENY: THE FOUNTAIN OF YOUTH IN OUR PETS' PAST, AND IN OURS

How did it come about that dogs and cats were domesticated? Stephen Budiansky provides a thorough examination of the origins of domesticity in animals.[7] Was it a matter of tender or lonely hearted humans taming and adopting some of the wild creatures? Although we may have tamed and adopted many wild young puppies and kittens, it was not those kinds of actions that led to the domestication of canines and felines. As evidence, note that Native Americans made pets of bears, moose, and raccoons, but none of these species became domesticated. Likewise, in Egypt's glory days, attempts to domesticate hyenas and antelope failed, although Egyptian cattle were a domesticated species. Although individuals of these undomesticated species may have been tamed when they were young, when they became adults most broke their bonds with humans and responded to the call of the wild.

Earlier humans failed with all these "refuse-nik" species because the latter were too fearful and not docile enough for domestication. Also, even though individuals were captured young and then tamed, they did not breed much or at all when they became captive adults. Then what was different about dogs, sheep, goats, cats, cattle, and horses, most of whom were probably domesticated in more or less that order between fifteen and five thousand years ago?

The answer can be expressed in one word: neoteny. The word "neoteny" refers to the biological retention of juvenile traits in adulthood, with regard to both appearance and behavior. It turns out that all domestic animals, and we ourselves more than most, are neotenic. Members of neotenic species are more appealing to us than individuals of those which aren't; their baby-faced youthful appearance makes us more friendly toward and protective of them. However, that's only one of the reasons that neoteny led to domestication. More important was what neoteny did directly for these species. It gave them, as adults, the behavior of the young. The young are always more flexible, more venturesome, more curious, quicker learners, and less fearful than know-it-all non-neotenic adults.

Cats and Dogs: An Odd Couple of Domesticated Creatures

But why, you may well ask, why did all these species (who eventually became domesticated) become guzzlers at the fountain of neotenic youth? What force drove them, and, decidedly, drove us, toward this evolutionary change?

For neoteny, the driving "force" was time, the quick-time of ice-age environmental change. The repeated rapid—by geological standards—rise and fall of ice ages, for a million years prior to their disappearance some ten or so thousands of years ago, made life difficult for stick-in-the-mud-and-melting-frozen-tundra species. Neoteny was a way to accommodate to rapid change in the physical world. Long before domestication entered the picture, creatures that chanced on mutations that made them neotenic were rewarded with higher survival rates. Neotenic species could also more easily enter into a later unspoken agreement with humans to form an aid society, one where all domesticated species profited from their cooperation with the sapiens sapiens subspecies. Need I say that we, too, benefited from the agreement? Likewise, our own neoteny, which gave us small faces in large heads with large brains relative to body size, contributed to the flexibility that allowed us to take advantage wherever we saw opportunity in the world around us. As Stephen Jay Gould remarks, flexibility is the hallmark of our evolution, and "we are, in a more than metaphorical sense, permanent children."[8]

NOTES

1. Elizabeth Marshall Thomas, *The Hidden Life of Dogs* (New York: Boston, Mass.: Houghton Mifflin Co., 1993).

2. Steven Mithen, *The Prehistory of the Mind* (London: Thames and Hudson, 1996), p. 224.

3. Myrna M. Milani, *The Body Language and Emotions of Cats* (New York: William Morrow and Co., 1987).

4. Desmond Morris, *Catlore* (New York: Crown Publishers, Inc., 1987).

5. Daniel C. Dennett, *Kinds of Minds* (New York: Basic Books, 1996), p. 115.

6. Daniel Rancour-Laferriere, *Signs of the Flesh* (Bloomington, Ind.: Indiana University Press, 1992), p. 40.

7. Stephen Budiansky, *The Covenant of the Wild* (New York: William Morrow and Co.), 1992.

8. Stephen Jay Gould, *The Mismeasure of Man* (New York: W. W. Norton and Company, Inc., 1981), p. 333.

PART TWO

LIVING APES, SOMEWHAT LIKE OUR APE FORBEARS

INTRODUCTION

"THE RICH," F. SCOTT FITZGERALD WROTE in a short story, "are different from you and me." This remark drew a cynical response from Ernest Hemingway: "Yes, they have more money." It's an interesting question: Does wealth change the nature of its possessors?

How do the rich themselves weigh in on this issue? In most times and places of our long past, the rich voted with their kidskin-gloved butcher fingers on Fitzgerald's side of the scale. In the past, when wealth and power were almost synonyms, the rich often viewed their own blood as regal, hardly diluted as it dribbled down to them from deities, and nothing like the thin stuff they bled from the common folk beneath them.

However, in the United States, the rich have become of two minds on this issue. In one mind, they have convinced themselves and many of us that Hemingway was right. In our democratic land there simply is no upper class. They view themselves as middle class, like most folks. The only other classes are the working class and—ugh, there should be a four-letter word for it—the nonworking class (which of course refers not to the rich but to the unemployed poor). In the other mind of the rich, they can't help taking a view close to Fitzgerald's: Their wealth is Social Darwinian evidence of their superiority. After all, free enterprise does allow natural selection to bubble the best to the top, doesn't it?

Putting the issue of the nature of the rich in a broader way, does the gradual increase in the quantity of some possession or capability reach a threshold above which its quality changes? As with the proverbial camel's back-breaking straw? Or is it the case that no matter how much ho-hum hay we have in our stack, we get possessed by a new and different spirit only when we acquire a qualitative difference, a needle?

We can transpose the question to a colloquy between two apes, linguistically gifted somehow with sign language better than what they normally learn from humans. Thus Fitz, the admiring first ape, says, "Yuh know, the sapiens are different." The less impressed Hem replies, "Yeah, they got more stuff to eat."

Paleoanthropologists address the same issue when they classify themselves as being either "splitters" or "lumpers." Fitzgerald-like splitters are inclined to see differences in fossils as important enough to suggest different species. Lumpers are like Hemingway. They more often see fossil differences as minor, as little more than divergences between individuals of the same species. In a broad sense, the ability to split and lump is what big brains are (not all, but partially) about. Millions of pattern-shaping neuron-networks make connections with one other, connections that allow us concepts about concepts, and thus allow us to experience concepts as related in "this" way, but unrelated in "that" way.

In Part Two, I will do both splitting and lumping as we explore the lives of apes. I must start by splitting with Terence Deacon. In *The Symbolic Species*, he makes a lumping and splitting that I can't agree with: "[There is] only one significant difference between language and nonlanguage communication . . . word meaning and [symbolic] reference."[1] By lumping word meaning with symbolic reference, he implicitly (and, in my opinion, mistakenly) denies or ignores the possibility of meaning and symbolic reference for gestures. In effect (even though they use gestures), he splits apes (and some birds) away from a somewhat humanlike life with meaning and symbolic reference.

With regard to concepts and gestures, I will split the apes from lower animals (other than some birds), and lump them above some threshold with humans. That means you'll see evidence that apes not only *act* on getting power and sex, they *think* (use concepts) about power and sex. With regard to morality, I'll split apes from us and leave them lumped with lower creatures. They lack the conceptual mental capability to think about right and wrong.

Implicit to this look at the great apes is the assumption that the common ancestors we humans share with them were similar to these existing apes. The activities and presumed attitudes of the modern apes may represent good approximations for those of our own long gone ape ancestors. In truth, it is not only the great apes, but also other living simians, such as baboons and macaques, that have advanced beyond lower mammals and seem at times almost human.

The great apes—the gorilla, orangutan, chimpanzee, and bonobo (pygmy chimp)—are species living now about whom we know a fair amount. I will examine four aspects of the mentality of the great apes, especially the more social of them, the chimps and the bonobos. First (in chapter 4) is family life, where we will concentrate on the at times tender, at times indifferent or brutal, chimps. Second (in chapter 5) will be a look at the expanded mentality of apes, relative to that of lower animals: The mental expansion of self, of perception of others, and of their conception of the space and time they live in. Some altruistic behavior, such as that of an ape risking its life to save another, is not a Darwinian adaptation, but an adaptation byproduct. It is based on the ability to identify with others and feel them to be a part of self. Also, ape awareness of others as distinct individuals makes their communities more like human societies. Third (in chapter 6), we will examine the perhaps at times titillating subject of ape sexual activity. We will see that apes are much freer sexually than lower mammals from nature's confining restraints. Nonetheless, ape sexuality is dissimilar from that of humans. In part that is due to the retention by most ape species of female sexual heat. Fourth (in chapter 7), we will examine ape communication as the use of gesture and utterance to convey concepts to others. The use of a communication system in this sense, largely lacking in lower animals, is something that apes share with us. For example, we will see how the pseudosexual bottom-presenting behavior of chimps is a way of using gestures to communicate the acknowledgement of dominance and thus to deflect aggression.

Note that the gestures of apes may differ from those of humans in significant ways. In "A Comparison of the Gestural Communication of Apes and Human Infants," Michael Tomasello and Luigia Camaioni suggest that even one-year-old children use gestures in ways that are beyond the capability of apes. "Chimpanzee gestures seem to be used for *imperative* purposes to request actions from others . . . not declaratively to direct the attention of others . . . for the sake of sharing interest . . . or commenting."[2] They also note that "chimpanzees use a number of gestures intentionally, that is, in flexible ways tailored for particular communicative circumstances."

NOTES

1. Terrence W. Deacon, *The Symbolic Species* (New York: Norton, 1997), p. 43.

2. Michael Tomasello and Luigia Camaioni, "A Comparison of the Gestural Communication of Apes and Human Infants," *Human Development* 40 (1997):7–24.

4

FAMILY LIFE
WITHOUT FATHER
GROWING UP CHIMPISH

DESPITE THE ABSENCE OF FATHERS from families, chimps have a rich and enduring family life. Most mothers have several young, with four to six years the usual spacing between them. Infants are of great interest to the rest of the community, but especially to their immature siblings. Unlike the half-grown and the adults, who must be carefully deferential to their social superiors, the infants and young juveniles (who have not yet mastered the rules of what passes for polite chimp society) get patted and petted by whomever they barge in on. Mothers of both youthful and adult children come to the defense or support of their young, unless the scrap is with someone from a higher ranking family. Likewise, children who are old enough come to the defense of their mother or siblings.

To refer to chimpanzees as "fathers," "mothers," and "children" is to use language to *humanize* these apes to some extent. This use of language was not only frowned upon but viewed as unacceptable not long ago by editors of scientific journals. Jane Goodall was one who not only pioneered the study of chimps but did the same for such humanizing language in science writing.[1] Her books about these creatures with whom she lived for so long, *In the Shadow of Man*, *Through a Window*, and *The Chimpanzees of Gombe*, are the source of most of the information in this and following chapters about apes.

Sexual interaction between adult male children and their estrous mother is rare. She wants none of it and her son's interest is half-hearted. Do we know why they avoid what, for humans, would be called incest? No, we don't.

I can infer, however, that each has developed a certain mindset that is based on the way they acted toward one another while the child was growing up. The son's new interest in sexual interaction provides a new mindset, one which is in conflict with the earlier one in his head, and with hers.

Chimp infants show an odd response when their mothers are copulating with suitors. Young chimps attempt to interfere by jumping on the engaged suitor and pushing him away. Perhaps they resent their mothers' momentary (chimps copulate for only a few seconds) higher priority for love-making. We know that the emotional bond between a mother and her immature child is strong. The mother of an infant who has just died may carry or drag the body around for a few days. She is often listless and disinterested, even in food, for a longer period. For a young child whose mother dies, the situation is worse. The child's depression and subsequent disinterest in food sometimes leads to death.

Let's look at a now well-known family, whose members' names begin with "F," whom we meet in primatologist Goodall's books. We meet them when the old, somewhat frail mother, Flo, is in her forties—a much more dilapidated age for chimps than for humans. Her youngest, daughter Fifi, is two; juvenile son Figan is seven, and older son Figan is almost mature at twelve. Flo is usually attentive and easy-going with young Fifi. Whenever (a couple of years later) Fifi whimpers and stretches her open hand out, a gesture we would name asking or begging, to her mother who is eating a bunch of bananas, Flo usually lets her take one. When the banana at issue is the old lady's last, however, the two get into a screaming, hair-pulling brawl. Most chimp mothers are less generous than Flo; their infants have learned that insistent open-handed banana begging often gets them nothing less than a "knuckle sandwich."

WHEN PROVISIONED WITH BANANAS, TOP CHIMP MALES FOUND NEVER-KNOWN USES FOR POWER

The consequences of bunches of bananas becoming available to chimps are significant for the light they cast on these apes. They also illuminate human behavior that started when occasionally plentiful produce reached us with the agricultural revolution.

For family F and their fellow chimps, the bananas were provided in increasing quantities at the human observers' camp site. Those observers slowly slipped from the straight-and-narrow path of ethological purity by providing more and more bananas to lure more chimps more often to where it was easier to observe them. Remember the old joke where the drunk looks for his lost keys beneath the street light? When asked, he admits losing them a ways down the dark street, but says he is looking where he is because the light is better. Were the drunk an ethologist or a primatologist, he might have done better. He would have used a huge electromagnet to attract the keys into the light, perhaps rubbing off some of their edges in the process, but coming up with precious metal for all that.

For the chimps provisioned with bananas, the intervention did not dull rough edges; instead, it sharpened them. It certainly achieved its original purpose, to bring chimps frequently to the observers. This made less necessary the taxing task of tracking the creatures over rugged terrain. It also turned out to be an unintended but instructive experiment showing how primate behavior changes in response to "unnatural" concentrated abundance.

Chimps normally seek and find fruit as it ripens on trees within their territory. Each individual fruit must be plucked by hand and popped into the mouth. Pulling rank to get good fruit from heavily laden trees does occur on occasion, particularly with the top males, but it is limited. Once a chimp, such as top ranked Mike, threatens and thereby succeeds in displacing another ape, he doesn't try to prevent the underling from feeding nearby or elsewhere on equally succulent fruit from other overburdened branches or trees. Put in terms of the first of Mother Nature's figurative three commandments for conscious creatures, "Look out for number one," Mike has acquired the feeding spot he sought. He can do no better, and so he has no reason to deprive others from finding and eating fruit.

When first the feeding station was started, it was no more than a place where large bunches of bananas, brought in from nearby towns, could be dumped over a broad area. The immediate response of the first chimps, and of the many others who soon showed up, was gluttony. Males in particular can and do eat several dozen bananas, hardly pausing for more than something like a belch. The increasing number of visiting chimps soon made it necessary to build steel-lidded concrete banana boxes that were sunk into the ground. To reduce the danger to the human experimenters from the sturdy, greedy chimps, wires were attached to the lids. This allowed the boxes to be opened from a distance. Note that even out-of-shape captive chimpanzees are four or five times as strong as an in-shape human athlete (as Richard Wrangham and Dale Peterson inform us in *Demonic Males*).[2]

HOW WE GOT TO BE HUMAN

The chimps soon make a mockery of the attempts to control them with boxes, lids, and wires.

More important to us here, the experimenters found it increasingly difficult and soon impossible to make sure the females and the youngsters got what the humans thought was a fair share. The powerful males no longer merely displaced and then ignored an occasional lower ranked ape, as they had in the forest primeval. In the presence of a never before known abundance of fruit, they acted out new rules for the new world they encountered. No lesser creature dared reach for a stray few bananas within the sight of the new order of bullies. Before provisioning, power was used, just as food was available, at the "retail" level. When provisioning provided the concentration of "wholesale" quantities of food, the apes expanded their natural desires for "more" for themselves, at the expense of others, to "unnatural" wholesale levels.

Look at *power* in the lives of communal creatures. I agree with Matt Ridley when he says, "Power-seeking is characteristic of all social mammals."[3] Power is primarily social. Its basis exists as a nonconscious neural-network pattern in all social animals that exhibit something like a pecking order. For lower social animals, power is a real thing in the same sense that sex and food are real, but it is not a concept that they can use. Dorothy Cheney and Robert Seyfarth show that vervet monkeys understand power in the sense of knowing the pecking order position of every other animal in the troop.[4]

Maxine Sheets-Johnstone examines the roots of human power: "Present-day accounts of power assume power and power relations to be . . . a peculiarly human idea. . . . In consequence, the accounts nowhere come to grips with the question of the genesis of power . . . a hominid evolutionary heritage."[5] Even her usage, limiting humanlike power to hominids, is too confining. Not only humans and prior hominids, but apes, too, have sufficient mentation for power to exist as a (conscious) concept. In chapters to come, we will see good evidence, in the form of gestures, that apes have the concept of power.

Mike, Goliath, and the other high-ranked chimps gave new meaning, in the land of the apes, to the notion of power. Before provisioning, power was certainly a valuable thing to have: it gave a better choice of food and females. It may also have provided a bonus: fear in the eyes of the weaker. With provisioning, chimps went big-time with power, which now meant gluttony, greed, and, perhaps, the satisfaction of depriving others.

Family Life without Father: Growing Up Chimpish

Another primatologist, Frans de Waal, describes the same power problem arising with captive chimpanzees in a twenty-five-acre enclosure provided for them at Holloman Airforce Base in New Mexico.[6] The absence of separate facilities for feeding the apes individually or in small groups resulted in violent fights at every meal. At the Dutch Arnhem Zoo, he tells us the problem was avoided by splitting the apes into small groups within separate cages before giving them their daily provisions.

Where did this simultaneous desire to satisfy the self and deprive others come from? It did not result from new mutations at the genetic end of the mentation of apes. Rather, it is something learned. What was so easily and instantly learned by these apes was that they could satisfy Mother Nature's first commandment—look out for number one—and her second (issued to communal species)—thou shalt make thy life with thine own kind—at the same time. The pecking order itself, which is seen in communal creatures as smart as hens, was the first way in which learning added to self-centered and social instincts to create something new, something called power. For the top dogs among communal creatures with concept-using brains like apes (and some birds), the pecking order was only the beginning of power. When it reaches the level of a (conscious) concept, power means two things at once. First, to do well for yourself with the things of your world, of which food is the first. Second, to do well in your own eyes and those of your own group—which meant raising yourself while diminishing others. Lower animals don't have the mentation to handle power beyond the bare bones of a pecking order. Given a suitable situation, chimps have enough conceptual ability to expand their ideas about power. That is why the apes "went ape" when provisioned with bananas. The higher stakes provided reason to raise the ante and every bet in the use of power.

On a far greater scale, our own twelve-millennia-old agricultural revolution provided a similar sudden availability of food at "unnatural" wholesale levels. In response, human Mikes and Goliaths, far more imaginative than chimps, also rose to the occasion. They not only moved to new quarters at the intersection of Egocentric Street and Communality Drive, they erected a control tower there, a tower where they, too, went ape with power. With brains and minds of human size, they went beyond wholesale power, they went to manufacturer levels with self-taught deprivation of others, and also with gluttony, greed, covetousness, pride, anger, and lust. The slothful were too lazy to move.

We'll look further at this expansion of human social power in chapter

13. Note here that human power, in a somewhat different sense, now enables us to bring other species, including that of the apes we are here verbally observing, to the point of extinction.

YOUNG CHIMPS' LEARNING: A LOT DURING A LONG CHILDHOOD, BUT NOT MUCH MORALITY

Return now to the chimp family F. We see that before long, Flo has a new son, Flint. His story is short and also sad. His earliest years go well enough, with good care from his mother. Most infants are weaned when they are about four. The infants respond to weaning by lashing out at their mothers with resentment, anger, and temper tantrums. While she is weaning, a chimp mother usually stops her child from getting his expected free bus rides on her back. She also stops the infant from sleeping in her nest, built nightly by bending leafy branches and twigs in a tree. Flo starts to wean Flint at about three. But even when he is five, he still has tantrums, he hits and bites his mother, he screams and flails on the ground. Old and tired now, Flo, who has always seemed to give wisely to her offspring, does not give him the back of her hand. Flint continues to ride his mother's back and sleep in her nest when he is long past six.

When, after a couple more years, worn-out Flo dies, Flint, now a juvenile at the age of eight and a half, cannot handle the loss. Depressed and without appetite, he dies soon after. Flint's story is a sad one indeed, hardly imaginable for any lower mammal. Should we draw from it the moral, "Spare the rod and spoil the child?" We could, but don't we already have, from our human condition, an overabundance of cautionary comments, many of them conflicting one another?

In contrast with Flo, some chimp mothers are intolerant if not indifferent to their young. Passion was a chimp mother who, far from indulging her infant daughter, Pom, acted with a neglect that seems, to us, almost malign. At two months of age, Pom hurt her foot badly enough to interfere with her normal ability to cling beneath her mother by holding onto her hair. Rather than constantly supporting the infant with one hand, as most chimp mothers would do, Passion keeps pushing the baby onto her (Passion's) back. This is when little Pom is only one-third as old as most infants who make this shift. Thereafter, even in pouring rain, Passion never allows the infant to return to the underneath clinging position. Indifference is Passion's normal attitude toward Pom, whether the infant wants to nurse or the mother wants to go elsewhere. The youngster learns to survive by devoting all her attention to her mother. When Pom is weaned at about four, she suffers a longer and deeper version of the depression common to chimp infants in this transition.

When Pom has a newborn sibling, Prof, she shows the usual interest chimps have in babies, but the interest does not last and Pom reverts to her depression. Surprisingly to those of us who start to think chimps are almost like humans, there is no sign of rivalry or resentment by her toward her younger sibling, nor is there any such behavior by other young chimps with younger siblings.

Pom (like other young chimps in that position) does not formulate the concept that the new sibling has, in one sense, taken her place. Yet that connection is something that young human children universally make. In fact, a recent book by Frank Sulloway makes a scientific study of how birth order affects the character and behavior of children.[7] The author shows there is a significant difference, in particular, between the first-born and subsequent children in a family. First-born children have no competition when they first start to grow up. Soon, however, there are others, younger and weaker, competing for everything wanted from the parents. Those born first tend to be conservative throughout their lives, which means mainly defending whatever they have. In contrast, a large fraction of those born after the first are inclined throughout their lives to get what they want with a shake-up of the existing power structure.

The question of interest to me here is this: Does this competition for power among children in the family have genetic origins? Or is it only a matter of concept learning? I could take a reductive position here and claim adaptive genetic roots for this behavior in humans, roots that did not exist in our ape ancestors. I could say we have a gene for "The Cain Mutiny" that makes siblings frequent competitors and sometimes murderers of their brothers or sisters.

A simpler explanation exists for this difference between humans and chimps. Assume you share egocentricity and communality with apes, and those two instincts form the basis for the learned concept of social power. You need nothing more than a bigger brain and a better-sized mind that forever push you to find some meaning in your situation in life. To generate competition with your siblings, it is enough to make self-taught conceptual sense of your world, sense that is congruent with your ancient instincts. This sense is not an objective truth about the world. It is a social truth that is created, a truth about your kind.

To get back to our chimp cousins, when Pom, about seven, is with three-year-old Prof one day, she hastens to climb a nearby tree after seeing a large snake nearby. When Prof does not react to the snake, Pom scurries down, scoops him up, and hurries back up the tree. Later, as an adolescent and as an adult, Pom has no friends except, oddly to us, her mother. When young Pom is in heat, she is not like the usual young female. Although she seeks the sexual response of males, unlike her peers, she runs away shrieking after she receives copulative males. The males, however, may not have the brains to think that Pom in heat is not a hot number.

There is doubtless, somewhere in our fables and cautionary tales, as good a moral for Pom's story as the one for young Flint's. But chimps have few morals, in the other sense of socially shared concepts of what is right and wrong, as we shall see. Nor can they draw child-raising precepts from their observations of mothers and children like those described here. As for us, surely we have an overabundance of human sources from which to draw rules for rearing our own young. Even though we often find those sources inadequate and conflicting, it is doubtful that our knowledge of apes can add more than a few crumbs to our mixed bag of moral guidance. In sharp contrast with that from parents who *demonstrate* moral behavior, there is only minor value in moral advice that is merely *voiced*. One exception is in the enjoyment of a play like Shakespeare's *Hamlet*, where Polonius starts to pontificate to his son:

> And these few precepts in thy memory
> Look thou character. Give thy thoughts no tongue,
> Nor any unproportion'd thought his act.
> Be thou familiar, but by no means vulgar.

With Laertes's eyes glazing, Polonius draws near the end:

> Neither a borrower nor a lender [be],
> For [loan] oft loses both itself and friend,
> And borrowing dulleth [th'] edge of husbandry.

Family Life without Father: Growing Up Chimpish

Morals apart, we can conclude that chimps, like us but unlike lower animals, have a long, impressionable, and sometimes flawed childhood that contributes to the shape of their adulthood.

The story of Passion and Pom has a supplement which is a shocker, when Pom is more grown up. One day, Passion, with Pom not far behind, steals a tiny infant from its mother's arms. The frightened mother flees, but this is no mere kidnapping, for at this point Passion sinks her teeth into the front of the infant's head, killing it. The mother returns, sees her bleeding infant, and departs again. Then Passion feeds for several hours, sharing the flesh with Pom and Prof, Pom's younger brother. They chew each bite with a handful of leaves, as chimps normally do with the meat of their occasional prey, young monkey, bushpig, or bushbuck. The next year, when the murdered infant's mother has a new child, the participants repeat the feast of flesh. Over several years, such raids by Passion and Pom account for the deaths of half a dozen infants. In "Comparative Socioecology of Gorillas," David Watts notes that both "male chimpanzees and gorillas are infanticidal."[8] So the killing of infants by apes (by male apes at least), is not extraordinary, but the eating of such babies remains decidedly unusual, if not abnormal.

It's hard to find appropriate words for this highly unusual behavior. Cannibalism is a term we use for creatures with a moral sense. In *From Lucy to Language*, Donald Johanson and Blake Edgar discuss evidence for human cannibalism. They note that a cranium "belonging to *Homo heidelbergensis* displays diagnostic stone tool cut marks" with no sign of chewing by carnivores.[9] "Around 600,000 years ago, this individual was intentionally defleshed. . . . But whether the butcher ate any of this flesh cannot be answered." They also note evidence that neandertals and early sapiens occasionally resorted to cannibalism.

The concept of cannibalism has overtones that are inappropriate for chimpanzees. I will distance us (from what we can't help regarding as horrors) by changing the subject and noting that chimps usually kill their monkey, pig, or buck prey before eating them. In contrast, there are accounts of baboons biting into the belly of prey without killing it. The difference might be, if I dare to so call it, an intellectual advance of a sort. That is, it is possible that chimps sometimes understand the distinction between living and dead prey. In contrast, baboons may distinguish only beasts in the bush from those in the hand, and pay no heed if the prey squirm during the first few bites.

Chimpanzee mothers are usually closely attached to their young, and seem to understand when one of their number has stopped living. Goodall writes about the mother of an infant who had lost the use of his limbs due to polio: "She supported him carefully and each time he screamed she cradled him at once. . . . The moment he died (or, at least, lost consciousness . . .), she began to treat the body in an extremely careless way."[10]

Why, then, would a chimp mother continue for days to carry her dead infant? Perhaps for one of the reasons we humans have burial ceremonies and wakes. We cannot fully accept death on the instant it occurs—we need time for its teeth to sink into us.

CHIMPS HAVE STATES OF MIND: THEY MIGHT AMOUNT TO CONFLICTING MULTIPLE MINDS

Mellisa and her near-year-old son, Goblin, provide us with another vignette about these barbarous creatures who make human barbarians look like high-society members nibbling tiny sandwiches at a tea party. One day, little Goblin wanders over to nearby top-ranked Mike, whom all knowledgeable chimps treat with kowtows and other chimp variations on toadying and trembling. Mike is contentedly munching on a mouthful of figs. He turns to little Goblin, reaches out, and pats him gently on the back a number of times. A touching example, is it not, of how understanding and accepting the adults can be toward an infant?

Not far off, sudden vocalizations announce the impending arrival of other chimps. Immediately, all the apes—except the infants—get excited. Mike himself arises, all his hairs rise and puff up his presence. He begins to vocalize. He stands erect, he raises his arms, he shrieks. He starts to move, he looks for branches to break and hurl. Every chimp who knows anything hastens to get out of his way. Unconcerned, little know-nothing Goblin again totters toward Mike. Vastly excited now, Mike begins to charge. Charge where? Charge forward, where whoever is there will learn who's in charge. Goblin, in his path, is some thing or other he seizes and drags like a small branch. The infant's mother, suddenly brave, throws herself at Mike. She gets swiftly beaten up as he passes, but she succeeds in freeing her infant, and hurries off with him.

What has happened to Mike? Does he have two last names, one of them Jekyll, the other, Hyde? Yes, something like that is what has happened. Mike changed his mind. He put his gentle avuncular teacup-taking mind back in

116

the cupboard and pulled out his big mugging mind, the one suited for the stir of exciting power. In chapter 2, we saw a bird changing its mind back and forth between defending its nestlings and defending itself from a predator. One mindset said, "My young are part of me, I must defend them," until the predator got too close to the bird. Then the other mindset took over with "The first commandment is to look out for number one" until the plover had safe distance from the attacker.

When I say that we humans frequently do the same kind of mind changing, what I mean is what Steven Pinker writes of in *How the Mind Works*: "Mental life often feels like a parliament within. Thoughts and feelings vie for control as if each were an agent with strategies for taking over the whole person."[11] Of course, we differ from the chimps. When we lose our temper and find our anger, most of us don't beat up babies, although some of us swat them. When we switch back from the unbalancing mug mood to that of the more level teacup, we often experience guilt, a feeling that is unknown to the uncultured and unreflective ape. But that is a different issue, one of morality again, to be explored later.

Before we leave Mike and his changing minds, two more issues must be addressed. I am not suggesting some kind of magical or science-fiction scenario, where, like mysterious creatures from outer space, apes and humans can change from one complete and isolated mind to another. Nor am I implying seemingly unconnected multiple personalities, although those might be an extreme and extravagant version of multiple mentalities. For all that we know with clarity and certainty about our multiple minds, we could as suitably refer to them as "mindsets," "states of mind," or "frames of mind." One advocate of regarding each brain as associated with a multiplicity of small subminds is Marvin Minsky, in *The Society of Mind*.[12] He notes that most of us frequently change our minds about whether we have a single self or multiple selves. In *Evolving the Mind*, A. G. Cairns-Smith discusses some unfortunate humans who are of precisely two minds in a more literal sense that is not pertinent here: those who had had cut the direct connections between the left and right hemispheres of their brains.[13] More specific about the *domains* of normal human mentation is Steven Mithen.[14] He finds four cognitive domains, language, psychology, biology, and physics, that were quite isolated from each other prior to a time about 40,000 years ago. After that time, he thinks a changed mental structure allowed more fluidity in the transfer of information between these realms.

These minds we change have some limited indirect connection with one another, via perception and memory. Even within an ape, there might be some understanding in each mind of the other minds' activities. More than with us, strong feeling associated with a particular frame of mind dictates actions despite any understanding.

The second issue is who or what in the chimp's head is the subject, "Mike," and what is the relationship between his minds and "him?" (To avoid clutter, the idea is abandoned here that only humans have "minds," whereas other animals have more limited "mentalities.") There is a multiplicity of mind-sets in Mike's head. We have seen a mind that is benevolent to young children. We have seen an excited assertive aggressive mind that wordlessly says, "What's going on here? I'll handle this! I'LL HANDLE IT!!" There is a family-oriented mind that lets him feel his siblings and mother are a part of him and his interests. As we will see, there is a sexual mind that is on the lookout for females in heat. Apart from these and other minds that occupy him, is "Mike" something different? He is little more than the sum of these minds, or rather, the totality, as there is no central "I" doing the summing in his head.

Well, that's easy for me to say, especially as neither the creature named Mike nor the unsummed-mental-sum "Mike" can argue with me. What, then, is to be said about each human and the "I" that seems to be in charge, the "I" that nakedly proclaims itself emperor though it may have as many minds as a Hydra had heads? (The Hydra, a mythical serpent that was slain by Hercules, originally had nine heads. However, when one of the heads was cut off, it was replaced by two others. So it's hardly easy to know how many heads a Hydra had.[15])

The "I" that claims to be ruler within each of us presents a tougher question than its like in Mike's head. It is not tougher because we are totally different from our ape cousins in that aspect of mental structure. Although we have more communication between subminds, we are still similar. Nor is the problem only that our vastly greater reflective consciousness has endowed us each with the concept of an "I." Rather, it's tougher because it is always one of those "I"s—one of mine right here—that must address the issue. The difficulty is due to self-reference or circularity, which usually becomes paradoxical in complex conceptual structures. For example, the paradox of the shaven barber who shaves every man in town who doesn't shave himself, and who does not shave anyone who does shave himself. Who shaves the barber? We cannot say. For another example, the paradox in a sentence that reads "I think that I am not real in the same sense that my several minds or frames of mind are real." When one "I" looks at the other in such a sentence, the effect is cockeyed.

I want to return our eyes to the apes and leave the "I" concept here, even though its paradoxical nature leaves me vulnerable to comments, within me or within you, from subminds we could call Dr. Heckle and Mr. Jibe.

LOSERS IN CHIMP FIGHTS:
THEY WANT BOTH TO GET AWAY FROM
AND TO GET CLOSE TO THEIR TORMENTERS

In the process of growing up, male chimps slowly make the transition from life with mother and siblings to life with adult males and females. That transition involves learning and establishing their place in their society. It is low, rather than high, society they are entering, so learning how powerful they and others are is a large and important task. Consider Flo and her son Figan when he is entering puberty. His hormones are beginning to flood him, and frequently he engages in a charging display like that of Mike. When he does so at eight, his mother ignores him. A year later, his physical and social power has increased enough so that she hastens to get out of his way. After another few years, Figan and the others near his age frequently contest each other for superiority, but the contests are mostly at a level approaching that of gestures, and not that of blows.

Figan and the other youngsters are usually of two minds about the powerful males near the top of the pecking order. One of those minds is fearful of the top apes, which is understandable enough, given the brutal power that the latter can activate so swiftly and suddenly. The other mind is attracted to the male leaders. This seems odd. What's going on here? Is this a form of male bonding, an ape version of the recent "men's movement"? Is it something most human males don't like to hear about, sexual attraction between males? Another anecdotal tale will steer us in a more fruitful direction.

One of Figan's fellow almost-adults goes, in primatologists' circles, by the name of Evered. On one occasion, Evered finds himself in the same position vis-à-vis powerful Goliath as was infant Goblin in the path of mighty Mike. As might be expected, charging Goliath brutally beats up youthful Evered. When the mayhem stops, does Evered, bleeding and at a loss for several handfuls of hair, run away? He may wish to for a moment but, no, he does not. On the contrary, he follows on the trail of galumphing Goliath. When the latter, in his majesty, deigns to stop, Evered, still screaming his anguish, creeps toward his tormenter. Shaky Evered gets close, humbly turns to present his rump (in a later chapter we'll see that to be a gesture of deference), and then, quaking, crouches low to the ground. After a few moments, Goliath, with the chimpish equivalent of noblesse oblige, accepts this touching acknowledgment of his grandeur. He changes to his benevolent mind, and pats Evered's back for a while until the youngster is calm.

Instead of simply scramming, why does Evered seek his tormentor? He does so because Goliath's fearless power and strength is also the source of potential comfort. Only if poor Evered gains, or regains, for himself good terms with the source of that frightening power can he feel safe.

Here are a few lines from Gerard Manley Hopkins's "Carrion Comfort," wherein a human protagonist talks about his antagonist, who is his God. He also talks about their struggle. I will try to show how the human protagonist's account closely parallels fearful Evered's interactions with powerful Goliath:

> But ah, but O thou terrible, why wouldst thou rude on me
> Thy wring-earth right foot rock? lay a lion-limb against me? scan
> With darksome devouring eyes my bruised bones? and fan,
> O in turns of tempest, me heaped there; me frantic to avoid thee
>   and flee?
>
> Why? That my chaff might fly; my grain lie, sheer and clear.
> Nay in all that toil, that coil, since (seems) I kissed the rod,
> Hand, rather, my heart lo! lapped strength, stole joy, would laugh, cheer.
>
> Cheer whom though? the hero whose heaven-handling flung me,
>   foot trod
> Me? or me that fought him? O which one? is it each one? That
>   night, that year
> Of now done darkness I wretch lay wrestling with (my God!) my God.

Evered, too, has wrestled with his powerful antagonist. He, too, has been rudely rocked by a wring-earth foot and a lion limb—Goliath's. Evered may not feel a tempest from his antagonist fanning him, but he too feels blown away by his opponent's power. Evered, too, is in conflict, frantic to avoid and flee his antagonist, yet not doing so. Instead, Evered, too, abjectly acknowledges the power of his opponent. His way of kissing the rod or hand is to present his bottom and prostrate himself. After demonstrating his fealty to a greater power, Evered, too, is rewarded for his obeisance, and he is calmed by the touch of his antagonist's hand; no doubt his strength and joy in life will also soon return.

Family Life without Father: Growing Up Chimpish

As for Goliath's rod, it is the same one that old Flo failed to use on young Flint, when we wondered whether "Spare the rod and spoil the child" was appropriate for the upbringing of young chimps. After all, it is parents who first wield the rod of power which is later shifted to the hand of a natural or supernatural authority.

Wordless Evered is no match for Hopkins's eloquent protagonist when it comes to providing an explanation for the dreadful blows that fell upon him. Hopkins's character confidently explains the reason was to separate his "grain" from his "chaff." Humans commonly can and do generate explanations for whatever has befallen them. In fact, we insist on having meaningful reasons from ourselves and others. If Evered had the advantage of more conceptual ability plus advanced training in sign language, he might have figured out a reason why Goliath with his rod so crudely caned him: "To teach me my proper place" or "So I wouldn't get spoiled." Without training in Latin, shame, and social concepts, Evered cannot be expected simply to have announced his guilt: "mea culpa."

My purpose here is hardly to compare (powerful chimp) Goliath with (Hopkins's protagonist's) God. Rather, it is to point out that apes and higher hominids are commonly of similar dual minds about the powers-that-be in their lives. One mind fears that its bones will be bruised as a wring-earth right foot rocks it, the other wishes to kiss the rod or hand and thereby gain strength and joy.

Apes exhibit the basic concepts about having power or about reacting to it. For humans, holding and responding to power has gotten complex. Once upon a time, our rulers had it all—political power, wealth, glory, fame, you name it. Now it's not just emperors, but presidents and other political winners; movie, sports, and religion stars with images on large or small screens; private bosses, billionaires, dominatrixes; there's no limit on the kinds of people who hold diluted power, people to whom we'd cozy up if we could, and avoid if we can't.

Sucking up to power or scramming are not the only choices available. Some apes (and some humans) have another option, as we shall see when we encounter (in chapter 7) braver chimps. There we also encounter Prometheus, the Titan who continued to defy the gods after he stole fire from them to give to humans. More commonly, for both apes and sapiens, one mind is in fear of the mighty antagonist, while the other mind is eager to bend the knee and beg for blessing in the hope of gaining "the peace that passeth all understanding"—and relieveth anxiety in passing.

HOW WE GOT TO BE HUMAN

Another example, less dramatic but no less interesting, of these two peculiar aspects of human mentation regarding a deity, can be seen by examining the philosophical argument in Blaise Pascal's wager that it is safer for humans to choose theism over atheism. Richard Popkin writes that Pascal tried to show it is preferable "to believe in God. . . . If there is a God . . . he is infinitely incomprehensible to us. But either God exists or he does not exist, and we are unable to tell which alternative is true. However, both our present lives and our possible future lives may well be greatly affected by the alternative we accept. Hence, Pascal contended, since eternal life and happiness is a possible result of one choice (if God does exist) and since nothing is lost if we are wrong about the other choice (if God does not exist and we choose to believe that he does), then the reasonable gamble, given what may be at stake, is to choose the theistic alternative. He who remains an unbeliever is taking an infinitely unreasonable risk just because he does not know which alternative is true."[16]

I am wholeheartedly with Pascal in preferring an eternal happy life, preferring it to an eternal unhappy one or no eternal life at all. However, I can't help noticing a flaw in Pascal's argument, a flaw that demonstrates that he was of two minds about the nature of the deity. First, Pascal tells us the deity is utterly unfathomable. He does no shilly-shallying there, right? Despite that, he also tells us there is a huge risk in not believing in the deity's existence. How, then, could we possibly assess what is an unknowable risk that is associated with nonbelief? Also, why is it a risk with such negative consequences? The answer is that the second of Pascal's minds regarded the deity as very much a knowable egomaniacal deference-demanding Someone. That is a Someone, who, somewhat like the God of Hopkins's protagonist, will reward obsequious believers and punish defiant nonbelievers.

How it came to be that many people have believed their god behaves like a sometimes-benevolent, sometimes-brutal, respect-demanding most-powerful-of-all chimp may never be clear to us. That they have such beliefs, however, suggests that the long-gone ancestors that we share with apes had similar attitudes toward powerful others. It further suggests that those ancient apes had even more distant animal ancestors who, like dogs, had no choice (because they lacked concepts) other than to lick the limb that battered them.

This long history of obsequious behavior toward powerful entities may suggest to some that it has a genetic component. I suggest otherwise. That capability is not, in itself, a genetic adaptation. Rather, it is no more than an adaptation byproduct, a byproduct whereby egocentric social concept-wielding creatures place a high value on both self-preservation and social power. It's clear that self-preservation is the core of the egocentricity that is our first law of being. On the other hand, power, whether it be one's own or that of one's antagonist, is primarily relative to that of others in one's group. Thus, it is central to the communality that is another law of our existence.

There are some situations where one of these laws pushes a creature in one direction, and the other in another. We saw this with the initial "uneducated" ground-nesting plover in chapter 2. At other times, such a bird may have the learning and conceptual imagination to make a happy marriage between these two parts of its mind; it can serve both of these parts of its mentation by dodging its predator as it saves its nestlings and itself. I'm sure that many apes and many humans have done as much. Indeed, let's hear a cheer for parts of mentality that work together rather than against one another.

NOTES

1. Jane Goodall, *In the Shadow of Man* (Boston, Mass.: Houghton Mifflin, 1971); *Through a Window* (Boston, Mass.: Houghton Mifflin, 1990).

2. Richard Wrangham and Dale Peterson, *Demonic Males* (Boston, Mass.: Houghton Mifflin, 1996), p. 6.

3. Matt Ridley, *The Red Queen: Sex and the Evolution of Human Nature* (New York: Macmillan Publishing, 1993), p. 195.

4. Dorothy L. Cheney and Robert M. Seyfarth, *How Monkeys See the World* (Chicago, Ill.: University of Chicago Press, 1990).

5. Maxine Sheets-Johnstone, *The Roots of Power* (Chicago, Ill.: Open Court Publishing Company, 1994), p. 43.

6. Frans de Waal, *Chimpanzee Politics* (New York: Harper & Rowe, 1982).

7. Frank Sulloway, *Born to Rebel: Birth Order, Family Dynamics, and Creative Life* (New York: Pantheon, 1996).

8. David P. Watts, "Comparative Socioecology of Gorillas," in *Great Ape Societies*, ed. William C. McGrew, Linda F. Marchant, and Toshisada Nishida (New York: Cambridge University Press, 1996), p. 25.

9. Donald Johanson and Blake Edgar, *From Lucy to Language* (New York: Simon & Schuster Editions, 1996), p. 93.

10. Jane Goodall, *The Chimpanzees of Gombe* (New York: Cambridge University Press, 1986), p. 383.

11. Steven Pinker, *How the Mind Works* (New York: W. W. Norton & Company, 1997), p. 419.

12. Marvin Minsky, *The Society of Mind* (New York: Simon and Schuster, 1986), p. 12.

13. A. G. Cairns-Smith, *Evolving the Mind* (New York: Cambridge University Press, 1996), p. 176.

14. Steven Mithen, *The Prehistory of the Mind* (London: Thames and Hudson, 1996).

15. *Webster's New Twentieth Century Dictionary,* 2d ed. (William Collins+World Publishing Co, 1975).

16. Paul Edwards, ed., *The Encyclopedia of Philosophy,* vol. 6 (New York: Macmillan, Inc., 1967), p. 54.

5

CHIMPS ARE NO CHUMPS

MENTALITY WITH ROOM FOR NEW CONCEPTS

HUNTING AND BASHING OTHER CHIMPS: MORE SPORT THAN WAR

CHIMPS HAVE CLOSE AND DIFFERING relationships with one another, both as individuals and as members of family groups, and not merely as undifferentiated others of the same kind. This is an expansion from a lower animal's categories, the kinds of things it differentiates in its here-and-now world. I'll use the word "social" instead of the word "communal" to mark this difference from lower mammals, who pay less heed to the individuality of another of their own kind. This social community in which chimps and bonobos live is in some ways remarkably like a human community. One of the large differences, however, is the dearth of group values driving individual and group activities for the apes. Members of one chimp community will differ from those of another in some of the things they eat, with no apparent reason except that social habit places value on certain foods. Thus, some chimp groups use stones to smash nut husks, and others simply do not. Although different communities of chimps may have different food preferences, there is no evidence that they have different concepts of appropriate behavior toward individuals.

Among the things that all chimp communities share is a pecking order that says who threatens and who feels threatened when two members of a group encounter one another. In all chimp communities, the basic

---

pecking order is expanded by means of behavioral gestures that show and acknowledge power.

To see how a scientist examines ape gesture, look at Kim Bard's "Intentional Behavior and Intentional Communication in Young Free-Ranging Orangutans."[1] Bard finds that food-sharing by a mother with a child is increased by the child's use of a communicative gesture: placing an open, cupped palm-up hand under the mother's chin. In another study, Bard and others find that laboratory chimpanzees use a gesture of referential pointing, directing an index finger to bits of food (that are outside the reach of the caged animals) while alternatively looking to the food and to a nearby human observer (who could get and give the food to the gesturing chimp).

These studies are painstaking efforts, with a number of apes and a number of instances with each one, that allow mathematical assessment of the confidence with which the assertion is made that the behaviors are indeed *gestures*. My own assertions about apes using other gestures are not based on such efforts, so I cannot provide means of associating them with a confidence level.

What distinguishes *gestures* from most ape behavior (and all lower mammal behavior) is that they are *about* something other than themselves. This quality of "aboutness" is discussed in *Human Evolution, Language, and Mind*, by William Noble and Iain Davidson. They describe three kinds of behavior that are about something: "icons, indexes, and symbols. Icons resemble what they refer to, . . . such as a pantomime . . . ; indexes are constrained by . . . what they indicate (for example, smoke as an index of fire). . . . Symbols stand in arbitrary relation to their referents."[2] These ideas arose with reference to *human* behavior, which may be why there is no mention above that *consciousness* is basic to all of them. It is also the neglect of this subjective aspect that allows some scientists to suggest that computers know what symbols *mean*. In my view, our country cousins, the apes, use *gestures* as both icons and indexes that refer to (conscious) concepts of power and other things. They use gestures as symbols, in the sense described above, only with the experimental languagelike behavior they learn from us.

Even when rare and aberrant behavior by individual chimps occurs, such as that of Passion and Pom, the mother-daughter pair who stole and ate infants, there is nothing in the way of shared group response to those events. Other chimps, observing such an event, seem simply not to know what to make of it; it is beyond their ken. And the malevolent pair, what went on in their heads? We simply don't know. What we do know is that Passion, the mother, in prior years seemed to have a remarkable absence of behavior toward her infant daughter that would be called empathetic or compassionate if she were human. Could the same flaw in her mentation contribute to her treating others' infants as prey? If she was so flawed, perhaps it was a flaw with partly genetic origins, a flaw that made her an ape version of a psychopath or sociopath.

We do know that after a while the aberrant infant-killing-and-eating behavior ceased. The mother-daughter pair, of course, showed no signs of remorse or guilt. How could they? Remorse and guilt, social concepts for humans, who know they and their peers have minds, are beyond the means of chimps. A simplified version of conceptual shame, in contrast, is something that perhaps even the mother, as well as the daughter, might have shown (with obeisant behavior) were the pair caught in the baby-stealing-and-eating act by a powerful and threatening Mike or Goliath. Cringing Evered (whom we encountered in chapter 4), whether or not he believed that specific acts of his induced Goliath's anger, might well experience something like shame. After all, we see what looks much like (nonconceptual) shame in the sheepish behavior of pet dogs that are chastised by their masters for the indoor deposition of excreta.

For shame to metamorphose into guilt, internal conceptual versions of the masters, with both their enforcing rods and some enticing carrots, are needed. The possession of such internal monitors is a human capability that we see no sign of in either dogs or apes. Primatologist Frans de Waal notes that both *desire to belong to a group* and *fear of punishment* are basic for the development of human norms and respect for rules.[3] De Waal comments that submission to higher authority is both fundamental to human morality and commonly seen among many animal species. He also notes that the more advanced stages of human morality are beyond dogs, although dogs are sensitive to blame and praise.

Consider what seems like a local group activity and value, but probably is not such. On many occasions a group of adult mostly male members of a chimp community engage in aggressive displays that signal their hostility to a group from another such community. The second group usually responds in kind, but physical battles normally do not take place between the two groups.

However, on occasions when a "task force" from one group encounters a single member of the other group, or perhaps two members if they are female or young (thus smaller and less of a threat), the result is usually a brutal attack. The attack may leave the victim—sometimes a female with an infant

—bloody, bitten, bruised, and broken-boned, but not dead. The severity of the attack, however, is such that the victim commonly dies soon afterward. In one case, the attacks from members of a larger community on individuals of a smaller one continued until the latter was, member by member, destroyed.

At first blush (*blush* if one identifies with the victims; *rush* if one identifies with the creatures that bash them), this looks something like human tribal strife. The differences, however, make that comparison suspect. The chimp "war" party may include estrous females. When fighting does break out, it is with many aggressors bashing a single opponent. That victim is often female. Whereas for small-scale human warfare, the battle is usually between groups of males on each side. Human females are commonly part of the spoils and not direct casualties of the fighting.

If this chimp activity is not war, then what is it? In Shakespeare's *King Lear*, Gloucester says:

> As flies to wanton boys, are we to th' gods,
> They kill us for their sport.

These words may be pertinent to the description of our contesting party of mostly male apes. Note that the "wanton boys" and "th' gods" are plural in number, such sport is often a social activity. Note also that the risks for these sport-enjoyers are not great, killing neither the flies nor us entails much hazard. The several-against-one battles involving members of another community might be described as chimp "sport," rather than as chimp "warfare."

We could say that the sport called *hunting* is what the group of aggressive chimps is engaged in. It is something like the male-chimp hunting for game such as the young of monkey, bushpig, or bushbuck. Hunting for game may be one form of chimp sport; hunting for an outsider, a member of a different group, to victimize may be another. Hunting for game is sometimes started by single males, with other chimps showing up later to join the hunt or to beg for a share of the prey. Hunting for a "foreigner" victim seems always done by a group, but the plurality of the aggressors may be because it is less risky that way.

The frequent presence of an estrous female in the aggressive group needs remarking. As excited and active as the others, she saves the group's activity from being describable as exclusively male aggression. Nonestrous females do not join these hunting parties; perhaps they understand that they do not have the puffed pink sexual parts that act like white flags to turn off the aggressive minds and invite the sexual minds of the foreign males they might encounter.

Apart from a name for their activity, it seems that the chimps who go off in search of such sport or battle all know what they are doing. The chimps play their war game very carefully. They maintain strict rules of silent stealthy approach to their sought targets. The normal loud utterances that are common to other chimp activities are suppressed by all. They are equally careful about avoiding noises from the breaking of twigs or branches with body or limb. Although they hardly behave like Boy Scouts are assumed to behave, they move like patrolling scouts are supposed to move.

They also behave, let me note, as though they had concepts in their heads. They must all have the same concept in mind, although primatologists have not figured out how the concept is transferred from one mentality to another by use of behavioral or vocal gesture. Also, as Jane Goodall writes, chimps expend "considerable energy in *creating* opportunities to encounter [foreigners]."[4]

Unlike the activity of ganging up on and destroying individuals from another community, male chimps often individually behave brutally toward females within their own community. Wrangham and Peterson describe how each young male chimpanzee, after entering adolescence, systematically starts the charging, hitting, and kicking that will dominate, one by one, all of the females. Each wanton boy chimpanzee uses as much of his newfound strength as needed to get his target to kowtow to him. A female sometimes resists the attempt of a youthful male to bully her. When she defies him, he batters her into submission. It's hard to believe the "wanton boys" don't experience satisfaction in these low-risk assertions of their newfound social power. In subsequent years, these males will "often attack females without apparent provocation and with similar ferocity."[5]

Richard Wrangham and Dale Peterson discuss theories that seek to explain ape forced-sex. One theory claims such behavior is directly adaptive; it results in more fertilizations of females and more progeny for the males. A less directly adaptive theory (that of Barbara Smuts and Robert Smuts) says the purpose of male coercion of females is not fertilization, but control—for future fertilization. Note that these are both theories of Darwinian genetic adaptation, and that neither theory shows any interest in what goes on in the mentation of these rambunctious males. About this, Wrangham and Peterson comment, "Even if we can't prove claims about what apes know and think and feel, though, we will make more mistakes by ignoring such signs of mental power than by taking them seriously."

Describing the group chimp activity of hunting down and bashing an individual of another community as either war or as sport may or may not be suitable. But using more moral terms such as murderers, though tempting, is not appropriate. No jury would consider convicting them, for they simply do not understand what they do, not in the way that humans understand. Apes don't have any moral sense of right and wrong worth speaking of. Such a moral sense comes from the understanding and acceptance by a group member of values that are held by significant others in his or her group.

We know, for example, that there have been human communities whose members felt obliged to do the right thing by hurling stones to crush a wrongdoer. In other moral communities, the group values are such that the awareness of less-than-flawless behavior within the group may inhibit the stoning of flawed outcasts. The pursuit and attack by a group upon an outsider chimp might be viewed as the bare beginnings of such moral motivation. As for such harsh behavior within a community, there is an account of unusual behavior that might be evidence that chimps do a diminished variation of stoning other chimps that were once friends and neighbors. That occurred, as de Waal tells us, when some chimps at the Gombe National Park suffered from an epidemic of poliomyelitis.[6] The partially paralyzed victims might have preferred some comforting; instead, they were treated with hostility and fear by their healthy neighbors. Otherwise, values in this sense are largely absent in chimps, as was evidenced in the group's lack of response to the infant-devouring chimp pair.

A case could be made that killing other chimps has an adaptive advantage; by removing potential competition, it increases the odds that the killers will have progeny that prosper. But that potential gain may be balanced by the increased risk of harm and death if the attackers themselves become targets in the future. It is likely that the infrequent hostile and successful pursuit of outsider apes is not so much adaptive behavior as example of an adaptation byproduct. Bigger brains that can handle more abstract concepts were themselves adaptive. The interest in the warlike or sportlike excitement of bashing outsiders may be merely an incidental byproduct. However, behind the wargame lies the understanding of "us," our group, and "them," not our-group. That, too, may be an adaptation byproduct wherein big-brained communal creatures expand on the smaller-brained nonconceptual feelings of comfort with familiar others and distress with strange others who come too close.

CHIMPS PROJECT THEMSELVES INTO OTHERS, PERHAPS INTO STORMS, BUT NOT INTO LADDERS TO BALANCE THEM

Although much of ape behavior is describable as that of narrowly egocentric creatures living in the local-here and immediate-now, some of it does not fit unless we expand the idea of such egocentricity. Take, for example, the altruistic behavior of young Pom sitting in a tree, who sees a snake near her baby brother on the ground. She quickly climbs down the tree, grabs the infant, and reclimbs to safety. All this is at some risk to herself. Likewise, a mother chimp will risk a beating and come to the defense of her infant when an overexcited male may harm it. This is even further removed from simple egocentric behavior than the example, in the previous section on lower animals, of the mother bird seeking to protect its nestlings. To accommodate it, we must conjecture that the young female chimp identifies with the infant, feels it to be, somehow, part of herself. This expansion from a mother animal's feeling for its young is not normally seen in lower mammals. This altruistic activity of Pom's is no Darwinian genetic adaptation. Rather, it is a byproduct of other genetic adaptations that led to larger brains that could entertain more complex social behavior. Chimps trying to do what Pom did might frequently suffer poisonous and fatal snake bites, so even if the behavior had direct genetic roots, they might not be adaptive ones.

Related to Pom's concern for her younger sibling is the general close "family" concern among chimps. Each chimp grows up in a family of mother and siblings. It will often provide direct support for family members when a conflict with other chimps arises. Such recognition and special behavior toward family members is also seen among monkeys. Monkeys share with apes an understanding not only of their families but also of everyone's place in the local hierarchy. Frans de Waal describes how a rhesus monkey mother carrying her own infant may pick up and briefly hold a stray infant, but only if the latter belongs to a high-ranked mother.[7] So it may be said that monkeys and chimps have expanded their self-interest to include family members, and to exclude others unless they are of high rank or happen to be "friends."

Another type of behavior that is too complex for the simple model used for lower animals can be seen in the male pecking order and its influence on access to females. In "Chimpanzees and Bonobos: Cooperative Relationships among Males," Toshisida Nishida and Mariko Hiraiwa-Hasegawa describe male competition for females.[8] The males in a chimp community commonly form a hierarchy that defines who gets to "do" a female.

In such a pecking order, I will call the top three positions "alpha," "beta," and "gamma." Let me name the initial top three animals One, Two, and Three. Sometimes in such a community, Three and Two will form a coalition and succeed in deposing One, so that Two becomes alpha. Sometime after this, Three may collude with One to depose Two and restore the original order. Also, at times Three may use his "swing" power between One and Two to gain sexual access to a female that would normally be denied him by one of those two. In this situation, Three and whoever is alpha have clearly gone beyond an interest in only the narrow here-and-now. Were there no threat of a future coalition to depose him, alpha would chase Three away from the female. So alpha's here-and-now has expanded to somewhere-nearby and the-near-future. Likewise, Three would not normally pursue a female when in the sight of alpha. He does so because he knows of the future-elsewhere threat he represents to alpha. So, he, too has moved to a larger "here" and a longer "now." The current alpha and Three have moved to a more encompassing situation or episode, however, only after Three's power was initially established by actually forming a coalition with Two and deposing One. That suggests that it was past experience and not abstract thought about the future—which humans on occasion use—that brought about the possibility of a new and different situation.

A protracted struggle (described by de Waal) occurred between three male chimps where the leadership changed several times.[9] These chimps were part of the Arnhem Zoo colony where the apes had a fair amount of outdoor space available to them. The apes were given food every evening after they entered the small low-occupancy indoor cages provided for them, so they did not have the need of wild chimps to find their daily bread. As a result, these unemployed apes had far more time than their wilderness brethren for social activities, including the struggle for power between the three adult males.

The first leader was the oldest male, Yeroen. Then a months-long struggle with the second-rank chimp, a struggle mostly of blustery gestures rather than blows, took place. When Yeroen lost, he gestured his acceptance of the change; he used the rapid deep bows called "bobbing" to acknowledge the ascendancy of formerly number-two Luit, who had formed a coalition with number-three Nikkie.

Chimps Are No Chumps: Mentality with Room for New Concepts

After Luit's ascension, Nikkie began confronting once-mighty Yeroen until Yeroen acknowledged Nikkie's second-place status and fell to third himself. Soon thereafter, Nikkie started to challenge Luit, with some help from Yeroen. In response, Luit sought friendship with his once-rival Yeroen. Nonetheless, power slowly shifted again; Luit's power was declining while Nikkie's was increasing with Yeroen's help. Finally, Nikkie acceded to the Arnhem Zoo throne.

Power had shifted from one chimp to a second and then to a third. There were only three adult males in the colony and all had now been at the top. One might have expected the last shift to be the end of the tale. But it wasn't. Perhaps these unemployed apes had too much time on their four hands. Nikkie's coalition with Yeroen did not hold; the two had a serious fight. That allowed the once-second king, Luit, to regain the top position. At that point, the chimps—who seem to understand immediate social power and its recent history almost as well as do humans—may have realized how unstable was the grasp of power by any of the three. At any rate, what happened next was tragic for one of them. One night, the two once-top apes ganged up on the other—the second and also the present king, Luit—and did him so much physical harm that zoo veterinarians could not save his life.

Female adult chimps often extract termites from a termite mound by inserting a small bare branch into a hole in the mound and, after termites have moved onto it, the chimp withdraws it. The chimp usually encounters the termite mound without such a "fishing rod." She leaves the mound to find a blade of long grass or to break a suitable branch from a nearby tree. If her tool is a branch, she strips it of twigs and leaves and returns to the mound to do her "fishing." To do this, it is clear the chimp's here-and-now is larger and longer than if she simply stood at the mound with no way to get at the termites. I suspect, however, that it is uncommon for the chimp to make the fishing rod first and then go in search of a termite mound. She is less likely to make a fishing rod, an activity hardly gratifying in itself, and then go looking for tasty termites. More commonly, she encounters evidence of the tempting but inaccessible insects and keeps her concept of them in mind long enough to find and return with a fishing rod.

Goodall's book, *In the Shadow of Man*, describes the downhill races during thunderstorms that are a form of behavior chimps engage in that differs from all others.[10] It differs markedly in that it has no practical purpose. At times, when a thunder-and-lightning storm starts, a number of adult males will hoot with excitement and charge at full speed down a slope, breaking branches from trees, brandishing them in the air and hurling them ahead. When each reaches the bottom of the incline, he plods back up the hill and then, with resumed vigor, charges down again. The activity is usually performed many times. The obvious question is, why on earth are they doing that?

Before I offer my answer, I will quote a question about humans posed by Paul Kurtz: "Is there deep within the human breast a transcendental temptation that reappears in every age and accounts for the ready acceptance of myths about the transcendental?"[11] Here's part of his answer: "The fundamental premise of those who believe in a magical-religious universe is their conviction that there are hidden and unseen powers transcending the world, yet responsible for what occurs within it. . . . Early anthropologists considered magical thinking to be a prelogical mode of response to natural events . . . they attributed them to animistic causes, believing that material objects and animals have an inner spiritual consciousness like ourselves, which was separable from the body and had causal efficacy."

The downhill-running apes are beginning to do what humans do all the time: They are finding personal meaning in the impersonal events of the inanimate world around them. The male apes are identifying with the storm by projecting their vaguely felt assertive feelings and concepts of strength and power into it. Then the storm's vivid activity makes a huge reflecting mirror of the sky. That thunder and lightning, seeming "motivated sky-agent behavior," is introjected back into the participating apes to color their conscious perceptions with feelings of high excitement and energy that elicit their downhill charges. In this view, the storm is viewed not as an index or a symbol but as an icon of strength and power for the apes. It lets them feel the vague nonverbal equivalent of "Some powerful agent is doing something exciting up there," plus "I'm part of it" or "I'm fighting it."

Pertinent in a negative sense to the idea of chimpanzee projection is something they—even the ones that have been constrained to live in human-controlled environments, where it might be more appropriate— cannot do. Yet it is something that even young human children can do. Langer discusses chimps and ladder-balancing. Chimps can't readily balance a ladder against a wall. Although they have excellent balance for themselves, when attempting to climb a ladder to get food in a laboratory experiment, they don't seem to get the idea—without being taught—that a ladder can be leaned stably with the tops of both verticals against the wall and the bottoms of the verticals away from the wall. "The reason for this is that any object an animal uses as an instrument becomes an extension of its own body, more like a prosthesis than a tool. . . . The chimp's balance is superior to man's but cannot be projected by him into an object."[12]

A young child with a toy ladder does the balancing fairly easily. The difference is that the child nonconsciously perceives the ladder as something that is in one way like himself, namely, a bilaterally symmetrical object. This he does by projecting his own sense of symmetry and balance into the ladder. This then permits him to place it vertically against a wall with relative ease. In the old film, *Singing in the Rain*, the hero, played by Gene Kelly, joyfully projects his feeling of being in love: "If I were a gate I'd be swinging." Similarly, the child nonconsciously projects: "If I were a ladder, I'd be balancing like this." But the ape, although somewhat able to identify himself with a fellow family member or even a thunderstorm, cannot intuit the likeness between himself and a ladder, and so cannot project himself into one in order to reliably balance it against a wall. This is not to say that apes could not be taught to balance a ladder against a wall. They probably can be taught to do so; having others around that make learning important can compensate for many things that don't come naturally, for apes as well as humans. That's part of why it is so devilishly difficult to distinguish nature from nurture in the affairs of apes and humans.

Before leaving this subject of expanded ape mentation, let's consider the extent to which apes share, in some measure, the reflective consciousness that we humans have. In his forward to *Self-Awareness in Animals and Humans*, Louis Moses writes, "It seems reasonable to suppose that any animal with even a minimal conception of self should be capable of recognizing who or what it is."[13] He then goes on to describe the technique (developed by Gordon Gallup Jr.) for assessing recognition of self in a mirror. Moses comments, "the animal might simply recognize its body in the mirror while having little in the way of a conception of its own mental life." An answer to the question "Do apes have some minimal conception of themselves?" is precisely what Gordon Gallup and other scientists sought to learn in "Self-Awareness and the Evolution of Social Intelligence."[14] They did this by examining the evidence that an ape is aware that the reflection it sees in a mirror is its own. For several, but not all, tested chimps, orangutans, and bonobos, and for only one tested gorilla, the answer was yes. Each recognized itself in a mirror, but what was the nature of the "self" that was recognized? What kind of "me" did each see? Gallup thought it likely that mirror self-recognition implies mental self-recognition. In "When Self Met Other," in Daniel Povinelli and C. G. Prince disagree with Gallup about the mirror-self meaning a mental self.[15]

I believe each ape saw itself as a body that sometimes engages in behavior, which is what we humans often see in a mirror. Sometimes we see more—sometimes we see ourselves as embodied minds, such as the embodied mind we see in Rembrandt's marvelous portraits of what he saw in his mirror. Of course, some of us, like Narcissus, find in our mirrored image mainly an opportunity for self-love.

NOTES

1. Kim A. Bard," Intentional Behavior and Intentional Communication in Young Free-Ranging Orangutans," *Child Development* 63 (1992): 1186–97.

2. William Noble and Iain Davidson, *Human Evolution, Language and Mind* (New York: Cambridge University Press, 1996), p. 64.

3. Frans de Waal, *Good Natured* (Cambridge, Mass.: Harvard University Press, 1996), p. 92.

4. Jane Goodall, *The Chimpanzees of Gombe* (Cambridge, Mass.: Harvard University Press, 1986), p. 528.

5. Richard Wrangham and Dale Peterson, *Demonic Males* (Boston, Mass.: Houghton Mifflin, 1996), pp. 141–43, 256.

6. Frans de Waal, *Good Natured*, p. 83.

7. Ibid., p. 101.

8. Toshisida Nishida and Mariko Hiraiwa-Hasegawa, "Chimpanzees and Bonobos: Cooperative Relationships among Males," in *Primate Societies* (Chicago, Ill.: University Chicago Press, 1986), p. 176.

9. Frans de Waal, *Chimpanzee Politics* (New York: Harper & Rowe, 1982).

10. Jane Goodall, *In the Shadow of Man* (Boston, Mass.: Houghton Mifflin, 1988), p. 52.

11. Paul Kurtz, *The Transcendental Temptation* (Amherst, N.Y.: Prometheus Books, 1991), pp. xv, 454.

12. Susanne K. Langer, *Mind: An Essay on Human Feeling*, vol. 3 (Baltimore, Md.: Johns Hopkins University Press, 1982), p. 47.

13. Louis J. Moses, forward to *Self-Awareness in Animals and Humans*, ed. by Sue Taylor Parker, Robert W. Mitchell, and Maria L. Boccia (New York: Cambridge University Press, 1994), pp. xi, xii.

14. Gordon G. Gallup Jr, "Self-Awareness and the Evolution of Social Intelligence," in *Behavioural Processes* 42, nos. 2–3 (1998): 239–47.

15. Daniel J. Povinelli and C. G. Prince, "When Self Met Other," *Self-Awareness: Its Nature and Development*, ed. by M. Ferrari and R. J. Sternberg (Guilford Press, 1998).

6

APE LOVE-LIFE

NEAR THE BORDERS OF SEXUAL EDEN

BONOBO SEXUALITY: LOTS OF IT, SOMETIMES WITH THE SAME SEX, AND IN WHATEVER POSITION WORKS

AS SEXUAL CREATURES, SOME APES are more active than humans, let alone lower mammals. Consider the bonobos (also called pygmy chimps, although they are not much smaller than ordinary chimpanzees), who are giants when it comes to sexual activity under captive conditions. In the wild, however, recent studies have shown bonobos to be less active sexually. One such study is "Comparing Copulations of Chimpanzees and Bonobos: Do Females Exhibit Proceptivity or Receptivity?" These data on chimps and bonobos in the wild "do not support the conclusion that female bonobos exhibit prolonged estrus."[1] For the populations observed, "the copulation rates of adult male bonobos . . . were lower than those of adult male chimpanzees." Thus, greatly increased bonobo sexual activity under workfree captive conditions suggests that idle bodies are put to use in the human "devil's workshop" by increased rubbing of sexual itches.

Much of the behavior described here is taken from studies of these apes in zoos. There, bonobo females have sexually attractive swollen genitals about three-quarters of the time, compared to half the time for chimpanzees. Even in the absence of such estrous swellings, they mate frequently. When, absent such estrous swellings, they *seek* such sex, they may be entertaining lustful *ideas, conceptual* prurience. That is beyond the range of lower mammals, who are limited to prurience in the form of the lustful *desires* that go with their estrus or rut.

In contrast with chimpanzees, a high-ranking male bonobo does not act as if sexual opportunity is a scarce commodity, one that must be monopolized by pressuring a female to go away with him on a long "date," or by denying females to lower ranking males. Males who encounter an "occupied" female look for sexual opportunity elsewhere. Or they more or less wait their turn, in a short line, as it were.

Bonobos are different from chimps about both sex and power. "The chimpanzee has an appetite for the second, the bonobo . . . for the first. The chimpanzee resolves sexual issues with power; the bonobo resolves power issues with sex."[2] Also, the bonobo male, unlike the male chimp, has little power over females, as they gang-up on any male who tries to throw his weight around.

With captive bonobos, the sexual "action" is pretty much continuous. Throughout the year males have sex several times—even ten or more times—a day, as do females, except for the few days in their estrous cycle when their interest may have dwindled to nought. Wrangham and Peterson note that bonobos can mate dozens of times a day. Also, a frequent part of a sexual encounter is the use of hands or mouth to manipulate each other's genitals. Nor, it might interest you to know, is bonobo sexuality limited to adults. Sexual activity starts "long before the onset of puberty—from about one year old."[3] (Perhaps we should remind ourselves that there have also been human societies in which children are sexually active.)

Unlike that of gorillas or chimps, copulation between female and male bonobos is sometimes in the orientation once regarded as exclusive to humans—face to face. Such face-to-face sex was found to occur much more commonly in compounds of captive bonobos than in the wild. Frans de Waal reports that zoo matings are about 80 percent face-to-face, much greater than an estimated 30 percent in the wild.[4] (That orientation was once thought by members of the clergy—at least by those devoutly devoted to saving souls in the South Seas—to be the only one appropriate to humans; presumably, all orientations except "the missionary position" were too beastly for us. Obviously, at the time it was not known that bonobos sometimes assumed the missionary position.)

Also like that of some humans, sexual activity among bonobos is sometimes with a partner—yes, sir, we should face it—of the same sex. A pair of females accomplish this by rubbing of genitals against genitals. "Hoka-hoka" is the name given to such rubbing by human communities living near the wild bonobos. Richard Wrangham and Dale Peterson inform us that typically a female pair who engage in hoka-hoka end with "mutual screams, clutching limbs, muscular contractions, and a tense still moment. It looks like orgasm."[5] A description of the sexual interactions of male pairs is provided by de Waal.[6] A male couple sometimes engage in mutual penis rubbing and thrusting.

Bonobos pairs often participate in prolonged mutual gazing into each other's eyes, not across a crowded room, but both prior to and during their engagement in sexual activity. Nor do bonobos break only our once-presumed monopoly in the animal kingdom on unusual sexual positions and partners. They also step with us onto the information pathway by using facial and vocal signals, (optional) gestures, to communicate concepts of desired copulatory position.

PRURIENCE IN MALES MAKES SEX OBJECTS OF NOT-HOT FEMALES

It has been observed for male and female bonobos that at times the male provides food for the female before or after copulation. Wrangham and Peterson describe such a seeming exchange by chimps of sex for meat.[7] A male who has meat will often be approached by a female. She turns and presents her private parts in a sexual invitation. He, still holding his prized meat in hand, accepts her offer; afterward, he shares some of his meat with her. Such providing of food might, conceivably, be altruistic generous-minded gifts. However, the males are not generous to females who are not venerious. Given that warm-hearted gestures and acts are less than universal in human affairs, there might be a more cool-headed explanation for such ape offerings. Perhaps it is an exchange of one thing for another. But what, if anything, is the female providing in exchange for meat and why?

In one sense it is obvious: the female provides the sexual opportunity that the male desires. Doesn't the male, however, provide the sexual opportunity that is desired by the female? So why is the male providing food as well as copulation? One answer is that the food exchanged is often meat, flesh that is hunted and found almost exclusively by the males. Meat is something the females have learned can be obtained from the males.

Something else may also be involved here. To see what it is, let's look more closely at estrus. Primatologists Sarah Hrdy and Patricia Whitten, in "Patterning of Sexual Activity," tell us that estrus is normally characterized by three related changes.[8] First is an increase in sexual swellings and aromas that provide *attractivity*, the female's value as a sexual stimulus to males. Next is an increase in *receptive* behavior, which acts to facilitate copulation by showing the readiness of the female. Third is an increase in *proceptive* behavior and appetitive activities by the female, such as going up to a male and "presenting." Presenting consists of the female turning away from the male and bending forward to present the attractive locus of her estrous itch.

It is attractivity, the first of what are called the three characteristics of (female) estrus, that is of interest here. For the bonobos and chimps, attractivity is provided to the males by increased female sexual swelling and odors. Among bonobos and chimps, males frequently examine and sniff aromas from a female's sexual skin. In Jane Goodall's *In the Shadow of Man*, a photograph pictures such male sniffing of female aroma.[9] Why do they do so? In part, no doubt, for the same reason dogs smell each other's genitals. Even when the female is not in heat, sexual aromas are interesting, as are all body odors, to these creatures with better noses for news than ours. Additionally for apes, the odor may be highly significant to males who are trying to anticipate the future attractivity, receptivity, and proceptivity of the females. This is in sharp contrast with female chimp and bonobo sexual attitudes toward the male. The female who is interested in copulation need never concern herself much with determining if the male might have such interest in the near future. Unless he's sick or dead, a bonobo adult male is rarely unavailable to the interested female. Nor, for that matter, is the situation different with chimps if no high-rank male is nearby to inhibit a lower-ranked male.

Now the bonobos "know," in a nonverbal way, much of what has been said here. Both males and females know that the males are commonly looking for receptive females. They also know that such females rarely need to look for males, since they are normally available. Goodall writes, "[The female chimp] does not need to work at maintaining the sexual relationship. . . . It is the male who must put effort into his relationship with her. . . . In one captive group the sole adult male groomed females significantly more [when they were approaching maximal sexual swelling.]"[10] They all know that there is, in effect, "something" that the females sometimes "have" that the males want and look for, and that there is *not* a similar sexual thing that the males sometimes have that the females look for. In other words, they all know that females in heat are sex objects for the males, but that the males are not, in the same sense, sex objects for those females. That social fact may be a second reason why bonobo and chimpanzee females trade sex for food from males. Frances White, in "Comparative Socioecology of Pan Paniscus," suggests something similar, that the sharing of meat by bonobo males with females may be in exchange for *future* sexual access.[11]

There's no reason to believe there is a *genetic* root to the behavior of the females. Like any animal (of either gender) that perceives an opportunity for gain, these females take advantage of their status as sex-objects to get a few morsels of meat from a male who has prey and wants copulation.

After getting his fill (or better, getting emptied of immediate desire) and without the female getting her fill (females in heat do not get "filled" until their estrous interval ends), the male goes off. Before long his postcoitus disinterested down-time ends and he is on a prurient quest for a new sex object. The female in turn—but with no disinterested down-time—soon finds another male to sooth her estrous itch.

Another variation on the theme that I am here suggesting—biologically based rut contributing to conceptual prurience in male apes—can be seen by looking at the sexual interactions of orangutans. Note first that, for two reasons, scientists have difficulty understanding orangutan sexuality. First, these apes live high in the trees, which makes it difficult to observe them. Also, females do not exhibit the genital swellings that clearly indicate an estrous condition of receptivity. Yet, at times, some males—those adult males, about the same size as females, who are called "small males"—force females to copulate, despite the latter's attempts to avoid mating. This suggests that these small-male adult orangutans may be driven by a conceptually prurient expansion of genetic-originated rut.

Carel Van Schaik and Jan Van Hooff discuss such forced sex from the female's viewpoint in "Toward an Understanding of the Orangutan's Social System": "Orangutan females could . . . minimize sexual coercion if they experienced . . . advertised receptive periods."[12] But then they say, "An estrous female attracts multiple males who compete vehemently for sexual access to her." To me, this suggests they may be unclear about what constitutes an estrous female if she doesn't have advertised receptive periods. If so, their problem is understandable: How is a primatologist to know that a high-in-a-tree orangutan female is sexually receptive if the swellings that advertise such heat are absent? For all that, the primatologist's problem absolutely pales when compared with that of the orangutan male.

HOW WE GOT TO BE HUMAN

In two respects, the sexuality of orangutans is somewhat like that of humans: The males have a prurient expansion of rut, and the females show no physical signs of receptivity. Unlike most societies of humans most of the time, however, orangutan societies do not put strong pressure on males to refrain from forcing sex on females. So, for orangutans, prurient males' uncertainty about females' receptivity, plus the lack of restraining social attitudes, may be among the reasons for what is, sometimes, forced sex. There must be more to it than that, however, as it is only the small males that are known to force sex on the females. Also, as Anthony Rose tells us in "Orangutan, Science, and Collective Reality," orangutan sex is mostly consensual, and that when forced sex occurs, it does so only with young adults.[13]

The smallness of the small-male orangutans is somewhat of a mystery in itself. "Ordinary" adult males are about twice the size of orangutan females. The small males, in captivity, remain small for many years before they too become big. Wrangham and Peterson describe the orangutan small male as "an adult male frozen in an adolescent's body."[14] They also use the word "rape" instead of the somewhat less demonic-sounding "forced sex" for the copulations that are sometimes initiated by these small males, copulations that are decidedly undesired by the females. About forced sex, Wrangham and Peterson write as follows: That its occurrence is an "*ordinary* part of a species' behavior implies that it is an evolved adaptation to something in their biology." But consider humans, where the removal of clothing is an ordinary part of our species' sexual behavior. Surely our sex-without-clothing syndrome is more in the way of an adaptation byproduct than it is an evolved adaptation. I think forced sex, by male orangutans or male humans, is more likely an adaptation-byproduct than an adaptation.

It is worth noting here that both negligible signs of estrus in the female and occasional forced sex by the male also coexist in the stump-tailed macaque monkeys. Frans de Waal tells us that the stumptails are studied for their very active sex life.[15] Captive stumptails, like zoo bonobos, don't have to go out and earn a living. This may contribute to their copulating several times a day.

The general pattern of male sexual behavior discussed here, perceiving females who may not be in heat as desirable sex objects, is an ape conceptual expansion of the never-ending male quest among lower mammals for estrous (thus attractive and receptive) females, a quest that has come to be called, for some lower animals, either the Coolidge Effect or rut. For apes, it is noted here as male prurience, where being prurient means "having lustful *desires* or *ideas*." Male and female lower mammals have only lustful *desires*; ape and human males and females have both lustful desires and lustful *ideas*.

Yet it must be said that the literature describing ape sexuality does not do so in terms of something that could be called male prurience or conceptually expanded rut. So I am under some obligation to explain this absence. I think the discrepancy arises, in part, because *animal* male sexuality is commonly perceived by ethologist and primatologists as "just" a response to the swellings and behavior of estrous females.

So in natural—nonexperimental and nonzoo—situations, it is not at all evident from observable behavior that male sexuality includes anything beyond a simple *response* to female estrus. The idea of male rut in lower animals as an active "thing" in itself—as an *endless* interest in new, "hot" females—that idea itself did not gain acceptance until an artificial experimental situation was used to examine male sexual behavior. And even there the males did not do anything other than what they do in the wild. The only difference was that they continued to do it, continued endlessly to do it— continued only because the experimenters provided the in-heat females the males needed to do it. The observation of this repeated doing led to the concept that males endlessly desire to "do" more receptive females, and that concept acquired a name from a slightly salacious and surely apocryphal joke about Calvin Coolidge and his wife.

It is conceivable that an experimental situation could be contrived to determine the extent to which apes are subject to such endless male prurience. Need I say that it is highly unlikely that such tests will be performed? Apes are too much like humans for their use in such experiments not to raise questions of propriety. Even though the experiments would be "harmless"—the worst consequence to the male apes might be sexual exhaustion—they are not likely to be performed. This is certainly so if, as seems to have happened with such tests on lower mammals, the experimenters themselves just might get exhausted before the experimentees do. Far worse, an experiment exploring the limitations of male apes in continuing to copulate with "fresh," "hot" female apes would surely get leaked to an ever prurient press and thus to a seemingly pubescent public, which together would make fools of the would-be experimenters. So unless someone develops a more subtle and ethical, as well as a less risque and thus less risky kind of study, male prurience may remain unproved in apes. But unproved does not imply untrue or unprovable.

HOW WE GOT TO BE HUMAN

Another point about primatologists' greater discussion of female than of male ape sexuality is worthy of note. Maxine Sheets-Johnstone comments about animal sexual studies: "In contrast to sexual swellings in females, virtually nothing is said of sexual swellings [erections, that is] in males."[16] I can think of a couple of possible reasons for such relative silence. First, social scientists, both male and female, belong to a species whose adult members are so universally aware of the frequent likelihood of such swellings as to take them for granted in a closely related species. Second, even scientists may, at least in part, nonconsciously acquiesce to a prurient and still-some-what-patriarchal society that prefers to peer at female private parts.

One more semipolitical point needs voicing here. In our considerably less than completely patriarchal society, ideological feminists may find sin in my suggestion that both humans and apes have *genetic* origins, as well as learned ones, for some of the prurient behavior of males. That's certainly understandable. I just hope they don't successfully invoke Hera to push Zeus aside and hurl his thunderbolts my heretical way.

Even if all this is belaboring the issue of ape sexual behavior, there are two reasons for so doing. First, to suggest that the nature of male simian sexuality is worthy of innovative further exploration by students of ape behavior, even though it may be risky to the explorers. The second is to prepare us for a look at some of our human sexual behavior later, after we visit our predecessor upright-ape and erectine species. The point need not be kept in breath-holding suspense. It is that, despite the many intervening evolutionary advances, human males may be similar to chimp and bonobo males in that they retain the biological roots of male prurience, roots that exists as rut in lower mammals. I will suggest that human males, like those of apes and lower mammals, are often on the lookout for sex objects, most commonly females, that are attractive and receptive. Human females with an interest in sexual activity do not normally do much searching for males, as several men are usually nearby, willing and eager to help out. Saying that does not imply that women cannot be prurient; they can—and often are. Women's prurience, unlike that of men, is mostly learned and not innate; it has lost most of its innateness as a result of the disappearance of estrus. Note, however, that some women report an increase in prurient feelings when they are ovulating; so all of estrus may not be lost for primate females who are orangutans or humans.

Chimpanzees are unlike bonobos in that the males are often possessive with regard to the in-heat females. This may be related to the fact that female chimps are in estrus for a briefer portion of the ovulation cycle. Thus, the males, unlike bonobo males, may feel there is not enough in the way of female sex-objects to go around. Chimp males often try, with partial or complete cooperation from their chosen targets, to get females to go away with them, to go somewhere where the consorting couple will be away from the potential competition from males in the remainder of the troop. For the males to do that requires *conceptual* mental activity. At times, a chimp male tries to get a female who totally lacks the puffed-up sexual skin of estrus to follow him away from the group. Even more clearly, that calls for lustful ideas as well as lustful desires in the mentation of the male. Sometimes, chimps are also choosy about their sex partners; attractiveness and receptivity have more reference (than exists for lower animals) to individual potential mates. Old Flo, who was introduced with her family in chapter 4, was said by her human observers to be a popular sex partner, eagerly sought by young and old males even when younger adult estrous females were around. This, too, suggests a conceptual component in ape sexuality.

SEX FOR GORILLA MALES AND ORANGUTAN FEMALES: THE FORMER ARE SOMETIMES INDIFFERENT AND THE LATTER ARE OFTEN DISINTERESTED

The gorilla is a less social primate than are the chimp, bonobo, and human. They live in families, but the families are normally separated from one another by a good deal of space and never united by shared activities. A family usually consists of one controlling male, the silverback (so called for the white hair that starts to appear when he is no longer a young adult), several females and their young, and several subordinate younger males. When, for whatever reason, a silverback dies, "his" females' infants are likely to be killed by the new male who takes over the family. This is explained as behavior that is adaptive in the normal Darwinian sense; females unencumbered by infants are more likely to have new young soon, young who have the new male's genes. That may well be the case, but it doesn't address the difficult question of how the new male experiences his activity. Of course, similar behavior occurs with lower mammals, such as lions who kill the cubs when they take over a pride. These killing males must experience something when they engage in this behavior; they are not mere nonconscious machines. Do they regard their own infants, who come from their females' bodies and may smell like them, as therefore okay? Do they regard other cubs as a threat, somewhat like the threat of strange males? It is also of interest that, once her existing infants are killed, the female normally stays with the new male, even though she could, without opposition, migrate to another silverback's family. Is this because, the past otherwise forgotten, she feels safer with a male who has shown himself to be powerful?

Gorilla silverbacks are at times unresponsive to the estrous allure of their females, and do not get enticed into performing the sexual act unless another male shows interest in usurping their position. It seems to me that the silverback's response might be far more positive if he were beckoned by an unfamiliar female who is not part of his family; were that the case, it might suggest male conceptual prurience in gorillas.

Anthony Rose describes some unusual gorilla behavior.[17] This activity suggests that male gorillas may have a concept-based learned expansion of innate rut. Captive male gorillas have been observed to assertively initiate copulation with *nonestrous* captive gorilla females. When such a female was in close proximity, the male engaged in a sexual chest-beat display and then copulated with her. Primatologists deduced that such behavior was similar to the wild gorilla's behavior when an *estrous* female came within a few yards of him. It seems to me that the captive male considered the female's close proximity as evidence that she was receptive sexually; he also ignored the evidence that was missing, the absence of the sexual swellings that go with estrus. It's true enough that the cramped captive quarters initiated the male's misreading of the female's message, but that's besides the point. The point is that there is evidence here of conceptual activity in the male gorilla's mentation; the subject male gorilla has a nonverbal concept (crudely: "Hey, she wants it, she's sexually receptive!"), a concept mediated and suggested by what, to him, seemed to be a gesture on the part of the female ("When she comes close, she's giving me the ooh-la-la, she has the hots for me").

As we saw above, orangutans have a pattern of sexual activity that is different than that of the other great apes, perhaps because they are less social and live rather solitary lives. The females have very little in the way of estrous sexual tissue swellings at any time. This may be one result of the solitary life. After all, there's no value in advertising in magazines that have almost no readers with the needed binoculars. Also, for most of the eight years between births of her young, the female has no sexual interest in males. For adolescent females, the situation is quite different. Wrangham and Peterson tell us that they are sexually curious and playful, they investigate the vagina with toes or objects, and they masturbate.[18] These young females sometimes seem to seek sex from adult males, who, in turn, are disinterested. Intercourse between females and big males is sometimes, like that of bonobos, face to face. Also, such matings may start with each creature mouthing or fingering the other's genitals; perhaps it amounts to "foreplay" that is related to the virtual absence of estrus for the females.

APART FROM INTRASEX STRUGGLES, LOWER MAMMALS LIVE IN SEXUAL EDEN, BUT APES RESIDE ASTRIDE ITS ROCKY BORDERS

What is to be made of all these great ape sexual activities? Compared with those of lower mammals, we can see that they are less automatic, no longer turned all-the-way on or off simply by female seasonal chemistry. They are more based on conceptual mentation that is mediated with languagelike gesture; they are more under the volitional control of the individual creatures.

Ignoring for present purposes the fierce sexual competition that is common between males and occasional between females, lower mammals can be said to live in a sexual Eden. That is, they live in an Eden solely in terms of relations between the two sexes. The females, by virtue of estrus, can always attract males when they want them. For the same reason, when the females don't desire males, they are never bothered by the latter. Likewise, males desire females only when the latter are sexually available. Nor are animal males ever under any experienced duress from sexually unfulfilled females. It's true enough that the powerful high-ranked animals commonly get a larger share of sexual activity than do the lowly, but we're talking here about sexual Eden, not about communist heaven.

Apes, however, although still close to that estrus-based sexual Eden from which humans—and most of our hominid predecessor species—were long ago evicted, nonetheless live mostly astride its rocky borders. There is anecdotal evidence that certain chimp males can hardly get the time of day, let alone a guaranteed lay, from many estrous females. These poor slobs seem to lack what was once vaguely identified among humans as "sex appeal." Some chimp females, such as old weary worn-out Flo, have so much sex appeal that, had they the mind for it, they could probably sell their services for fresh meat from every male; whereas other females almost have trouble giving it away. Also in the latter category are female in-heat gorillas in a "harem," who sometimes can't get a rise out of their male "keeper" until a would-be usurper appears. Some female orangutans, who must cope with small males forcing sex on them, can see the tough twisted terrain where humans reside outside sexual Eden. Clearly nature's firm genetic shaping of sexual behavior for lower mammals is much less in play for our great ape cousins.

At times this results in behavior that may not fit Darwinian strict constructionism, behavior that does not seem to be adaptive in itself, such as female-female or male-male bonobo sexual activity, or bonobo sex several times a day most of the time. All such behavior is either underkill or overkill if viewed as having the purpose of leaving progeny in the world. Of course, an adaptive purpose may be conceived for the behavior. For example, bonobo frequent sex has been said to make the creatures more social, which in turn may mean improved survival prospects. That may well be true, but I suspect that if bonobos had infrequent sex, an adaptive value could be conceived for that behavior. (Think of the mad general in the old movie, *Dr. Strangelove*, whose central personal concern was to retain and not expend his presumed limited supply of precious bodily fluids. His belief in the value of not parting with his seminal fluid is shared by many real people today, such as the villagers in India described in chapter 14. Perhaps that idea originated by analogizing semen with blood, the expenditure of which is usually risky.) The central concept—that there is a random mutational genetic origin for all significant behavioral patterns, and that adaptation selects the behaviors that survive— has been treated as a Darwinian dogma to be leaned on and to lend support to unproven ideas. A glance at many aspect of human life, such as written language (unavailable to the young) and music-making (of doubtful help for having more children), or such as having a lifespan long enough for us to see our grandchildren grow up, suggests it is not always true for humans. So why should it always be true for apes? Apart from how frequently they have sex, and with whom, some ape activities (like racing downhill in a thunderstorm) are more likely to be byproducts of adaptation than adaptive.

NOTES

1. Yukio Takahata, Hiroshi Ihobe, and Gen'ichi Idani, "Comparing Copulations of Chimpanzees and Bonobos: Do Females Exhibit Proceptivity or Receptivity?"

Great Ape Societies, ed. William C. McGrew, Linda F. Marchant, and Toshisada Nishida (New York: Cambridge University Press, 1996).

2. Frans de Waal, *Bonobo* (Berkeley, Calif: University of California Press, 1997), p. 32.

3. Richard Wrangham and Dale Peterson, *Demonic Males* (Boston, Mass.: Houghton Mifflin, 1996), p. 213.

4. Frans de Waal, *Peacemaking Among Primates* (Cambridge, Mass.: Harvard University Press, 1989), p. 198.

5. Wrangham and Peterson, *Demonic Males*, p. 210.

6. De Waal, *Peacemaking Among Primates*, p. 190.

7. Wrangham and Peterson, *Demonic Males*, p. 11.

8. Sarah Blaffer Hrdy and Patricia L. Whitten, "Patterning of Sexual Activity," in *Primate Societies*, ed. Barbara B. Smuts, Dorothy L. Cheney, Robert M. Seyfarth, Richard W. Wrangham, and Thomas T. Struhsaker (Chicago: University of Chicago Press, 1986), p. 370.

9. Jane Goodall, *In the Shadow of Man* (Boston, Mass.: Houghton Mifflin, 1988), p. 169.

10. Jane Goodall, *The Chimpanzees of Gombe* (New York: Cambridge University Press, 1986), p. 394.

11. Frances J. White, "Comparative Socio-ecology of Pan Paniscus," in *Great Ape Societies*, ed. William C. McGrew, Linda F. Marchant, and Toshisada Nishisada (New York: Cambridge University Press, 1996), p. 30.

12. Carel P. Van Schaik and Jan A. R. A. M. Van Hooff, "Toward an Understanding of the Orangutan's Social System," in *Great Ape Societies,* ed. William C. McGrew, Linda F. Marchant, and Toshisada Nishida (New York: Cambridge University Press, 1996), p. 10.

13. Anthony L. Rose, "Orangutan, Society, and Collective Reality," in *The Neglected Ape*, ed. Ronald D. Nadler, et al. (New York: Plenum Press, 1996).

14. Wrangham and Peterson, *Demonic Males*, p. 132.

15. De Waal, *Peacemaking Among Primates*, p. 145.

16. Maxine Sheets-Johnstone, *The Roots of Thinking* (Philadelphia, Pa.: Temple University Press, 1990), p. 94.

17. Rose, "Orangutan, Science, and Collective Reality."

18. Wrangham and Peterson, *Demonic Males*, p. 133.

7

APE GESTURE AND UTTERANCE
STEPS ON THE INFORMATION HIGHWAY

WHY CHIMPS PRESENT THEIR RUMPS TO AGGRESSIVE OTHERS

ANOTHER FASCINATING ASPECT OF APE sexuality is that (like humans) they seem to be able to divert sexual instincts and activities to nonsexual purposes. I am referring here to the practice (widespread among chimps and bonobos of all ages and both sexes, and also among other apes such as baboons) of presenting the rump as a seeming sexual offer to a more powerful threatening animal, with the consequence that the threat does not lead to a hostile act. This behavior appears to be identical to the sexual bottom-presenting performed by a female in heat. The offer of what seems like sexual interest is made by an ape who is being pursued, where the pursuer is often a hostile male and sometimes a hostile female. As a consequence, the pursuer stops the hostile pursuit and mounts the offered rump—be it that of a juvenile or an adult, that of a male or of a female—for a few moments (without seeming to have a serious interest in copulation). Then the sideshow of apparent sexual activity ceases, and more to the point, the aggressive behavior of the pursuing animal also ceases.

What's going on here? And why does it work? Maxine Sheets-Johnstone agrees that something is going on: "Presenting is of course a standard non-human primate way of both acknowledging the threats of another and calming them. In Gombe Stream chimpanzees, for example, Jane van Lawick-

HOW WE GOT TO BE HUMAN

Goodall observed presenting to be "one of the most frequently observed elements in submissive behavior patterns."[1] However, neither author explains how it is that presenting "acknowledges and calms." It shouldn't surprise you to hear that I will make the case that such presenting works because it is a *gesture* that represents a *concept* in the gourds of the apes involved.

In *Origins of the Modern Mind*, Merlin Donald decidedly denies that "apes in the wild . . . possess even a rudimentary system of voluntary gestures or signs."[2] He also states that "no animal except humans has ever invented a symbolic device in its natural environment." Regarding this last statement of his, I find it more than interesting to hear Donald discuss the behavior of the chimp called "Mike," who figured out how to intimidate his fellow apes by banging two kerosene cans. "[Mike] had to generalize from a number of event perceptions, and then interrelate them." Certainly, it is difficult to classify the behavior of members of even our own species, let alone that of individuals in another crowd, such as apes. Nonetheless, I think banged cans were a rudimentary symbolic device that Mike invented to generate intimidating gestures.

About the general issue of whether or not apes engage in gesturing, there have been studies by scientists. In "Indexical and Referential Pointing in Chimpanzees," David Leavens, William Hopkins, and Kim Bard have some data.[3] This is data about chimps' pointing, not in a natural environment, but in a laboratory. In the presence of a nearby human observer, a chimp would point to accidentally dropped food-items outside his home cage, then gaze alternatively at the human and at the food. Their research allowed the authors to conclude that such pointing behavior is intentional and communicative gesturing that occurs in the absence of language training or explicit shaping of behavior.

Now let's look at several answers to the presenting problem, none of which may provide a complete explanation by itself, but all of which may contribute to some understanding of both ape sex and ape communication.

The first explanation says somehow underneath such rump offering and rump taking there is real sexual desire. It is conceivable (barely so) that a young male, novice to the sex game, his blood a-boil with newfound hormones, might regard the rump presented to him as a real opportunity for coitus. Would he not, however, quickly wise-up? Before long, he would ignore offers from females who are neither in heat nor hot to deliver the promised pleasure, let alone have any interest in false offers from pipsqueak half-grown males. Please don't give a thought here to buggery. Chimps do not engage in anal intercourse, nor do bonobos (who do engage in male-male sexual activity). Evidently, neither ape can conceive of such a thing. So the pursuing ape cannot regard the offered male rump as providing the equivalent of "any port in a storm." Also, the presentation is sometimes made by a female to another (often higher-ranking) female in hostile pursuit. Female apes have not been observed to mount other females for sexual gratification. (Female bonobos do have sexual interaction with other females, but its form is face-to-face rubbing of genitals against genitals, called "hoka-hoka" by humans who live nearby, not mounting and thrusting.)

So it is quite unlikely that real opportunity for sexual activity lies beneath what seems to be a disarming pseudosexual rump presenting. Old-time comedian Henny Youngman for decades disarmed his audiences with his one-liner: "Take my wife. Please!" Truculent chimps are, in fact, disarmed by what seems like an ape parody of Youngman's joke: "Take my rump. Please!"

A second explanation suggests ape male behavior here may be somewhat similar to certain human male behavior. For example, there are the always reliable bait-and-switch television beer commercials. There, the attention of human males is transferred from some prior interest to images of sexually attractive, friendly young females in the midst of cans, bottles, and glasses of beer. (The word "attractive" is used here in the broad sense, not the narrow, sexually heated one.) The import of the images is that buying and drinking beer will increase the odds of the likes of such females entering the lives of the ever-hopeful watching males. Impossible dreams perhaps, but the ads presumably sell beer.

It might be that for a male ape, an offered rump is like the image of a shapely young female to most male humans. It is something to which attention simply must be paid, no matter how unlikely it is that a real sexual opportunity is present or will ensue. But this explanation would hardly explain females offering their rumps to hostile other females, who in turn briefly mount them and then abandon their anger.

Rather than pursue this connection with male human sexuality here, let's pursue the opposite one, that we are falsely projecting human male motivation into the lives of male apes. Thus, a third explanation says that maybe male or female ape responses to phony offers of sex are simply what students of animal behavior call a "ritual," by which they mean a formalized set of acts that may have little or no meaning to the agents. These acts are somehow, via biology or learning, entrained in each agent's mind. That would suggest something like one of those old Charlie Chaplin movies where, when the pursuing villain gets close, Charlie suddenly stops, turns, lifts his hat, and bows. The villain, despite his desire to pounce and pound on Charlie, finds that, like it or not, he, too, must pause and go through the bowing and hat-lifting formalism. In those old movies, the villain and Charlie resume the chase at that point.

Although our chasing ape does not resume the hunt—rather, it abandons its anger and its pursuit of the offender—we could simply assume a slightly different structure to the "ritual." Interesting though this explanation may be, it's really kind of a cheat. It offers a name, an intriguing name, "ritual," but no real explanation. (By the way, ethologists made a poor move by borrowing from anthropologists the word "ritual," when what they mean is "formalism." Animal "rituals" have none of the underlayer of high import and solemn significance that is always attached to human rituals.)

To find something better than amazing ignorance, graceless beer commercials, or empty canned ritual to explain the pursuer ape's response to a pseudosexual offer, let's again look at the idea of gesture. We humans use gesture much like we use language, as a means of volitionally conveying concepts to one another. Whenever we use our body, in whole or in part, to refer to (rather than directly to interact with) some part of our world, we are gesturing. One of the simplest human gestures is pointing, wherein we extend a finger in a certain direction so that if another person were to use the extended member as a gun-sight, that person would look at what we want him to see. No lower animals use anything like finger pointing to convey information to one another.

If you are thinking of the more or less "human-made" subspecies of canine called *pointers* as a related counter-example, think again. It is the human hunter, not the dog, who conceives of the dog's behavior as pointing. The dog holds its disarming still pose in the presence of prey as part of its genetic hunting behavior, not to get other dogs or a hunter to look at the prey it has in its sight. Nor does a dog pointer have the mental option to point or not to point.

By the way, although apes rarely point with their fingers, they more commonly point with their arms, and at other times they do "peer" in a direction of what seems to be interesting to them. Donald discusses the ability of human children to engage in pointing behavior and to direct their gaze to a place where their mothers seem to be looking. For him, however, "Chimpanzees lack this central component of intentional gazing and pointing: the ability to realize the intentions of others."[4] For me, Donald's views are less credible than those of primatologists who see some ape behavior as *pointing gestures* and other as *deceptive peering*.

Ape Gesture and Utterance: Steps on the Information Highway

About animal deception, Cynthia Mills supplies a nice example in "Unusual Suspects." She tells us of the consequences when "a baboon, chased by a group of older males, suddenly stopped and stared at the horizon. The attackers stopped and stared, too—though there was nothing to see—until they forgot why they had been running and settled down for a grooming session, saving the fugitive from a beating."[5] I find it interesting that, although scientists sometimes explore animal *deception*, for the most part they do not examine anything named animal *gesture*. Perhaps this is because deception can more readily be thought of as no more than objective *behavior*, whereas gesture carries the implication of subjective *mental* activity behind it.

When chimps do point or peer, their fellow apes will usually look in the same direction to discern what was of interest to their comrade. As for peering, if the first ape is looking at something that really interests him, he is not employing a gesture. However, there is anecdotal evidence that on occasion an ape will "fake" such an activity and successfully divert the attention of higher ranking neighbors for a few moments—during which the possessor of the "lying eyes" seizes a nearby desirable bit of food. To deceive in that way is to employ peering as a gesture, just as intentional arm (or finger) pointing is a gesture.

Gesture differs from language in two ways (among others). First, gesture commonly uses only a single integrated body movement to symbolize what is to be conveyed to others. This contrasts with the structured set of meaningful but usually arbitrary throat "movements" used in vocal language. Second, gesture is always a "natural" icon or index that is appropriate to the concept being symbolized, and not arbitrary as are the components of verbal language. Examine two gestures, finger pointing and winking. It would not do, say, to wink an eyelid, instead of pointing a finger, to get someone to look in a given direction. Nor would it do to point, rather than wink, to convey the notion that things may be other than they appear to be.

Although apes rarely use the gesture of finger pointing and never that of winking, we can expect that apes—who live in a mental world containing concepts, which are largely absent in that of lower animals—do need and use gestures to convey some of the concepts that are important in their lives. In chapter 2, we examined and rejected several possibilities of lower animal gesturelike behavior. However, apes have much more in the way of (conscious) access to social concepts than do lower mammals, and the use of gesture is part of how that greater mentality is expressed.

HOW WE GOT TO BE HUMAN

An idea which was rejected for wolves—the use of gesture to convey acceptance of inferiority to a superior—warrants endorsement for apes. When two male chimps have had a brouhaha, the loser frequently shows his wish to be friendly with the winner. The winner rejects attempts by the loser to embrace and hug him. The victor rejects them, that is, until the loser overcomes his reluctance to perform what looks to us humans like a gesture of fawning, bending the knee, butt-kissing of a sort. The loser's actual behavior is one of making himself small by standing down on all four limbs, while the victor stands tall on two legs and even erects his hair (with the thrill of it all) and looks still larger. Sometimes the loser adds a series of bobbing bows to his self-belittling gesture of standing low. Only after this peculiar exchange does the victor accept and return the loser's attempts at embraces. Can you do other than believe that the loser is gesturing to show his deference to the winner?

That said, let's look at how well suited to its purpose is the loser ape's gesture. He literally makes himself smaller, thus conveying the concept "you are larger, more powerful, and superior to little weak inferior me." The notion "bigger is better" is acted on throughout the animal kingdom: birds and hairy beasts all fluff themselves out and stand tall in contest situations, while butterflies develop patterns on the backs of their wings that look like large eyes that scare off onlookers. Making oneself large is usually a smart move. For an animal to go counter to the natural enlarging impulse by making itself small is highly unlikely unless doing so allows it to get something important that it wants, in this case the friendship of what might otherwise continue to be an enemy. For young Evered, whose beating by Numero Uno Goliath was described in chapter 4, you may recall that prostrating himself gave him peace through strength—Goliath's strength.

Less brainy social animals, such as wolves and dogs, certainly suck up to their leaders. But they don't employ (optional) gestures to do so, they always suck up. That suggests that, for them, social power is unquestionably a real thing, but it is not real as a concept in their mentation.

Another seeming gesture commonly used by chimps is that of holding out a palm-up open hand to another chimp. It is used for more than one purpose. If the second ape has food, the gesture asks for a portion, as described by Kim Bard.[6] If the gesturing ape is in the midst of a struggle with an opponent, it may hold its hand out to a nearby friend, asking for support in the altercation, as described by Frans de Waal.[7] If the requestor gets support, evidenced by the friend's approach, it will turn back to its struggle with new vigor as it now expects help from its supporter.

156

Ape Gesture and Utterance: Steps on the Information Highway

Making oneself small, enlarging oneself, and requesting something are all behaviors that are in wide use in ape communities. This suggests the likelihood that they are learned gestures rather than simply spontaneous ones. However, if an ape (it may take a smart one) understands the concepts that apply to its immediate situation, it can invent an appropriate gesture. An account of just such an invention is provided by de Waal.[8] He describes the behavior of a chimp whose hand was hurt in a fight with a rival. For a few days after the fight, the chimp with the injury limped when it moved (on all fours). That was certainly not a gesture. Days later, however, the animal did not limp at all, except when its rival was in sight. Such limited limping is something in the way of a generalized gesture, or at least a deceptive one. Furthermore, it may have been a gesture that was invented on the spot. Primatologists, however, would not presently use the word "gesture" for this deceptive pseudolimping behavior. As scientists, they would have to find, somehow, many almost identical behaviors. Then the behaviors would need to be evaluated by more than one observer before they felt free to rely on that word. To do all that, they would need lots of time and perhaps some luck.

An example of deceptive behavior that is broader than simple gesture is seen in the following account of the trickiness of Rico, abstracted from *Primate Behavior*, by Duane Quiatt and Vernon Reynolds.[9] Rico is a subadult male, one of a group of rhesus monkeys who live in a controlled setting. Rico was consistently successful in getting four other monkeys to accept his grooming. This occurred despite the fact that each time a groomee started to relax and enjoy the services, Rico would suddenly nip his victim. The authors of this paper, I must note, describe Rico's activity in terms of cajolery and submission rather than deception.

Although it seems that Rico's intentions were to deceive, what went on in the mentation of his victims is less clear. Were they deceived each time? Surely after a couple of nips, they would have learned that Rico was setting them up for a sting. Perhaps, each time they allowed themselves to be "talked into" the grooming, which they knew would be pleasurable, even though it might, later, end unpleasantly. (In the comics, Lucy-the-deceiver regularly persuades Charlie Brown to attempt to punt the football that she pulls away just before he would have kicked it.)

APES, HUMANS, AND PROMETHEUS: FACING UP TO A POWERFUL FOE

Although appeasement of a powerful foe may be the safest move, it is not the only response taken by apes. Ethologists have noted examples where apes show themselves to be attempting to disguise and overcome their fear of an antagonist. Primatologist Frans de Waal provides an anecdote that I shall reexamine here. In the midst of a struggle with an opponent, one chimp suddenly found itself with what is called a "fear grin"—lips pulled wide showing clenched teeth—on its face. The chimp turned from its foe and used its hand to first hide and then reshape its facial features. After three attempts, the ape succeeded in removing the fear grin from its face. Then it turned to confront its antagonist once more. So there are times when simians face down their fears and resume the agonistic struggle they came close to giving up. This, too, can fairly be compared with human behavior. Gulping mouthfuls of humble pie in the face of overwhelming power is doubtless the norm for both apes and humans. Being the norm does not make it an adaptation; it is only an adaptation byproduct. The fact that apes, as well as humans, sometimes overcome their fears rather than succumb to them is more surely a byproduct of adaptation.

Such humans may not respond to great power, human or superhuman, by fleeing. They may not respond by bending the knee, torso, or head, as Gerard Manley Hopkins' "Carrion Comfort" protagonist does in response to the might of his God, as described in chapter 4. We can see a perhaps romantic but surely inspiring model for these courageous individuals in Prometheus. In the ancient story, he took fire from the gods and gave it to humans. The reaction of top-ranked all-powerful Jupiter to this offensive insult to his majesty was to nail the Titan to a mountain, and to send sharp-beaked raptors daily to tear at his liver. The response of Prometheus to raw rapacious power was not at all like the craven collapse and mea culpa of Hopkins's protagonist. That is evident, at least in Act 4 of Percy Bysshe Shelley's "Prometheus Unbound," where his version of the Titan is addressed and described as follows:

> To suffer woes which Hope thinks infinite;
> To forgive wrongs darker than Death or Night;
> To defy Power, which seems Omnipotent;
> To love, and bear; to hope till Hope creates

From its own wreck the thing it contemplates;
Neither to change nor falter nor repent;
This, like thy glory, Titan! is to be
Good, great and joyous, beautiful and free;
This is alone Life, Joy, Empire and Victory.

Wonderful words, aren't they? Of course, Shelley's Prometheus, the creature described above, is a half-god Titan, whereas apes are only simian, and Hopkins's protagonist is only human.

RUMP-PRESENTING AS A PSEUDOSEXUAL GESTURE OF APPEASEMENT THAT ALWAYS WORKS

We have seen that, after losing a fight with another male ape, making oneself small can be regarded as a useful gesture of appeasement. Going back now to chimp pseudosexual behavior, we can view rump-presenting as another gesture of appeasement. Presenting certainly prompts similar results: if not the embrace of the more powerful would-be aggressor, then at least the pursuer's abandonment of hostility. There remains the question of why the particular gesture of rump-presenting works successfully. It certainly is not as obviously suitable as is kowtowing by making oneself small.

I suggest that it works because the chimps have sufficient minds to abstract (from what they regularly see of ordinary male-female sexual interaction, which includes females presenting their rumps) the idea that the females are at times also reacting to the greater physical strength and power of the males. At times, the males do use their physical superiority to aid them in getting sex from the females. Indeed, as Richard Wrangham tells us, primatologists have observed chimp males forcing sex upon females in their natural habitat; the same has also been observed with wild howler monkeys and with captive gorillas.[10]

This is not to say that the presenting females are always trading sex for peace. On the contrary, those presenters are commonly trading sex for sex. However, on frequent occasion, a male presses, pressures, or forces a reluctant estrous female to present her bottom and copulate. Sometimes he induces an estrous female to go away with him from the other males who might successfully compete with him for her receptive body. (Note that, if the female is reluctant to go along, such a male uses another gesture to enact his pressure—he shakes a tree limb. Why does he use that particular gesture? He makes himself more intimidating by enlarging his behavior to include the activity of the aerial branch.)

It is likely that all the apes, male and female, young and old, understand that there is a link between male interest in having females present their bottoms sexually and male threatening power. From there, it is only a small further step to link female presenting with female obeisance. To generalize from the frequent use by estrous females of obeisance (in the form of accepting otherwise undesired copulation) as a successful technique for warding off male aggression, the chimps developed the additional concept that anyone, in heat or not, female or not, adult or not, can present a rump to disarm an aggressive ape.

This may explain the behavior of the rump presenters, but what about the behavior of the aggressors, the presentees? Be they male or female, they, too, have learned that, in their society, presenting is often used as a gesture of appeasement. So they take satisfaction when homage to their power is shown, and that gratification satisfies the prior angry impulse and thus dissipates it. For a male, there may be an additional factor: the sight of a female ape offering her rump may be a learned signal of the latter's receptivity, one that is almost as effective as the perhaps part-learned, part-instinctively understood signal of the sight or scent of an aromatic puffy pink rump that represents female receptivity. In part, he may respond to an offered rump with a new frame of mind. By mounting the offered bottom, he abandons his prior frame of mind and his previous hostility. He realizes the offer was not real, dismounts, and goes about his normal everyday business of seeking new opportunity. (All this, I'm sure, without the least likelihood of blushing over his faux pas. Embarrassment is a social feeling that requires others who share the idea of what is embarrassing. As such, it is not within the reach of apes.)

When the rump-presenting gesture is offered by a pursued ape and accepted with a pseudosexual mount by the pursuer, the two animals are reconciled, they are friends again. The use of these gestural behaviors, the rump-offering and the rump-mount, are found not only among apes, but by stump-tailed macaques, as described by de Waal.[11] The bottom-mounter clasps the hips of the presenter in a reconciliation behavior called a "hold-bottom." Sometimes, the supporters of the dominant animal in such a reconciliation get in line to form a several-member train of hold-bottoms.

APE UTTERANCE: IT HAS A VARIETY OF FUNCTIONS, INCLUDING PROTOLANGUAGE

Ape vocal utterance, like that of humans, can be divided into two kinds, depending on whether or not the utterer intends to influence one or more others. For the simpler and more common kind, there is no intent to communicate or otherwise influence anyone. These vocalizations are directly related to the state of emotion, feeling, or mind of the vocalizer. For humans, they can be grunts, groans, ga-gas, even the sounds from a mutterer, and those between syllables from a stutterer-utterer. Of course, the sounds that start in a gut or a stomach may not end up as vocalizations.

Not quite as simple, but still not directed at others, are the voiced sounds that can be understood by the hearer as showing knowledge about the physical or social environment. These include sounds always emitted by lower animals when they encounter food or they sense an approaching predator. Why does a vocalizer who is not trying to influence someone vocalize? There is usually no reason, none at all, known to the mind or mentation of the utterer. The self-expressive urge is common to all of us lower mammals, apes and sapiens. It may be an adaptation byproduct of the biological inheritance of these sound-makers. Lower animals and apes partly inherit and partly learn self-expressive vocal behaviors that are used by others as signal that a predator is near, or those that announce the finding of a supply of food. These may help the survival of others, even though the issuer has no such intention in its head. Of course, such utterances, when heard by prey or predator, may diminish, not to say extinguish, the prospect of a particular utterer's survival.

Less common are the second kind of utterances, that is, vocalizations that are issued with the intent of influencing others. Trying to determine which creatures, other than humans, do this intending-to-influence requires getting into the mentation, the thinking in particular, of the creatures so as to learn their intentions. That is a task that scientists who study animals, and even many who study humans, have tried to stay away from. Their reluctance is understandable.

The (subjective) mental realm, which happens to be the main subject area of this book, is hardly accessible in the same easy way that the physical realm is. A certain amount of projecting (often with more pejorative intent called anthropomorphizing) is inevitable. Error becomes easy; finding evidence is difficult; and proof may be almost impossible. However, much of what is interesting and some of what is important, about lower animals and apes as well as about humans, lies in this seemingly nonphysical mental realm. So I'll press on with these words, my own making of recorded unvoiced vocalizations that are intended to influence others.

Such intentions to utter sounds that influence others come in two varieties. The first kind, which tries to affect the behavior of others, is found not only in humans but in apes and probably monkeys. Apes normally give calls announcing the presence of predators when other apes are nearby, but not when they are alone. So it seems that the calls may be more than mere announcements; some of them may be intended to get other apes to act appropriately. This would be consistent with other attempts to influence behavior, such as presenting to induce in others either sexual behavior or an end to aggressive behavior.

The second kind of intention for utterance seeks to change, not the behavior, but the mentation of others. However, changing mentation is usually an indirect way of changing behavior. Humans often attempt to influence other minds. I know I do. I believe you do, too, although I doubt that I could prove it. Apes, however, do not go beyond attempting to influence behavior. In fact, apes don't even know that other creatures have mental activity, or that other apes are anything more than bodied bags of behavior. That would explain why these often warm and friendly creatures never seek to supply a comforting embrace when another of their kind is in despair from the loss of a mother or an infant. If apes do not know that other apes have any mental activity, that, in turn, is because they don't know that they, themselves, have anything going on in their gourds. They don't have brains and minds that are capable of engaging in consciousness of the reflective kind.

Apes do show themselves as capable of perceiving that they exist as bodies. Gordon Gallup describes the careful methodology he used to test an ape's response to its image in a mirror.[12] An ape (that had been exposed to a mirror for several days) was marked on its face (while under anesthesia) with a dye that was free of tactile and olfactory ques. After recovery from the anesthetic, the ape's response to its image was noted. An ape that saw the image as its own would reach to touch the mark on its face while watching its reflection, and then smell its fingers. Apes with no previous exposure to a mirror, when similarly tested, reacted to the marked image as if it were another creature.

Mirror recognition or not, apes are of course conscious in the basic sense that I think is suitable for animals. They experience pain, rage, fear. They experience the desire to get what they want and avoid what they don't. They feel hungry, wet, hot. They perceive objects, other creatures, parts of themselves, perhaps even their own reflections in a mirror. Apes also have concepts such as those for classes of predators and for classes of other apes. These things suggest themselves to me by knowing about some of the things they do, and by projecting myself into these low-brow cousins of ours. However, apes are not aware that they themselves have these emotions, thoughts, feelings, and perceptions. Despite all this, it may be that apes who have mastered a bit of language could learn that they have ape versions of minds.

Communicative vocal language, as humans practice it, is an expansion of the two kinds of ape intentions to influence others. We try to influence behavior directly with vocalizations that include words. We also try to influence minds and, indirectly, behavior, using both language and simpler vocalizations. Influencing behavior by first influencing minds is far more effective with us, but it is often much less certain what behavior will ultimately result. My words here, for example, may result in some disagreements that quickly induce book-closing, or in some interest or agreement that results in further reading, perhaps even some changing of minds, with surely unpredictable eventual behaviors.

The ability of chimps and other apes to learn some of our language skills suggests that they are able to use language symbols to influence behavior much like we do. Douglas Candland, in *Feral Children and Clever Animals*, examines the limited nature and extent of ape use of language.[13] A chimp called Washoe learned about 150 words of American Sign Language, and other apes have shown similar skills. But no ape has ever learned syntax, nor is it clear that their language use goes much beyond getting things to happen immediately. Their use of sign language is better described as limited protolanguage than as the complete language that speech or American Sign Language is for humans. More of the nature of protolanguage and complete language is examined in chapters 11 and 12.

In a recent study by several primatologists, it is suggested that bonobos in the field communicate—apart from the use of bodily gesture and vocalization—by intentionally leaving a "record" of flattened and broken vegetation to enhance the trail taken by one group between feeding sites. Such an enhanced trail lets subsequent groups follow earlier groups and meet at the next feeding area. The authors of that study suggest—not just that a new means of bonobo communication by leaving such record exists—but additionally "the signaler must, in a sense, enter into the mind of the potential receiver at the time of signal creation."[14] For this, they suggest the apes must have a "theory of mind" for other individuals. I think the apes do, indeed, have something far-out in their heads. Not as grand as a theory of mind, but still grand, is a theory of behavior.

For country cousins, apes are pretty slick. First, they took away our claim to fame as exclusive makers and users of tools. Then it was communicative gestures and utterances. Now, danged if it ain't communicative records. Well, I hope we can be confident of no ape leaving foot and hand prints on the moon, and of no apes swinging on the Internet.

DARWIN AS DEITY: ADAPTIVE EXPLANATION IS SWELL, BUT USING PROJECTION TO EXPLAIN CAN ALSO BE USEFUL

Before leaving these explanations of ape gesture, communication, and mentation in general, it may be useful to confront further the issue of the propriety of explanations that are not based on adaptation, on Darwinian genetic-originated survival value. Social scientists who study the lives and mentation of apes generally seek and find explanations based on adaptation. Who can fault them for that? On the one hand they are trying to contribute to a science. Unfortunately their science has, as yet, little in the way of scientific principle or foundation at all comparable to the towering contribution provided by Charles Darwin. On the other hand, many of the things we'd like to know about apes are not to be learned directly from the concrete behaviors of these astonishing relatives of ours.

The only other major tool that occurs to me—controlled projection into ape lives of what we humans might find meaningful—has two ends, one blunt and the other sharp, like an Acheulian hand-axe. The end that faces the apes is presently the blunt end, even though at times it seems capable of dissecting their behavior to yield valuable knowledge of the mentation underneath. The other end, the sharp one that faces the explaining axe-holder, sometimes can slip into that user's flesh even when the tool is not used naively. So it is not surprising that most students of ape mentation, as well as many students of the human mind, cleave to Darwinian adaptation as if it were the rock of ages. Or they act out the notion that there is no scientific salvation save explanation based on survival value. In so doing, they may forget that maintaining overly narrow scientific standards also has risks. There is a risk of inadvertently advocating a scientism that suggests idolatrous worship of narrowly "pure" science. That is a risk that psychology fell prey to a half century ago by trying to avow interest only in human response to stimulus. Cognitive psychology, one of the heirs to that largely disavowed ancestor, is currently tempted by a related over-reductionism to assume that inside the human mind is mostly algorithmic process without large other factors based on the consciousness that lets lives be experienced. Evolutionary psychology seems often to be overly determined to depend only on the good news (part of the modern-day gospel of "science") of Darwinian adaptation. There is also a more pedestrian risk with insisting on too much scientific purity, that of posturing with results whose intrinsic worth may be little more than that of the paper needed to print them.

Social scientists know they are in a tough field of endeavor. According to economist Samuelson, there is "only one proposition in the whole of social science that is both true and nontrivial. It is David Ricardo's Law of Comparative Advantage."[15] (That law gives one country the possibility of comparative advantage in some product even if it is less efficient at making the product than its trading partner.) Perhaps this is more than a bit excessive, but it should give us pause about social laws.

NOTES

1. Maxine Sheets-Johnstone, *The Roots of Power* (Chicago, Ill.: Open Court Publishing Company, 1994), p. 39.

2. Merlin Donald, *Origins of the Modern Mind* (Cambridge, Mass.: Harvard University Press, 1991, pp. 3, 126, 156.

3. David A. Leavens, William D. Hopkins, and Kim A. Bard, "Indexical and Referential Pointing in Chimpanzees," *Journal of Comparative Psychology* 110, no. 4 (1996): 346–53.

4. Merlin Donald, *Origins of the Modern Mind*, p. 171.

5. Cynthia Mills, "Unusual Suspects," in *The Sciences* (The New York Academy of Sciences, July/August 1997), p. 35.

6. Kim A. Bard, "Intentional Behavior and Intentional Communication in Young Free-Ranging Orangutans," *Child Development* 63 (1992): 1186–97.

7. Frans de Waal, *Chimpanzee Politics* (New York: Harper & Roe, 1982), p. 39.

8. Frans de Waal, *Good Natured* (Cambridge, Mass.: Harvard University Press, 1996), p. 44.

9. Duane Quiatt and Vernon Reynolds, *Primate Behavior* (New York: Cambridge University Press, 1993), p. 168.

10. Richard Wrangham and Dale Peterson, *Demonic Males* (Boston, Mass.: Houghton Mifflin, 1996), p. 138.

11. Frans de Waal, *Peacemaking Among Primates* (Cambridge, Mass.: Harvard University Press, 1989), p. 163.

12. Gordon G. Gallup Jr. "Self-Recognition: Research Strategies and Experimental Design," in *Self-Awareness in Animals and Humans*, ed. Sue Taylor Parker, Robert W. Mitchell, and Maria L. Boccia (New York: Cambridge University Press, 1994), p. 35.

13. Douglas Keith Candland, *Feral Children and Clever Animals* (New York: Oxford University Press, 1993).

14. E. Sue Savage-Rumbaugh, Shelly L. Williams, Takeshi Furuichi, and Takayoshi Kano. "Language Perceived: Paniscus Branches Out," in *Great Ape Societies*, ed. William C. McGrew, Linda F. Marchant, and Toshisada Nishida (New York: Cambridge University Press, 1996).

15. Matt Ridley, *The Origins of Virtue* (New York: Viking Penguin, 1997), p. 207.

PART THREE

OUR UPRIGHT PREDECESSORS, THE EARLY HOMINIDS

INTRODUCTION

OUR OWN SPECIES, HOMO SAPIENS, was preceded by a series of breeds which, as we move forward from the common ancestor we share with the great apes, looked and acted less and less like apes and more and more like us. What was once conceived as "the missing link" has turned into a series of found links, species whose linkage with one another, however, is still less than clear. The missing link was once expected to be a creature whose main difference from ancient apes would be its large brain, foreshadowing the brain for which we are famous among ourselves. That expectation was unfulfilled. Instead, an upright posture proved to be the first major change from prior apes. The brain got significantly larger only for the last two among the eight-or-so found links. But our human vanity regarding our powerful brains contributed to the names we gave these linking species. The first six (perhaps more) of them are named "australopithecines"—which means "southern apes"—and are also called "upright apes," or "woodland apes." They were the ones without much increase in brain size and with little or no making of stone tools. The remaining two (or more) species, with increasing relative brain size and increasing complexity of stone tools, were given our own family name, Homo, or human. This was done despite the evidence of genetic similarity that suggests it is also reasonable to regard all Homo species, ourselves included, as part of the clan of great apes. Never mind; if we don't indulge ourselves, who will?

Whatever we call our predecessors, there are some real differences between apes and humans. As elsewhere in this book, we will seek evidence here more of differences of subjective *minds* than of the objective brains and bodies that maintain those minds.

The six (some identify even more) species of upright apes lived between 4.4 and 1 million years ago. The two species (again, there's evidence for more than two) of early Homo lived between 2 and 0.2 million years ago. Some of the basic approximate data about them all are tabulated below, most of it coming from David Lambert's *Field Guide to Early Man*.[1]

Name	Years Ago (millions)	Height (feet)	Brain Capacity (cubic centimeters)
Upright ape:			
A. ramidis	4.4 to ?	?	?
A. afarensis	4 to 2.53 to 4	410	
A. africanus	3 to 1	3 to 4	450
A. robustus	2.5 to 1.5	5 to 5.5	500
A. boisei	2.5 to 1	5.3 to 5.8	500
A. garhi	2.5 ?	5 ?	?
Early human:			
H. habilis	2 to 1.5	4 to 5	650 to 800
H. erectus	1.6 to 0.2	5 to 6	880 to 1100

Brain capacity is the expression used to refer to the size of brain that will fit in a bone or fossil cranium. It's significance is relative to the overall size of a creature. Thus the last two of the southern apes may not have had more brain *capability* or mental ability than the first three, as *both* height and brain capacity increased. For comparison, note that our own brain capacity averages about 1,400 cubic centimeters. It also varies greatly, from about 1,000 to about 2,000 cc., and those variations have not been found to correlate with intelligence. So the organization of our sapiens sapiens brain may be more important than its size.

About Homo habilis in the table above, Maitland Edey and Donald Johanson say (in 1989) that the species may have been preceded by a still more primitive Homo species.[2] In 1996, Roy Larick and Russell Ciochon describe more recent confirming evidence that a species (Homo rudolfensis) emerged a half million years before habilis.[3]

These eight species, together with sister subspecies Homo sapiens neanderthalensis and Homo sapiens sapiens (us self-appointed "wise-guys"), are called hominids. As both the neanderthals and our own subspecies are quite recent, the above eight predecessors are referred to as early hominids. (The *number* of prehuman species is often out of date by the time it gets into print. In a 1996 book, Donald Johanson and Blake Edgar show *seven* pre-Homo species followed by *eight* Homo species.[4] With luck and the diligence of fossil finders, their numbers will be out of date before you read them here.)

Before examining what little is known about these early hominids, it is worth asking why the change from bent to upright apes came about. It is always tempting for us to think the cart is pulling the horse toward some future goal. Thus, before any early hominid links were found, it was expected that a large brain in the "missing link" would pull its near-ape owner on a path toward future human beings.

It is not at all certain, however, that a large brain would add survival value to members of an ape species, either now or way back then. Too large a step in size might not be a step forward. Over the many-million-year history of apes, nature may well have tried it (by random mutation) many times, and found it not worth the additional investment of blood and oxygen plus other modifications needed to sustain that enlarged brain. Countless mutations continue to occur within the stock of a species when a good deal of time passes without major environmental change. Most of those changes in the genetic plan fail; some continue to shape the species for its world. As more and more time passes, fewer and fewer of these mutations are adaptive. It makes sense, but no more than pedestrian sense, to suggest that such a stable environment might sometimes approximate the best of all possible worlds for a species. There is also a sense, equally unprofound, in which nature provides something approaching the best of all possible animals for each of the niches of the world.

Scientists are by no means in agreement about why and when adaptive change occurs. Eldredge calls those on the other side of the argument "ultra-Darwinians." He suggests their credo is, "extrapolation of generation-by-generation change under natural selection is all we need consider when framing a theory of adaptive evolutionary change."[5] He finds that ultra-Darwinians (such as Maynard Smith and Richard Dawkins) see competition for reproductive success the only real issue, and they find competition for food and other economic resources as insignificant.

The largest horse that pulled our evolutionary cart—and took a fork onto a road untrod by any other creature—may well have been (as is frequently the case) a change in the environment. In "East Side Story: The Origins of Humankind," Yves Coppens describes this change.[6] In this case, the force that shifted our ape ancestor species to a new path would have been nothing less than the slow shifting of the giant tectonic plates beneath the crust of central Africa. About eight million years ago the clash of hidden shifting plates started two related slow changes along a north-south line of east-central African geography: The gradual sinking of a long and narrow strip of land produced the great Rift Valley, and along the western edge of that rift, a line of mountain peaks was slowly thrust upward. As a result, the rift and the area to its east became ever less humid. It gradually lost its forests and woodlands, and evolved more and more into grassy plains and open savanna.

Millions of years after the African earth started to shift, on that east side of the rift mountains, "The East Side Story" began. Eldredge says that a global climate change, about 2.8 million years ago, underlies the dramatic turnover in African biota as habitats changed from tropical woodlands to more open savannahs.[7]

On the east side of the great African rift, the sun shed its first light on a new creature, a hominid, an ape who stood upright on the occasionally wooded and fruited plain. With thick forests fewer and farther between than across the mountains to the west, the first hominids had an adaptive advantage over their unmutated relatives. They could stand, walk, and run better than could their equally smart, stooped relatives. They could also make greater use of their freer hands. All this helped them find more food with less effort, especially because they had previously developed bigger and better molar teeth that could crunch the roots that were more common in the woodlands. And their tribe increased, even as that of their "parents" (the local apes who lacked the upright structure and better teeth) diminished and then disappeared.

That first hominid species, A. ramidus, "begat" (perhaps via an intermediary "tribe") the tribe of A. afarensis. In turn, afarensis begat africanus, who may or may not have begat the two tribes of larger-molared upright-apes, robustus and boisei. Latecomer garhi (who seems to have butchered antelope for meat and crushed their bones for marrow) or another of these upright-ape tribes in Africa begat the first tribe that are deemed "humans."

Those first humans were soon followed by, or became, a tribe named

"Homo habilis" ("handy man") by the first famous anthropologist Leakey who lived in Africa, Louis Leakey. Habilis, or perhaps another habilis-like tribe, begat erectus, "upright man," a larger brained and bodied human. This erectine tribe not only increased but left its home base in Africa. Mary Leakey, the second famous African Leakey, with Louis begat also-to-become-famous Richard Leakey (may that tribe, with all ours, increase at only a moderate rate).

Erectus (and/or another erectuslike species) spread out from Africa and begat some creatures with bone shapes between its and ours, creatures who —perhaps for lack of sufficient samples to warrant a better name or names —are called "archaic sapiens." Then either erectuslike creatures and/or archaic sapiens begat Homo sapiens' sister subspecies, neandertalensis and sapiens. This last guy's tribe (our very own) really increased. Far from first in that same subspecies was Adam—Abou Ben Adam, may his tribe (ours) cease its recent rapid increase—but I seem to be getting ahead of myself.

NOTES

1. David Lambert, *The Field Guide to Early Man* (New York: Facts On File, Inc., 1987), p. 98–117.

2. Maitland A. Edey and Donald C. Johanson, *Blueprints: Solving the Mystery of Evolution* (New York: Little, Brown and Company, 1989), p. 353.

3. Roy Larick and Russell L. Ciochon, "The African Emergence and Early Asian Dispersals of the Genus Homo," *American Scientist* (November/December 1996).

4. Donald Johanson and Blake Edgar, *From Lucy to Language* (New York: Simon & Schuster Editions, 1996), p. 38.

5. Niles Eldredge, *Reinventing Darwin: The Great Debate at the High Table of Evolutionary Theory* (New York: John Wiley & Sons, 1995), p. 6.

6. Yves Coppens, "East Side Story: The Origin of Humankind," *Scientific American* (May 1994).

7. Eldredge, *Reinventing Darwin*, p 149.

8

THE UPRIGHT APES, OR AUSTRALOPITECINES

A CLOSER LOOK AT THE theories offered for the momentous change in our ape ancestors from quadrupeds to bipeds is revealing—revealing, that is, of the frailty of human thought, even scientific human thought. Two hypotheses about this shift have been put forward by scientists in recent years. Richard Leakey and Roger Lewin describe them in *Origins Reconsidered*.[1] The first theory (that of Owen Lovejoy) compares the efficiency of locomotion of humans with that of four-legged beasts like horses and dogs, and found our mode of moving *inferior*. This implies that a two-legged stance, *by itself*, was not adaptive. Perforce, the hypothesis leaped far ahead. It suggested that bipedal apes got *swift* Darwinian advantage by having more frequent offspring than the four-legged fellows down the street. Then, to have more babies, it seemed to follow that the females would have needed help from the males. No, not copulative help, but help in the form of a *swift* shift to handouts from the males' hand-held collections of food.

Wait, there's more. Males who were willing to share their food with females would not gain the adaptive benefit of greater numbers of their *own* offspring, not unless the females abandoned estrus and formed ongoing close bonds with those same males. So, "naturally," the femmes both dropped heat and cozied up to the males. Of course, this abandonment of estrus and the establishment of male-female ties would also have to have occurred *swiftly* after the mutation that resulted in an upright stance that in itself was adaptively disadvantageous. Might this theory be, in a "woid," less Darwinian than it is "darwinoid"?

HOW WE GOT TO BE HUMAN

Oh, what a tangled web we weave when we think we can understand how we evolved. I am not pointing any fingers or toes at the scientist who built up this hypothesis. Rather, my digits fan out to any and all of us saps—that word is anthropological jargon for "sapiens"—who try to develop ideas about, say, whose hog ate the cabbage. All too often we later find ourselves eating our words encrusted in stale cabbage pie.

It wasn't long before this fully educated semi-guess about our origins was seen to be more than a bit flawed. Others (Peter Rodman and Henry McHenry) were in a position to propose a better hypothesis about why an upright stance was adaptive. This one started and ended with the idea that the immediate change itself, bipedalism, had to be of relative advantage. Again, the hypothesizers decided to compare the energy needed to walk on two legs with that needed to walk on all four. This time, however, they did it with an animal capable of both modes of movement. They did it with your everyday garden-variety chimpanzee. They found that it took neither more nor less energy for a chimp to walk on two legs. The authors of this theory next found that *human* bipedalism is considerably more energy-efficient that *chimp* quadrupedalism. This suggests the same was true of the first upright apes compared with their quadrupedal relatives. Given the likelihood that food was increasingly widely dispersed in increasingly fragmented forests, upright apes (who walked on two feet like us) would have used less energy, giving them adaptive advantage over their relatives who walked on all fours.

Note, however, that there was a helpful mutation that occurred prior to the one that made these apes stand upright. As Richard Wrangham and Dale Peterson tell us, this prior mutation produced thickly enameled massive molars that proved to be adaptive for the consumption of the roots that had become central to the apes' diet.[2] Notice that *each* mutation was of value—not in the future, but in the present—and thus adaptive. Viva Darwin!

THEY STOOD TALL, BUT THEIR BRAINS AND MINDS WERE SMALL

Other than to have gotten up on their hind legs, the several species of upright apes did little to merit their distinguished position as early hominids and more-or-less predecessors of the first human species. (However, if I'm right in the speculations we're coming to, an early upright-ape species initiated the change that led to our human kind of sex lives.) One of the few areas of knowledge we have about them results from fossil finds of jaws and teeth. Canine teeth, which are useful for aggressive purposes, were mostly smaller for upright apes than are those of present apes. Ancient upright apes also had molars, cheek teeth, larger than those of present apes or subsequent humans. This suggests their diet was mostly plant food that needed a lot of chewing. The last two of those australopiths were bigger and had even larger molars, relatively, than their predecessors, the better to munch and crunch larger quantities of roots and lower-quality tough greens when choicer fruits and tender shoots were not to be had.

There is good evidence from fossil size that the australopiths were dimorphic, that is, the males were a good deal larger than the females. This suggests a couple of things. One of the uses for this greater bulk, judging from present apes, was not just pushing females around but intimidating them into sexual acceptance of the males that were "hitting on them." Lest I raise some feminist or masculinist hackles here, let me hasten to add the obvious: greater size among males is useful for throwing one's weight around at anyone who is smaller, especially other males. In fact, it is the widespread belief among ethologists that most species that engages in intermale competition for sexual access to females end up with males considerably larger than females. Donald Johanson and Blake Edgar provide estimates of male/female body weight ratios. For afarensis, the ratio is 1.52; for africanus, 1.35; for modern humans, 1.22; for chimpanzees, 1.37; for orangutans, 2.03; for gorillas, 2.09. They also say that "in most but not all primates there is a correlation between body size dimorphism and canine [size] dimorphism."[3] They speculate that "the smaller canines in *A. afarensis* males may therefore suggest that, for whatever reason, there was reduced male-male aggression." Of course, such male competition sometimes leads to a different result, such as preposterously elaborate peacock tail feathers, presumably the better to catch a peahen's eye.

All of this male competing and posturing goes by the name of sexual selection, in contrast to natural selection. As is plain to see in a peacock's fancy feathers, sexual selection, as such, may not always result in higher survival value in the sense of better adaptation to a natural ecological niche. Pity the poor peacock, made into a mockery as a bird by the peahen's attraction to feathery elegance. On the flip side of this flip snide sneer at the peahen is the likelihood that the size and quantity of peacock tail feathers correlates well with size and strength of the male. Mating with upscale males, as any peahen that quotes Darwin on natural selection will tell you, increases the odds for her progeny to survive. That's true enough, but the peahens neglected to read on in Darwin, about sexual selection. Again, the point is that sexual selection—if and when other things are equal—may loosen, to its peril, a species' ties to improved survival via natural selection.

It is sometimes surmised that the upright apes used their newly freer arms and hands to collect food for future use, perhaps to bring it to a home base, perhaps to share it with others of their kind. It is sometimes suggested that these early hominids made tools of stone, bone, or tree branch, perhaps to help scavenge meat from carnivores and to defend such booty against other scavengers.

In *Blueprints: Solving the Mystery of Evolution*, Maitland Edey and Donald Johanson comment (1989) on the oldest (2.5 million years old) stone tools in the world. "Who made them? It had to be a human being. No australopithecine ever made a stone tool; there is no evidence for it anywhere."[4] About the social life of upright apes, paleoanthropologist Mifford Wolpoff found some evidence.[5] He saw a *healed* break in the fossil femur of an upright ape, suggesting to him that the individual was taken care of by others. This contrasts with the observations of primatologists that apes with injuries or illnesses are often shunned by their fellows. With these exceptions (and also *sharing*, to be examined below), it is unlikely that upright apes had the conceptual ability to do much more of the above-suggested things than do present-day apes.

UPRIGHT POSTURE,
THE ABANDONMENT OF FEMALE HEAT,
AND THE EXPULSION FROM SEXUAL EDEN

With so little known about them, there has been an ongoing open season for more-or-less enlightened opinionating on upright-ape activity, including their sexual behavior. Lovejoy's couple-decades-old views—about estrus ending and pair-bonding beginning—are of interest here, not just because they're now regarded as incorrect, but because they show how easily even a trained scientist's mind is seduced into something like wishful thinking:

> A pair-bonding system was arising . . . as a way of keeping a male attracted
> to a female and ensuring that she be impregnated by him through the
> strategy of fairly continuous mating instead of a frenzy of it at the peak of

her ovulatory cycle. . . . This is a gradual thing. . . . It starts when a female continues to look sexy for a week or so at the end of her estrus cycle. . . . Finally the estrus flags don't count; she has permanent ones that keep her man—her hominid—interested in her all the time. . . . And she'd better keep him interested, because she's fertile for only about three days. If she copulates once every two weeks, her chance of getting pregnant is pretty low. She can't afford that. It's her job to get pregnant quickly, as soon as she can handle the next infant.[6]

Note that the purpose of this new strategy is to get a male to stay around and make himself useful for something besides partnering the conception of babies. After all, the old animal strategy of estrus did that pretty well. It's clear that Lovejoy is not suggesting that this slowly changing-over-millennia "she-ape" is following a deliberate conscious learned conceptual strategy of displaying receptive behavior to get and hold a male. So it must be a (per-haps gradually) acquired, innate strategy. However, that strategy can't be one of being sexually "hot" all the time; that would bring too many males, all of whom would be too busy fighting to be of any help to her. So perhaps it is selective heat: When you find a male you like, that makes you hot, you "put-out" plenty; putting-out keeps him interested.

May I ask, however, what happens to her hots for him when, before long, she happens to get angry with him, or tired of him? She's still an ape, not smart enough to pretend (for the children's sake) to deliver hot sex, not smart enough to gesture "not tonight, you big ape." So, to take a less than wild guess, her sexual desire decreases, even vanishes. Then what happens to him, the big ape? He's no smarter than she is, why should he hang around? So he doesn't—he vanishes. Then where is she, wandering around with a broken pair-bond? This two-decade-old theory shows us how seductive it was to view ultimately advantageous changes as contributing to the *initiation* of the changes.

In a recent book, anthropologist Helen Fisher gives us another demon-stration of the cart-before-the-horse syndrome. She suggests the following theory about pair-bonding as a response to the presumed increased burdens that children were to females with an upright stance: First, there was the change to an upright posture. Then, because the young could not be easily carried on their mothers' backs, the females were greatly burdened by their infants. "So as pair-bonding became the *only* alternative for females—and a viable option for males—monogamy evolved."[7]

This suggests that the change to an upright carriage, in itself, was not adaptive because mothers had to schlep babies on their backs. So the second change—which somehow caused those creatures to live in male-female pairs that made life easier for the mothers—must have occurred immediately after the first one. Otherwise the overburdened upright apes might soon have dwindled and disappeared. I'd say this is not a compelling theory.

Before I go on, I'm obliged to point to a bit of cold water that might get sprinkled on the idea that estrus was abandoned by the upright apes. In *The Neandertal Enigma*, James Shreeve suggests that neanderthal women "may have shown visible signs of estrus."[8] Were this the case, then it is not likely that estrus was abandoned by either the upright apes or their successor erectines. I hardly need to say that the solid facts of this matter are not known.

Any thoroughly enlightening discussion of hominid sexuality should attempt to fully distinguish between "primitive" and "derived" aspects of such sexuality. These two terms are used by Edy and Johanson with reference to the study of fossils: "You trace the primitive features down until you lose them. You try to follow and lump together the derived ones. You note what's shared and what isn't."[9] Primitive features of a fossil "refer to those it shares with an ancestor and carries down relatively unchanged." Derived features "are evolved, altered. If two fossils share several derived features that are not seen in any other closely related type, it is an indication that the two are extremely close and one may descend directly from the other."

In *Why is Sex Fun?* Jared Diamond describes a theory of the evolution of female ovulatory signs, where the presumed common ancestor of chimps, gorillas, and humans had only slight signs of ovulation.[10] Then, only the gorilla retained this "primitive" condition. The chimpanzees evolved bold sign of ovulation, and the humans evolved concealed ovulation. By this theory, our predecessor upright apes would have either retained slight signs of ovulation or evolved concealed ovulation.

In *The Evolution of Human Sexuality*, Donald Symons says (in 1979) that in his own loss-of-estrus theory, "humans are derived from a chimpanzeelike ancestor, and the increasing importance of cooperative male hunting is considered to be the key to the loss."[11] In my view, however, cooperative male hunting would not have achieved high importance for our rather dim-witted upright-ape ancestors. Nor would extensive food-sharing, which Symons seems to regard as a precondition in his loss-of-estrus scenario, have been the likely case with the upright apes. So, if his scenario were viable, loss of estrus would not have taken hold until the erectines came into existence.

The Upright Apes, or Australopithecines

Sheets-Johnstone offers us an explanation of why it was likely that estrus disappeared after our ancestors switched to an upright stance. "Loss of estrus—understood with respect to its visual manifestations—is in consequence explainable on morphological grounds."[12] About the upright apes, she writes, "[Female] genital swellings and colorations would no longer be immediately visible advertisements of sexual readiness. . . . [They] would indeed require a shift to a quadrupedal posture every time the females wanted to approach a male sexually."[13] However, she suggests that such loss of female sexual advertisement was *not* the main sexual change that occurred then. "Loss of estrus—physiological and behavioral—can be explained in the light of continuous and direct male genital exposure. Typical primate estrus cycling was replaced not by year-round female receptivity, as is so commonly claimed, but by year-round penile display."

Sheets-Johnstone has two separable ideas here, both of which I am obliged to disagree with. The idea that male displays of penises replaced estrus cycling will be discussed below. The claim (by others) of *female receptivity* that she refers to is a claim, she seems to think, that the *important* thing about females (in particular, human females) who had abandoned estrus is that they were possessors of a sex organ that, as it is an orifice, made them continuously copulable. For her to suggest that serious people (who are not ideologues) hold such a weird notion is disappointing. Most ordinary postadolescents living in Western society would not take it seriously.

I, for one, do not know enough about hominid sexuality to provide anything like thorough enlightenment. It hardly behooves me to pretend to *know* what is primitive and what is derived about the unknown sex lives of the upright apes. That would be hubris or hutzpah, which might offend someone "up there" in the heavens. To be candid, it's not only offending one or t'other deity that I deeply desire to avoid. Also, it's that one of them might take action. Definitely, I do not want to be "off'd" if one of those Patriarchs thinks he's been "dis'd." Nonetheless, with or without the neutrality of Patriarchs, I hope to shine a bit of light on the sexuality of hominids from upright apes to us sapiens sapiens.

I will expand on the suggestion that estrus disappeared because female genitals were less visible in upright apes. Upright posture was followed by the abandonment of estrus by the females in one of the first species of these upright apes. However, the step to pair-bonding, to forming somewhat stable male-female pairs, did not occur until much later, when those upright apes were succeeded by the larger-brained erectus predecessors of ours.

Furthermore, this male-female pair-bond was not a biological adaptation, but rather a byproduct of the adaptation that produced bigger brains and thus better minds in erectines, erectines whose females had inherited freedom-from-estrus from their upright-ape predecessors.

The recently discovered fossils of what may be the first upright ape, tell us only a little about that species. In 1996, it was "still not clear if this small creature, *Australopithecus ramidus*, could walk upright or not."[14] So I'll assume it was the second upright ape, A. afarensis, whose females were no longer subject to periodic sexual heat.

As with their remote later relatives in the Asian jungles, the widely separated orangutans, broadcasting invitations for amorous activity by displaying swollen sexual tissue would have been ineffectual for upright-ape females. Whatever its other advantages, an erect stance was a hindrance to the transmission of female bodily signals in the form of estrous swellings suggesting sexual availability. Standing upright obscured such signals—the swollen state of female genitals—that were originally sent to males. It concealed them almost as effectively as does a half mile of intervening high jungle for the orangutans. That does not at all imply that regular sexual hunger, as such, had become a disadvantage to those upright apes. Rather, the adaptive value of estrous *swellings* was vitiated, diluted to virtually no advantage. So a random mutation—one that removed not only the swellings, but the related *periodic* itch for sexual gratification, and also the added burden to the body of regularly swelling the signaling genital tissues—was able to succeed.

Of course, a case could be made that the upright stance of those upright apes would have increased the visibility of penises, drooping or erect. Such an argument suggests in particular that easily visible erect penises would have provided females with a clearer sign of male interest in sexual activity. Present-day male apes, however, often deliberately display erections. They do that either by sitting with their legs apart or by standing and "swaggering" on two legs when they are "pumped up." Also, the upright-ape males had *innate* origins for the rut beneath their prurience, as did their predecessor ape and successor human males.

That suggests those ancient upright males were broadcasting their interest in sexual congress on all of the then known frequencies. So an additional paging channel in the form of more easily seen erections would not have added much to their prospects of finding sexually receptive females. This would have been the case because the upright-ape females (who were no longer subject to innate periodic sexual heat) were often *indifferent* to the messages on their pagers that proclaimed male interest. There would not have been any adaptive advantage, in the usual sense of Darwinian natural selection, for males who could display more outstanding penises. Note that "despite popular belief, the size of [human] male sexual organs has nothing to do with virility or potency. . . . Almost all orgiastic sensation in women is concentrated in the clitoris, labia, and related areas, within easy reach of practically any penis no matter its size."[15]

Another line of evidence, however, suggests that upright-ape penises, flaccid or upright, may have indeed become larger than those of the apes that preceded those australopithecines. The Homo sapiens sapiens penis is larger, relatively, than that of any current ape. This suggests genetic changes in the way of a larger male organ may have started with its increased visibility in one of our predecessors, perhaps in the upright apes. Does this contradict the remarks above? Recall again that Darwin suggested sexual selection as another kind of selection mechanism in addition to natural selection. We have another case of what seems like mind's power over matter: as with peahens' interest in peacocks with large fancy feathers, the upright-ape females' idea of what is sexually attractive may have resulted in an enlargement of this part of the males' bodies.

Caution is in order with regard to what seem to be thinly Darwinian notions—the coined word "darwinoid" may be better here (it has the right ring to it)—notions that hover between seeking-to-explain and seeming-to-explain. Note for example that, complementary to the idea that females selected males for large penises, there is a similar idea—theory would be too inflated a word—that human males have selected females for large breasts. Similar to the penis proposition, it is noted by some astute observers that human breasts are large relative to those of apes, and reasons for that are sought. Anyone can offer one simple reason: "bigger is better" is an ancient principle among animals. In *Our Kind*, Marvin Harris says that large pendulous breast would have signalled to men that the women were able to withstand the extra burdens that pregnancy and lactation produce.[16] Do you suppose he's on to something? (Sometimes this explanation is enlarged by claiming that bountiful breasts were selected to provide males with a frontal view of females that somewhat matched a rear view of bulbous bottoms. Presumably, the bulging behinds themselves interest men because they are so close to the scene—just around the bend—of female sexual action.)

Rather than let these ideas of the evolution of the human breast rest, we should note that such explanation is normally found in discussions of relatively large-breasted Western women. Explanations of why Asian females usually are small-breasted go in a different direction. They point out that such less-outstanding female bosoms are themselves attractive to males as they represent a more youthful, indeed, a more childlike form, one that suggests the greater innocence of the young. The idea that smaller breasts suggest *younger* women could be conscripted to serve Darwinian duty: Younger females would have more years during which they could have more babies. Why should the presumed *innocence* of young females be sexually attractive to some men? Perhaps such men find the thought distasteful and threatening that adult females might be lustful and desirous of more in the way of sexual gratification than they (the men) can supply. This idea—the sexual attractiveness of young females—has some shadows that are more black than shady. A century ago in Victorian England, many impoverished girls were pressed into prostitution. It is in Asia now, more than elsewhere, that large numbers of once-innocent female children live in brothels that are highly popular with fly-in-from-elsewhere male vacationers.

A darwinoid explanation of human sexuality—before its also somewhat-explanatory opposite is also found—is clearly tempting to anyone, including me, who would play detective about our long largely unknown evolution.

If, as I have suggested, females of afarensis and its successors among the upright apes were no longer slaves to estrus, how might their sex lives have been structured? The small stature of members of the first three hominid species suggests that, for better protection, these upright apes lived in large groups, of the order of fifty or a hundred, like the chimp and the bonobo, rather than small groups of five or ten like the gorilla. If they lived in large groups, they would have developed well-tuned social skills, as have both chimps and bonobos.

Bonobos (relative to chimps) have both more frequent sexual activity and less frequent social dispute, and the presumption is that the former somehow contributed to the latter. But would it make sense to take the sexuality of modern apes, in particular that of chimps and bonobos, as models for that of the first upright apes to abandon estrus? In *The Moral Animal*, Robert Wright thinks an ape-human sexual comparison is of value: "Lots of traits are shared by us and the great apes. . . . In the case of human mental traits whose genetic substratum is still debated—such as the differing sexual appetites of men and women—this comparison can be useful."[17] However, he reduces its usefulness by neither mentioning nor exploring the significance of estrus in apes and its absence in humans.

The Upright Apes, or Australopithecines

We know what estrous experience means for animals lower that apes (on the tree we use to symbolize evolution while ignoring branches other than our own). During the brief portion of the egg-readying cycle most likely to result in fertilization, the female is actively seeking relief from (what can somewhat dissociatively be called) an itch, an itch who's disappearance, or rather diminution, is enjoyed only when her sexual parts are actively rubbed. Female apes, most of them at least, have the swollen sexual tissues that show their receptivity to males, and they have them for many more than the few days that are optimal for fertilization. They are also less pressed by their instinctive need for gratification. This lets them be more selective about the males they are willing to accept as dispensers of quick, if all too brief, relief.

For its *experienced* component, there are three possible meanings for free-of-estrus. The first meaning is that these females are *never* receptive (or proceptive, which I will here include within the former term) to their males; they want no part of sexual activity, ever, although (if I am right about genetic male prurience) the males sometimes will still find the females to be attractive. This meaning is unlikely. It implies that only the pressing of sex on the females by the males will produce progeny; we see such forced sex only rarely with stump-tail macaques, orangutans, and sapiens, all three of which species seem to be largely free-of-estrus.

A second possible meaning for "free-of-estrus" is that the females in question are *always* receptive or proceptive; they are itching for relief in the form of sexual rubbing by a mating part, at virtually all times. Here again, a look at the orangutans shows no evidence that there is truth in this second meaning.

We are left with an obvious third meaning for "free-of-estrus," one that is somewhere in between the other two: The females are *sometimes* receptive and proceptive, and sometimes *not so*. It is this third meaning—which implies that the individual females have a measure of *choice* about their lustful desires and ideas—that I believe to be appropriate, partly from the evidence from stump-tails, orangutans and humans. Such estrus-free female sexual choice comes into being for species, monkeys, apes, humans, that can have the needed experiencing of concepts, and thus can choose to initiate the gestures to others that represent concepts, prurient concepts like "Come up and see me sometimes. What's wrong with, like, right now?"

Fisher seems to agree with me about what it meant for a female upright ape to be estrus-free: "Of all the payoffs of [being free of estrus], however, the most staggering was choice. . . . Lucy could finally begin to *choose* her lovers."[18] However, earlier on the same page, she writes, "With the loss of estrus, a female mate was continually available sexually." In my under-standing, taken together, the two statements suggest the possibility that muddy thinking about female sexuality is not restricted to males.

Sheets-Johnstone, in *The Roots of Power*, expands on what she refers to as the "reigning Western biological explanation of human sexuality."[19] It's true, as she says, that social scientists who write about sex sometimes say that women have "year-round receptivity," women are "continuously copu-lable," and that sexually, a woman is "a vagina . . . a mere passageway, an opening." Scientists who use such phrases are trying too hard to be scien-tific. Sometimes, they are also succumbing to a mindset about women that is common in cultures, even Western cultures, that are patriarchal.

In *Why is Sex Fun?* Diamond gives us an example of a scientist's prob-lems. He refers to females as having "constant receptivity." He also writes that "the peculiar sexuality of the human female forces husbands to stay at home . . . more than they would otherwise."[20] Shouldn't I inquire: What is this *peculiarity* that forces these husbands to stay home? Is it that their loving wives have holes where sex can be forced on them by other men? Or is it that those wives would constantly seek other men to fill their sex-holes?

Nonetheless, for Sheets-Johnstone to elevate such use of language to the status of *reigning Western biological explanation* is to adopt an overly ide-ological stance herself. Most people in Western cultures know that there is something in the way of a rough balance of power in the real sexual rela-tions of men and women who live together, a balance that comes about because they care for one another.

Sheets-Johnstone also refers to Sartre's famous statement, "sex is a hole." His statement is obviously ideological rhetoric. For her to refer to it in order to support her case about what reigns in Western thought, that is also exces-sively ideological. As for Jean-Paul Sartre himself, who was known for his belief that both a terrible freedom and conflict were common to the human condition, I must ask the following question: If, unlike the rest of us, he had but a single frame of mind, one that also believed that women, as sexual crea-tures, were simply copulable holes, what freedom and what conflict would he have believed his consort had? Would generous venereous Jean-Paul Sartre have granted Simone de Beauvoir the freedom to resolve or relieve her not-unlikely sexual conflicts with him by, say, turning the other hole?

I sought to bolster my notions of what it means to be free-of-estrus by making a sort-of-scientific survey. I sought a survey, not of overburdened-by-mixed-meanings humans, but of relatively simple-minded orangutans. So I looked through the World Wide Web's directories, hoping I could make a telephone survey of the orangutans in our Western society.

Unfortunately, I could find only one listing that seemed like it might do, that for an "Oran G. Utan." A sample of one might seem insufficiently scientific, but, hey, one is a lot more than none. So, rather than abandon the quest, I called what proved to be Miss Utan. After establishing a code (one grunt or two, for yes or no) for her unwordy responses, I finally gleaned the information that, at times, she did feel "horny" (her grunt, not mine) when there was a "hunk" around, particularly when she was ovulating.

Apart from whatever clarity the above provides, describing life without estrus for those early hominid ancestors of ours takes both guesswork and backward extrapolation from human females and males. I suggest it had two components for the females and one for the males. By definition, the females had little or no periodic desire for sexual activity. Second, when the females did have some such interest, it was a contingent interest. It was dependent on the presence of one or more familiar friendly males.

There is reason to believe early hominid males, in contrast with the females, expanded their sexual nature. Each maturing male was initially subject to an apish version of genetic-originated lustful desires. Those desires were soon expanded with learned lustful ideas, expanded into a search for sexually attractive and receptive females. To put it both figuratively and literally, assured easy entry to females was gone with estrus. So the males were obliged to add something new, something that was learned rather than genetic, to their pursuit of sexual pleasure.

The first thing the males would have learned would have been to be always on the lookout, to extend their prurience until, more pressing needs absent, it was virtually perennial. That follows because the females they sought were no longer clearly receptive by virtue of scent or sight of "heated" sexual parts. They learned to engage in more of what might be called a courtship. Part of that courtship was the development of a close friendship with the female in question. Also likely was an increase of what for humans goes by the name of foreplay: precopulation hugging, mouthing, touching, and rubbing. This was so because, at the initiation of a sexual encounter, the free-of-estrus females would not have had much of an "itch" for sex. Rather, at the start of a sexual interlude, these more-often-than-not unaroused females were able to entertain no more than the possibility of a rub down there. The conversion of that possibility to a probability was the task of the males, an obligation from which they perhaps did not shrink.

HOW WE GOT TO BE HUMAN

If these changes were indeed what took place, then the sex lives of upright-ape Lucy (the name given to one of the early important upright-ape fossil finds by Donald Johanson, the leader of the team that discovered her) and her friends were not like those of either chimps or bonobos. Neither Lucy nor any of her male or female friends would have come near matching the frequency of sex for chimps, let alone the bonobo standard (at least, in captivity) of several times a day on most of the good-weather full-belly days of their young adulthood. On the one hand, when Lucy was with a familiar male, she might have found the male refractory period after sex shortcoming enough to discontinue her dalliance with him. On the second hand, if another male friend was not waiting nearby, her interest in sexual activity might soon fade away. On the third hand (let's assume here the slow-witted early hominids may not have realized their hand count went from four to two when they mutated to an erect stance with genuine feet) even if a number of males who were her friends were standing by, as if she were the sperm bank in front of which they waited in line to make a deposit, and she chose to exhaust them, she would have been disinterested in the rest of the male community. So frequency of sexual interaction among apes and hominids may have reached its maximum with modern bonobos. Our entire side of the great divide between apes and hominids—thanks to the disappearance of estrus—may have been, as sexual athletes in frequency contests, no better than bush-league would-be bonobos.

A second divergence, from chimps and bonobos, in behavior for Lucy is that most of the time she may have done little signaling of receptivity. Such signaling is something female apes and monkeys do by bending over and facing away from a male to present the puffed up and pink state of their private parts. Without estrus, Lucy's private parts would not have been periodically swollen; any message readable by males would have been in much smaller print. More important, with no estrus, she would have infrequent reason to present herself as a receptive female. Of course, we shouldn't need to remind ourselves that a young female upright ape would have quickly learned that sexual congress can provide pleasure.

Something sizable in the way of an exception to the absence of sexual signaling would have come into existence as follows: If and when Lucy and her kindred realized that, to their males, there was at all times a major shortage of clearly receptive females, they, the females, could gain some advantage. The females controlled a scarce commodity, the receptive sex objects they themselves potentially were to the males. (By the way, with that realization there would also have come into being a new field for later sci-

188

entific study. Sprung from the womb of the libido and sired by scarcity economics, let's name it "scarcity libidinomics," and note that there's plenty of it around currently for study.) Present-day chimp and bonobo females sometimes act as if their sexual receptivity is related to a handout of food, preferably meat, from a male companion. For them, however, the game is only a marginal one; estrus tugs on them much as its masculine counterpart of rut pushes their males. Thus, males among the bonobos and, to a lesser extent, chimps, face no real scarcity of sexual objects, although they may have a perceived one that contributes to their handouts of food.

When the upright-ape females did realize what leverage they had, they would have learned the habit of trading sex for food from males—security, wealth, and fame had not yet been invented. I am not suggesting anything so complex as the notion, "Why give it away when you can sell it?" As far as we know, only a human animal can generate the concepts needed to make a good living by prostituting itself.

Return now to the situation of estrus-free females sending signals of receptivity to males. Lucy may have found an open stare or moving close to a male a gesture both more effective and less effortful than presenting her bottom. In these and other ways, the females would have learned that their sexuality, in itself, was valuable to males, and thus convertible into something of value to themselves.

We have seen that chimps present their bottoms not only to signal receptivity but also as a gesture of submission to disarm an aggressive pursuer. Would the upright apes have done the same? Isn't sucking up to power universal among social creatures with the mental stuff to do it? What particular gestures became custom for them may be forever a drop in the bucket of their unknowable behavior.

Look back again at the sexual interactions of lower mammals. They are certainly not based simply on the free will of the individual creatures. Nonetheless, in a way, they could be described as idyllic. The Bible tells us that once we all lived in the Garden of Eden. As we saw in chapter 6, the lower mammals still dwell (apart from the often fierce competition within each sex) in a sexual Eden. There is doubtless little in the way of total freedom there, but also little need to sweat over any puzzle about how to get or avoid a mate. We also saw that modern apes, because they are somewhat freer than lower mammals from nature's no-brain rules of sexual engagement, live astride the rocky borders of sexual Eden.

HOW WE GOT TO BE HUMAN

The Bible suggests that it was Eve and Adam that were expelled from Eden. Indulge me in the fanciful suggestion that the names were changed to protect the innocent upright apes. Then it was not Eve, but rather someone like Lucy who initiated the change. The corrupting fruit indeed came from a tree of knowledge. But the nature of the tree or the knowledge got jumbled during the long millennia of gestural transmission before speaking and writing developed. It wasn't knowledge of good and evil. To the females, the fruit of their tree gave freedom, freedom to be or not to be part of a copulating couple. Nor was it Adam near the tree, but rather several of Lucy's nameless consorts with whom she offered to share the fruit. It turns out they didn't care for the taste. "Freedom to not pursue females," they might have thought if they could, "who needs it? The idea leaves a bad taste in the mouth." So they spat out their share of the fruit.

The serpent told Lucy, "One bite and you're free of the curse—the real curse, the sex drug: itching for copulation whether all of you wants to or not." So she bit. Sure enough, driven from the garden of brainless sex were Lucy and the other upright-ape femmes. The males were sandbagged, they couldn't do anything other than follow the females. Of course, the snake in the grass didn't tell Lucy she'd still have to deal with the males, who were gobsmacked by the switch. Nor did the serpent say that the males would continue to throw their sexual weight around when they thought it might gain them to do so. Outside the Garden, on the dry plains of East Africa, the males among Lucy's tribe did not mutate to solve their problem. Rather, they learned to deal with their females' newfound freedom, learned to accommodate to their own social reality.

Some of Lucy's male cohorts, perhaps prompted by a simian Lucifer who switched from his serpent suit, occasionally forced sex on the females. This could not have been the common male strategy. As is the case with present-day apes, except for low-ranked males who are sometimes obliged to seek it covertly, upright apes normally engaged in sexual activity in public, that is, in a location within the sight of other upright apes. That means the male who forced himself on a female might lose some measure of his friendship with other females who witnessed the event. Not that the watching females were morally outraged; they had not eaten the fruit from the tree of knowledge of good and evil. (That may have come later, with an erectine Eve and Adam, as we shall see.) The upright-ape females saw a threatening male, and thus a creature to be avoided. Thus, forcing sex on females was probably an uncommon strategy of ignorance or desperation, engaged in by males who were young or who lived apart from the group.

So it was that one thing led to another with the upright apes. Upright posture minimized the effectiveness of swollen female bottoms as a signal of estrous receptivity. That led to the abandonment of estrus by the females, which in turn led to the males seeking friendship with the females and courting them with gifts of food.

Did that in turn lead to wider sharing of food? Greater intimacy and more time spent together might well have resulted in the females sharing their food with their males. The females would easily have extended their sharing to include their young. Did sharing lead to the deliberate collection of food to be consumed hours later or on another day? I don't think these upright apes, with brains little larger than those of modern apes, had the mental capacity for such an abstract idea, for such a stretching of their sense of the present time. What about near-monogamous enduring sexual pairings, and the two-parent families that might develop from them? Here again, the mentalities of upright apes could not make that much of a stretch, could not bridge the disputes and temptations that inevitably and frequently separate a pair, could not bridge them with recollections of a happier yesterday or with expectations of a better tomorrow. So the upright apes continued the simian way of life: families without fathers. Only with one of their successors, probably with the erectines (as discussed in the next chapter) rather than the sapiens, did fathers become part of families.

NOTES

1. Richard Leakey and Roger Lewin, *Origins Reconsidered* (New York: Doubleday, 1992), pp. 86–90.

2. Richard Wrangham and Dale Peterson, *Demonic Males: Apes and the Origins of Human Violence* (Boston, Mass.: Houghton Mifflin, 1996), p. 49.

3. Donald Johanson and Blake Edgar, *From Lucy to Language* (New York: Simon & Schuster Editions, 1996), p. 73.

4. Maitland A. Edey and Donald C. Johanson, *Blueprints: Solving the Mystery of Evolution* (New York: Little, Brown and Company, 1989), p. 353.

5. Maxine Sheets-Johnstone, *The Roots of Thinking* (Philadelphia, Pa.: Temple University Press, 1990), p. 213.

6. Donald Johanson and Maitland Edey, *Lucy: The Beginnings of Humankind* (New York: Simon and Schuster, 1981), pp. 309, 337.

7. Helen E. Fisher, *Anatomy of Love* (New York: W. W. Norton and Company, 1992), p. 153.

8. James Shreeve, *The Neandertal Enigma* (New York: William Morrow and Company, 1995), p. 335.

9. Edey and Johanson, *Blueprints*, p. 349.

10. Jared Diamond, *Why Is Sex Fun?: The Evolution of Human Sexuality* (New York: Basic Books, 1997), p. 81.

11. Donald Symons, *The Evolution of Human Sexuality* (New York: Oxford University Press, 1979), p. 130.

12. Sheets-Johnstone, *The Roots of Thinking*, p. 172.

13. Ibid., pp. 91–92.

14. Jean Aitchison, *The Seeds of Speech* (New York: Cambridge University Press, 1996), p. 50.

15. Tom Burnam, *The Dictionary of Misinformation* (New York: Thomas Y. Crowell, Publishers, 1975), p. 149.

16. Marvin Harris, *Our Kind* (New York: Harper and Row, 1989), p. 187.

17. Robert Wright, *The Moral Animal* (New York: Pantheon Books, 1994), p. 49.

18. Fisher, *Anatomy of Love*, p. 187.

19. Maxine Sheets-Johnstone, *The Roots of Power* (Chicago, Ill.: Open Court Publishing Company, 1994), pp. 85, 88, 91, 355.

20. Diamond, *Why Is Sex Fun?*, pp. 70–73.

9

OUR PREDECESSOR SPECIES, HOMO ERECTUS

THERE CAME ANOTHER TURN IN the weather roughly three million years ago. An arctic ice cap formed around the top of our globe. East of the mountains on the west edge of the African Rift Valley, the climate became still drier and cooler. In response to this slow shift, it is likely that the last two of the upright-ape species came into existence, as did the earlier of our human predecessor species. Presently, there is nothing like a consensus regarding which species among the early hominids was ancestor to the other.

When the weather started to shift this second time, there was a call put out to attend a two-day Emergency Congress of Early Hominids. The subject was "Natural and Sexual Selection for Coping with the Worsening Weather." The old mimeograph machine—there were neither faxes nor e-mail back then—cranked out a few thousand notices. Most of the upright apes threw them away without a glance, or they made paper gliders. Among those who did look at the notices, a couple dozen afarensis and a similar number of africanus—there were no other prehuman species around at the time—decided to show up. Like our chimps, the upright apes were capable of understanding a few words of language. At the idea of sentences, however, they threw up their hands, and the mere mention of syntax made their eyes glassy and their guts gassy.

So those who came were not completely clear about the subject of the meeting. Considering what the males among them often thought about, and occasionally did when they got together, it's not surprising that most believed the subject was "Natural Ways to Have Sexual Congress in Bad Weather."

At the start of the first meeting they made friends with one another.

HOW WE GOT TO BE HUMAN

They made friends, that is, after they worked out who would kowtow to whom. They were mostly leaders; politics was not as ideological back then, so they got friendly fast. As political types did in those days, they debugged and they hugged. They swapped, rather than stories, spit. On the second day of the meeting, when rest periods were a good deal longer—remember, these creatures were not sexual athletes like bonobos—they grunted and groaned about the weather, and whether or not to do something about it. The majority, as might be expected, thought the so-called weather crisis was nothing to worry about. Then, after more bowing and big-dogging with one or another of the new friends, most of them went home.

Before very long, afarensis died out. Africanus, however, gritted his worn-down teeth and gummed his way around for a million or two years before dying out.

A few at the conference who stuck around decided to respond to the worsening weather by developing larger teeth to chew tougher roots and more hard nuts and seeds. "Because teeth are capped by enamel, the most durable biological substance known, and because they possess a core made of a very hard mineralized tissue called dentine, they constitute the majority of fossil hominid specimens. . . . Robust australopithecines, which are thought to have subsisted on a diet of tough, low-quality food, had the thickest enamel of any primate."[1]

Some of those remaining upright apes tried, but not very hard, to grow larger bodies so the big teeth wouldn't look too goofy. (Johanson and Edgar tabulate estimates that disconfirm previous data suggesting robustus and boisei to be significantly larger than their predecessors, afarensis and africanus.[2]) After some parting intimacies with one another, these few, in turn, went home. They lived another million or two years as bigger-toothed but not specially larger-bodied robustus and boisei. Then they, too, died out.

A remaining few at the conference were fed up to their eyeballs with the thought of chomping still more tough dry veggies. They got the idea for a long shot. They would grow larger brains. They figured larger brains would help them make stone axes. Stone axes would help them get a little more meat to eat instead of those grisly greens and ratty roots. As I said, it was a long shot. After farewell formalities, they, too, went home.

These last few, plan-ahead types, lived another couple or three million years as humans: habilis and her relatives; erectus and his cousins; various no-name archaic sapiens; sapiens neanderthalensis; and eventually us, sapiens sapiens. In what follows, after a glance at habilis, a closer look at erectus is taken.

194

THE ACHEULIAN HAND-AXE, A TOOL
THAT WAS SYMMETRICAL, LIKE ITS MAKERS

How hominids began to plan ahead, we may never know. We do know, however, that habilis made simple stone tools and that erectus made more complex ones. "Stone and symbolic tools . . . [ultimately made their users] adapt to a new niche. . . . These tools became the principle source of selection on our bodies and brains. It is the diagnostic trait of *Homo symbolicus*."[3] So says Deacon in *The Symbolic Species*.

The first stones used, by habilis and perhaps by an upright-ape predecessor, were pebbles, smoothly rounded stones found in or near riverbeds. A pebble's life (to engage in the favorite human pastime of projecting ourselves into things) is not a happy one. It gets all its rough edges pounded and rounded in the course of wandering down a river. After using rounded stones for a while, some enterprising early hominids discovered that pebbles which still had a jagged edge or two did a better job of penetrating whatever needed smashing.

Some time after that, some early habiline (there may have been several habilislike species, called habiline) geniuses made a remarkable deduction. Using a pebble to bash not the usual animal or vegetable but instead a mineral (another stone) produced a more useful smasher. Like all new ideas, this one took time to catch on. After all, bashing a small animal or a big root, that put food in the old belly right now. What more could anyone want? Bashing a few chunks off another basher did absolutely nothing to satisfy a growling gut. A smasher that might or might not be useful later, tomorrow, or who knows when—what good was that? "Promises, promises, we've heard them before," that's what those who watched the creation of the first bashed bashers would have thought if they could have thought.

The darned fool things, for all that, worked, and worked well. After a few of the young switched to the new technology, not more geniuses, but tradition served nicely to carry on. That kind of tool, found in countless copies, is called the Oldowan axe, after Olduvai in Tanzania where some were first found. This axe is made by striking off, with another axe, a few flakes near one end of the pebble. (No, Maude, this ain't what them philosopher folk call an infinite regress; the first one was mashed with a natural-born pebble.) As knives and scrapers for working meat, plants and wood, the flakes were of as much use as the body of the axe.

HOW WE GOT TO BE HUMAN

To make what is called an Acheulian hand-axe, a tool developed by erectus, is another matter. This axe, named after the region in northern France where it was first found, called for yet another application of genius. We tend to think of geniuses in the singular, but if the right one is not around at the right time and place, another usually shows up before long or nearby.

The Acheulian hand-axe has been skillfully chiseled everywhere by another stone axe. It has no rounded surfaces left. More significantly, all such axes have been shaped to roughly the same contour. That contour is like that of an open human hand with the fingers together. The Acheulian tool is thin from front to back, like an open hand held vertically. It is roughly symmetrical left-and-right around a midline. Its top-to-bottom dimension is greater than its left-to-right width. Like a human hand, one end (call it bottom) is more or less rounded, and the other end (the top) is sort of pointy.

It is conceivable that the inventor who created the first one was a weirdo who intuitively created an axe in his hand's image, or even in his own bilaterally symmetrical body's image. Unless that peculiar shape had an enhanced utility, however, a tradition of such axes would not have worn well. With time, that is. The silent majority would soon have spoken up: Genius, shmenius, why make it so fancy? In fact, not only is there a question of why the particular shape of the Acheulian axe survived, but a prior question: Acheulian, shmacheulian, why not stick with the good old Oldowan axe?

The latter question is by no means an idle one. Erectus expanded out of Africa to the Middle East, the Far East, and Europe. The evidence is that, in east Europe and east Asia, the simpler Oldowan axe was the tool that everyone used, whereas the complex Acheulian axe was used in Africa, western Europe, and the Middle East.[4]

Why this difference? Was this a cultural matter, a matter of learning, habit and tradition? We know there were two or more genetically different types of erectuslike creatures. We don't know if they had differing inheritances of mental ability. Some have given the name, Homo heidelbergensis, to the slightly different fossils first found near, you got it, Heidelberg. Also, some give the name, H. ergaster, to the "Turkana Newcomer" fossils. (Richard Leakey and Roger Lewin, in *Origins Reconsidered*, describe the process and thrill of uncovering a complete fossil skeleton of what they called "The Turkana Boy" that was uncovered near Lake Turkana. They classify the creature as erectus rather than as ergaster.[5])

Were there erectines (as plural erectuslike species are called) whose different mental abilities resulted in the use of different stone axes? Perhaps the axe itself, of either type, was less important than the flakes that were chipped from it. Perhaps axes and flakes, the only part of what erectines made that survived for us to find, were but a minor part of the brave new world they erected. At present, no one can say.

An odd fact, kind of a stunning one, about Acheulian axes has come to light. In Kenya, a "kill" site about the size of a good-sized house, 40 feet by 65 feet, has been excavated. In it were remains of hippos and other large mammals, as were remains of 63 giant baboons. What is bewildering to my mind is the more than 10,000 hand-axes that were found in the site.

What sense can be made of 10,000 axes together in a small area? When an axe "found blood," was it thought thereafter to have lost its magic? Was it put to rest, then, with its blood brothers in a comfortable kill site, like an elephant's graveyard? Or was it the loser axes, the ones not made with the proper magical utterances from their limited lexicon—the ones that did not find their enchanted way into flesh for all to sup—that were tossed into the pit? In *The Origins of Virtue*, Matt Ridley presents another idea that might explain these tools. They were "mining stone tools at specialized quarries, presumably for export."[6] Perhaps a scientific team with great patience and good funding might learn something about erectus by examining the edges of some of those 10,000 tools, to see what if anything they once cut. Maybe they will find that those ancient stones never cut anything. They might be merely the junk that remained when useful tools were hammered from what originally were larger stones. In his discussion of much-later neanderthal worked stones, James Shreeve notes that one theory suggests that such a piece of stone "might be the only part of the original stone that couldn't be used as a tool."[7]

A MILLION YEARS BETWEEN UPRIGHT APES AND SAPIENS, FILLED WITH A LOT OF GESTURES AND A LITTLE LANGUAGE

The concept of an egocentric lower mammal living in a here-and-now world was stretched in both the time and the space dimensions (in chapter 5) to characterize the social great apes of our own time. The upright apes (who comprised the earliest hominids) had brain size not dissimilar to that of now-existing apes. That suggests their concepts about their world were not unlike those of present-day apes. This implies that the upright apes intuitively regarded those they were close to as a part of themselves, and they had notions of space that expanded from *here* to *nearby*, and of time extended from *now* to *the near future and the near past*. We must ask if those ideas can reasonably be stretched further to accommodate erectine activity.

Look again at the principle erectine tool, the carefully worked and shaped Acheulian axe. We have seen that, for termite "fishing," a chimp makes a tool by selecting a small branch from a tree and stripping it of leaves and twigs. The procedure starts when she is at the termite mound and, eager to get a tasty snack of fresh termites, goes off to make and return with a tool. Seldom if ever does she, when near a tree with suitable branches, make such a fishing tool and then go in search of a termite mound.

Examine erectus' tool-making situation with regard to the hunt for live or dead meat. Like the chimp, he wants to get something to eat, and a suitable tool will be of great help. What erectus wants is either alive and capable of rapid movement, or dead and scavengable only if he acts quickly before competing scavengers show up. It makes no sense, however, to suppose that the sight of potential prey or of a dead carcass sends him off to the time-consuming task of making a hand-axe. A chimp cannot use the conceptual apparatus of a slightly larger "here" and a somewhat longer "now" to select a branch and make a fishing pole to save for a future termiting expedition. Neither could erectus spend hours of effort making an Acheulian axe to use in a vague future elsewhere. He could not, that is, unless he had a better apparatus for conceiving such things in his mind. The erectine mind had additional means, means well beyond those of a great ape, for expanding his notions of his world's space and time, and of the nature of the creatures therein.

If we look at erectus' migration to environments far different from those of Africa where he started, we find the same problem. Erectus did not migrate because he was possessed by wanderlust. The migration took place over a period of tens or hundreds of thousands of years. During the lifetime of most individuals, the movements were a few insignificant and unnoticeable miles. Erectines migrated because they were a highly successful species, which means they increased their numbers rapidly. As such, and especially if meat was a large part of their diet, they sooner or later ate everything nearby. They simply had to move on. If erectines were creatures with only the expanded-here-and-now-ish conceptual apparatus of the ancient upright apes or the existing great apes, they could not devise the new ways of making a living needed in the new lands with the new climates. They had a better way of thinking.

Our Predecessor Species, Homo Erectus

Erectus did not have a mentality like that of the apes. Well, then, let's go to the other extreme and ask, why couldn't erectus have been like us? Not as good to look upon, and they were undoubtedly uncouth. Mentally, however, could they have possessed our kind of stuff? Did they have what's near the heart of our kind of stuff, language, the formally structured means to represent, think, influence others, socialize, and communicate about things and events using discursive verbal symbols?

The idea of erectines with a more or less modern language-using brain has its own problem: they weren't successful enough to suggest they had it. The following may sound like bragging, but that's hardly my purpose, as those guys were not in our league. They were around for 1.3 million years, compared to our one, maybe two, hundred thousand. During all that time, not only did they not go from hand axes to space probes, they made only slight changes in the shape of their axes. So we are forced to think their mental capability, although superior to that of apes, was well short of that of modern humans.

To start searching for erectines' in-between mentality, look at a modern human who was known as "Brother John." Brother John was a fifty-year-old man who worked as editor of a periodical for a religious order. (There is an extensive account of him in Merlin Donald's *Origins of the Human Mind*.[8]) Brother John suffered epileptic seizures for twenty-five years, short spells of a few minutes several times a day, and long spells of several hours every few weeks. During a seizure he remained conscious and active, but lost all ability to understand, use, and think with language. Surprisingly, his seizures did not otherwise incapacitate him. During each seizure he was wordlessly aware of what was happening and still able to cope with his situation. In *The Seeds of Speech*, Jean Aitchison notes that even though Brother John lost the use of language during a "spell," he was able to record his own largely senseless speech.[9] When he was speechless, the monk also kept a small radio at hand to determine when his own understanding of language returned. So it was clear that, speechless or not, Brother John had his wits about him.

On one occasion he was in the midst of a long seizure when he arrived by rail at an unfamiliar town. He was able to wander about in search of a hotel until he found one. His attempt to use gestures to get a room did not work; he was turned away. Still knowing what he was after without being able to name it or anything else, he wandered further until he found another hotel. There he was able to communicate well enough by mime to get a room. Too hungry and miserable to sleep, he went to the hotel restaurant and ordered food by pointing randomly to items on a menu whose words and letters meant nothing to him. He then ate (although the food was not to his liking), and finally went to his room and slept. When he woke, language had returned, so he could explain his seizure to the hotel people before he departed.

During his seizures, Brother John is an example of something highly unusual, someone who seems to be totally without language but who could still understand symbolic things and employ symbolizing gestures to function in a complex human society. It is conceivable, however, that he was deprived only of the conscious use of language and that he retained and used language in his head nonconsciously. So, rather than relying on Brother John alone, let's look for confirmation, confirmation of people who use symbols but not language.

Look at the congenitally deaf before sign language was in use. They were a group of ill-treated and often abused people, without language, who functioned in a complex human society at a level clearly unavailable to apes. Despite their handicap, they understood symbols and used abstract concepts well enough to be employed as domestic servants and in manual trades.

In *Human Evolution, Language, and Mind*, William Noble and Iain Davidson imply otherwise.[10] The absence of language "limits the consciousness of nonhuman animals to that of sensory sensitivity (conscious sensibility). . . . Human mindedness is essentially marked by the conscious attentiveness and articulateness that language enables." I can't accept that. It implies that languageless deaf people lack human gesture, conceptualizing, and attentiveness.

What erectus had was an expanded use of ape verbal utterance and ape ability to make and use nonverbal gestures, somewhat like the abilities of Brother John and the congenitally deaf of old, but without the huge knowledge base that the latter could learn by living in a fully human world. In *Language and Species*, Derek Bickerton goes further by suggesting that erectus may have had a limited ability to use "words," without having syntax and other properties of a complete language.[11]

Symbols and gestures allow creatures to refer to things and processes, to think about them, and to communicate with others about them. Symbols are to be contrasted with signs or signals, which directly stand for things and processes. A field mouse who hears what sounds like a snake slithering in the grass near its feet has received a signal, in this case one calling for immediate rapid movement elsewhere. Likewise, Pavlov's dogs were trained to expect food immediately when a bell was rung. They reacted to the bell by salivating as if it were food; the bell had become a signal of food for them.

These examples of signs or signals can be contrasted with human symbolic use of and response to, for example, the simple gesture of pointing a finger. In such pointing, someone exercises the option to refer to something without necessarily wanting it. Even young children can use and respond to the gesture of pointing a finger, but this is a behavior unseen in wild apes. However, neural tissue being so plastic, it probably could be taught to them.

Symbolizing gestures, and language as well, can also be used as signals. A young child may point to candy or use the word to show it wants candy right now, rather than to show its awareness of the existence of candy. Brother John used gestures as symbols to indicate his interest in getting a room to sleep in and a meal to eat. If one gesture did not get through, however, he could switch to others. This contrasts with, say, the male chimpanzee use of the activity of shaking a tree branch near an estrous female to highlight his demand that she follow him for a few rolls in the hay in the outback. The ape's gesture is less a symbol than those of Brother John. But it is not simply a signal demanding sexual acquiescence. The male may repeat the branch-shaking or he may perform the activity more vigorously, but he has no alternative gestures. The chimp female understands the male's gesture, but on occasion she may ignore it. There is a continuum between signal and symbol, and ape gestures lie somewhere between the extremes.

Although we sapiens have language, we still employ nonlinguistic symbols and gestures in activities like trades and crafts, games, athletics, art, dance, social formalism (like handshaking), and, of course, religious ritual. We also use and interpret socially what is called body language. Using only nonverbal symbols to communicate can be extremely limiting. As anyone who has played the miming game of charades knows, abstract concepts are extraordinarily difficult to convey with gestures. Charade players know this well. They beat the rap by breaking words into syllables and then acting out simpler concepts with names that "sound like" the syllables.

To say erectines did not use language as we their successors do is not to say that they did not use symbolic utterance at levels beyond those of apes. Erectines must have gone far beyond apes and had many sequences of utterances at their disposal, but perhaps not even separable nouns and verbs, let alone sentences with syntax. In fact, for the first few tens of millennia of our own stay on earth, even we Homo sapiens sapiens may not have had complete language, as we will see in chapter 12.

PROJECTING AND IDENTIFYING HELPED ERECTINES TO SHARE FOOD AND TO HAVE TWO-PARENT FAMILIES

Consider again the shape of the erectine hand axe. Its symmetry could hardly be accidental; it was undoubtedly achieved only by carefully chipping away at an originally unsymmetrical hunk of stone. Nor would the shape have lasted over the hundreds of millennia if it didn't have some significance (in addition to practical value).

Here are two questions about the axe's symmetry: (1) How did erectus conceive of making the axe symmetrical? That is, how did it come to be that he understood and thus could see symmetry? (2) Why did he make his axes symmetrical?

The "how" answer: He projected his own symmetry, or that of his hand, into his axe. This is much like the ability of a human child, as we saw earlier, to project his inarticulate knowledge of his own symmetry into a ladder, enabling him to place it stably against a wall.

Consider now the "why" of symmetry. For erectus, the axe was, among other things, a symbol of his own power and strength. In reading about the modern Neur people in Africa, we learn that a Neur man's spear was regarded (by an anthropologist) as "a projection of the self and [it] stands for the self."[12] Without any thought, erectus "naturally" made his axe's shape roughly symmetrical like his own body or like that of his hand. Then it would have vitality and strength, as he does. The axe's symmetry and dissymmetry also had practical purposes. Grabbing it by the broad end gave you something like a knife for stabbing or cutting. Grabbing the pointier end gave you a hammer for pounding or breaking. To handle such an edgy weapon, erectine hands would have been hardened with heavy callouses.

Now look at erectines' move into new environments, particularly into lands where the temperature was much colder than their native Africa. To do this, an erectine covered herself in animal hides, something no ape could do. Apes are too literal-minded. To others, an ape inside a antelope skin would seem either both creatures at once or one alternating with the other. Either view might spook out neighboring apes. To clothe themselves, erectines used their ability to project themselves into other creatures and to conceive them as something like themselves—perhaps to conceive animals as spirits living within bodies. A further step let them regard an animal's furry skin as something apart from both the body and the mysterious self within. Then they could more easily regard the fur as a symbol of protective warmth, warmth to be realized after they converted the skin to a hide. To do this last part, they used chips struck from an axe to shave off the fat and tissue from the skin's inner side. Perhaps the hide was also made by dint of laborious use of teeth and saliva, as was done by Eskimos to soften skins.

Erectines, unlike their upright-ape predecessors, made a living as both gatherers of local edibles and hunters of dead or living meat. A gatherer collects, but does not live hand-to-mouth as do modern apes who are not provisioned by humans. So how were erectines able to collect for future use?

Let's face first the question of the division of labor by gender. The erectines continued their predecessors' custom of (larger) males doing most of the hunting. What is the reason for this division? Genetic hormone-related differences between the sexes, as well as genetic size differences, make male apes (and, doubtless, made male erectines) not only larger but also more capable in hunting than females. In contrast, the gathering of food items was probably an activity of both genders, the males were neither smart enough nor rich enough to make wage slaves of the females.

To collect—beyond immediate need—food items such as tubers, roots, fruit, nuts, and beetles, the erectine (I'll presume a female) first had to conceive the need to do so. For whom was she collecting? First and foremost, as with all egocentric creatures, she gathered goodies for herself. Also, as we really don't see with modern apes, she collected for those she felt to be in some way a part of herself, mainly her young. To collect at all, she had developed the difficult concept of the future, when her own belly as well as those of her family would again need food. For such a concept, the sun with its daily rising and setting, or a mental image of crying hungry children, may have served as a symbol. To get the task done, symbols are as necessary as the concepts they represent. A late afternoon, wherein she relaxed with no thought of tomorrow, would soon be followed by the dying of the light, which, as a symbol of the day's coming end, would quickly prompt her to resume her gathering. After a while, community custom and individual habit would come to her aid, as would the concept of going out for the express purpose of collecting food for tomorrow.

Lastly, in order to be effective with her collecting, she needed a container. She had to conceive of one or more animal hides as a bag, and then use a coarse "string" of vegetation or animal tissue to "sew" it or to tie it together as a container. The possession of abstract concepts and symbols that represent them would have helped in an additional way. The ability to use symbols meant, for example, that an erectus adult leader could suggest to other erectines, with miming gestures and with a couple of utterances, that they take axes and go out to scavenge or that they take hide bags and go out to collect edibles.

The ability of the hunter to deliberately cooperate with his fellows is based on his identifying with them and regarding them as like himself. That ability also includes his knowing that he can influence them with gesture or utterance. This social capability contributed greatly to erectine success. As a creature with neither fangs nor claws, he needed his hand axe and its flakes that served as his tools for killing, opening, and tearing apart his prey. He also needed his fellow hunters.

An erectine could seldom run as fast as a live target could, so how did he get his axe close enough to it? One theory suggested by William Calvin in *The Ascent of Mind* holds that he threw his axe, not at an individual creature, but into a herd of closely packed animals, where accuracy was not needed.[13] When he threw it spinning through the air, its eccentric shape would not merely hit a creature, but tear through skin and flesh and perhaps wound it badly enough so that the team of hunters could chase and catch it on foot. Another theory holds that erectus developed into a long distance runner (like some modern men and women), superior in this ability to all his potential prey. Hunters could run after a beast for hour after hour, especially if they could conceive of themselves as a relay team replacing one another, until the exhausted creature was within reach of their axes. The notion of erectus as runner has been further expanded by suggesting that running may have been central to his life over enough millennia to permit mutation and selection to remove the thick hair from his body and to add sweat glands, like the ones we have, to cool him.

You've been hearing about a good deal of heads-up thinking in the heads of those erectines. Although such work had to be done, that doesn't mean each of our erectine predecessors had to think it all out about providing for the future and for others each time they sought food. Social custom and learning from others would have served them, as it does us, with the habits that make heavy mental lifting rarely necessary.

With their mental ability to move beyond an ape's world, the erectines became the first in our line to live in two-parent families. The larger heads that housed their increased brain size made it necessary that their infants were born long before they were as ready as are newborn apes to fend for themselves. To accommodate the growth of their brains and minds, their childhood was long by ape standards. Without the pressure of estrus regularly sending the females to the plurality of males needed to supply relief, and with minds such that each could feel the other as part of a "we," both sexes would have found comfort and advantage in maintaining a relationship and in sharing food. Unlike the upright apes that preceded them, early human species had the abstract concepts and symbols needed to think of both the past and the future. This would have helped them to bridge the frequent chasms created in their lives by the slings and arrows of outrageous fortune. That was part of what was needed to form the two-parent families that were invaluable for the survival and growth of a species whose individuals came into the world helpless and grew slowly for many years before they were of use to their fellows. I must note that, like most significant things about our predecessors, the existence of two-parent families—even for the neanderthals who followed the erectines—is disputed by some scientists, as Shreeve tells us in *The Neandertal Enigma*.[14]

Erectine grasp of concepts and symbols gave them a greater ability to understand a mate and to comprehend a past and envision a future together. This permitted many of the erectines to live in somewhat monogamous two-parent families. It was also the source of the desire, not just to share food, but to seek an abundance and save it. It also enabled neighbors to trade a meal today for one in a day or so, or for a hand-axe yesterday.

ERECTINE SEXUALITY: MALES SEE FEMALES AS SEX OBJECTS AND SIRENS

What was the nature of sexuality for the erectines? In its basics, it was similar to that of the upright apes, who (in my fanciful presumption) were driven out of the no-brain sexual Eden where estrus ruled lower animals and even some quadrupedal apes. In the new world where the upright apes like Lucy were driven, females often could choose whether or not to participate in sexual congress, and males (who retained genetic-originated rut that originally complemented female estrus) had to figure out when which females would copulate. Erectines' broader grasp of concepts, symbols, and gestures expanded their sexuality beyond that of their upright-ape predecessors.

I'll look here at two of the many routes such an expansion might have taken. The first has to do with the occasional male view of females as *sirens*.

HOW WE GOT TO BE HUMAN

That is, males viewed them as enticing, seductive, and even bewitching sex objects, but also as deceptive creatures who might not deliver the sexual responses they seem to promise.

The second avenue of conceptual expansion is that of privacy for sexual couplings. We'll examine the significance of sexual privacy, and see where erectines and upright apes stood on this issue that divides humans and modern apes.

Might it be that all this is a tad speculative? Perhaps you'd prefer golden silence. I doubt that you would rather have me preface these words by recounting the erectine experiences of my tens of thousands of previous incarnations.

The erectines had the ability, far beyond that of modern apes or ancient upright apes, to project some aspect of themselves or their inarticulate knowledge into things or creatures in the world outside. Recall from chapter 5 that male chimps may project their vague feelings of their own strength and power into a lightning-and-thunder-storm. Compared to us sapiens, chimps are rank amateurs at the projector's game.

With the arrival of the erectine creatures whose family name, Homo, was the same as ours, the first semiprofessional projectors—let's call them "projectionists" here—came on the scene. In fact, the erectines came into the booth behind the audience and behind the blank screen in the theater of the mind. Of course, using their multiple minds, they were not only the projectionists but also the performers on the once-blank screens, and were also the audience. As they were not the equal of the sapiens that followed them, the erectines could project only the bare outlines of what was to come when we "wise guys" finally arrived.

On occasion, to get out of the rain, erectines had run into caves. Sitting there bored, they sometimes decided to turn on the projectors and look at the short subjects that suited their limited attention span. These they followed with brief previews of coming attractions. Hardly the equal of television situation comedy, I know. It was nonetheless a lot better than sitting cold and wet outside with the few remaining apes in the midst of the thunder and lightning whose import of power those ancient apes projected onto the cloudy skies.

The pop-eyed erectines in the caves first saw coming-attraction snippets of a future tale of sirens. For a tantalizing few minutes they, especially the young males, could see brief projections on the blank screen. They saw projections of mythic human Odysseus as he stood, lashed to his ship's mast, listening to the song of the sirens. Recall that Odysseus desperately wanted to hear the dangerously appealing sirens' song. Yet he despaired over his own ability to resist its dire lure. So, wily as always, he had his marines lash him to his ship's mast (after he washed his ears). Then, before they manned the oars and slid through the water past the singing sirens, his crew stuffed cotton into their own ears.

The erectine audience saw Odysseus as he lustfully listened to the glistening sirens, and increasingly strained against his constraints. They heard him demand of his unhearing mariners that they free him, even as he remained bound. The young males in the erectine audience were beginning to comprehend how confusing it was to live in a world where male certainty about female receptivity was gone, divorced as it was from female attractivity in male eyes. So they cheered lustily when they saw the bodacious sirens and also when they saw the wily warrior himself beneath his upright mast, even though—remember, this was long ago—the film snippets in the projector were scratchy and had bad sound after the long journey backward through time. They cried because they could not clearly hear the sirens' song itself.

There is one thing, by the way, that should be clarified. There's no need to be concerned about the so-called translation problem. It's true that whatever Odysseus said was not actually what the erectines heard. But rest assured, they did not have any Universal Language Translator sent back to them from some presumed Star Trek future of ours. That would have been absurd. As everyone knows, those things can't be made, and besides, the erectines had an extremely limited language. What they did have, sent to them through negative time with the film coming-attractions, were, within what looked like hearing aids, Universal Language-to-Concept Translators.

There were hardly any old men, forty-ish was old back then, but, after all, they were the management. So for themselves they followed the Odyssey's siren scenes with a preview of an avant-garde film adapted from T. S. Eliot's "The Love Song of J. Alfred Prufrock." The images showed old Alf on the shore reminiscing about sirens-past as he ogled mermaids in the surf, sea-sirens with their upper (human) halves mostly above the surface of the water and their scaly fish tails, not cheap cloth but quality computer simulation—occasionally flipped out of the water before they climbed onto a rock to rest.

The import of some of Eliot's time-garbled concepts penetrated the projectionists as they watched the images of lusty mermaids in the surf, and saw the moving image of beached old open-mouthed Alf with his hand on a listless little thing that might have been a piece of limp seaweed.

The tanks of these almost-frail-at-forty erectine males were already more than half-empty of their youthful male hormones, so they were moved by the images of old Alf, himself hardly softly lured by the singing mermaids. Many of these older males in the audience were close to tears, seemingly because, like Alf, the sea sirens were no longer singing to them. The younger old men, in their thirties, tried to comfort those who despaired by reminding them that they, the old-old, were the leaders and could still get young females for themselves. The old-old curtly told the young-old to shut their crunch-holes.

To put it mildly, the above projections are more spoofs than speculations. But they are not just spoofs. If erectines could live in full family units—as their spreading successfully over half the globe suggests—their females had abandoned estrus. If erectine brains and minds were greatly enlarged from those of apes, they had dim intuitions about sexuality that foreshadowed those of their successors. Modern sapiens males have genetic roots in rut for their learned conceptual prurience (as chapter 15 will show). Our predecessor erectine males had some measure of a similar prurience.

SEXUAL ACTIVITY NO LONGER IN PUBLIC LIKE APES, BUT IN PRIVATE LIKE YOU KNOW WHO

Consider now that erectine sexual interaction differed in at least one additional respect—other than the male anxiety about sirens—from that of the upright apes that preceded them. In this second aspect, erectine sexuality was also like that of us, their sapiens successors. I am referring to the privacy with which sapiens usually conduct their sexual activity.

Why do sapiens humans always seek and find privacy for sexual interaction, even if it is only that of a sheltering blanket or hide in a corner of a crowded hut? Why do they never copulate where others are gathered, never in the clearing midst the circle of huts, never on the village green, never in Times, Red, or Tienanmen Square? (Ignore here public sex acts such as those performed for communal rite, for private gain, or for the modern communal right of advancing scientific knowledge, purposes beyond the basic one we all share.)

The answer is the question restated. We jolly well don't want someone watching us. Why don't we want them watching? The answer is as obvious.

We don't want them getting their jollies by watching us. To use a more modern figure of speech, we don't want them getting off on us. To use no figure of speech, we don't want them acting as if they are or might become part of or party to our sex act. Erectines shared enough of our mental "talents" to have also shared our desire for sexual privacy.

We normally want sexual privacy; the erectines may have wanted the same. Yet, modern apes have no interest in it (ignoring here privacy from the presence of a high-ranked male), and the upright apes would likewise have been disinterested in it.

To see why, transfer attention from the sex-act participants to the presumed observers. Look first at modern chimps, as rough models for the ancient apes from whom sprung the hominid line. Female observer chimps are not aroused by the sexual behavior they see. Rather, regardless of what they see, those in estrus are aroused and those not in estrus are not. In contrast, male chimp spectators (rightly) take a participating female to be in heat. Barring the often inhibiting presence of a high ranked male and given the absence of other females in estrus, observing male chimps, who are always pressed by prurience, start to get aroused by the performing female, and might as easily get aroused even if she were not performing. They act as if they have a good chance of mating soon with the participating female, and that is indeed the usual case.

Switch back in time now to participating and observing hominid upright apes, with the assumption that the females have no estrous cycle and the males continue with a learned expansion of genetic ape prurience. There is no reason for a female upright ape to become aroused by observing the participants in sexual activity. She doesn't have the mental ability to identify closely with the participating female, the ability that might enable her to get aroused. However, the observing male upright ape, pressed by extended prurience, acts much like the modern male chimp, and for a similar—but in his case, mistaken—reason. The sexual part of his mentality believes that a female's participation in sexual activity is a sign that the female is receptive in a general way—that she is, in effect, in estrous heat. Thus the male observer, or rather, the sexual part of his mind, thinks that after the act with her present partner is finished, she is apt to be receptive to an encore with him. So, as an observer who expects to go on-stage soon, he gets aroused. If, after the initial participants have completed their "performance," the female turns him down—assume here she is not in the mood—he takes it as bum luck. Possibly, another part of the male's mentality may have gained some understanding that females are sexually receptive on a case-by-case basis. That understanding gets easily overridden when he finds himself observing a female engaged in sexual activity. He reverts to construing a female so engaged as a female in estrus who will take on additional comers.

Now throw the time machine into forward for the moment, past the erectines, to participating and observing humans. We have switched, like it or not, to a scene where the watchers are mostly males who have paid to see a pornographic performance, although the scene might be a sexual rite or a scientific study. If it is other than a scientific study, the males mostly, like their upright-ape ancestors and like modern apes, get sexually aroused.

Human males, however, with a far greater understanding of their social and physical world, have options unavailable to those upright apes and modern chimps. A male could decide it is inappropriate that his most intense feelings were invoked and manipulated by the situation he had placed himself in. He could resent that enough to inhibit arousal. His social feelings about what is right and what is wrong could come to the fore and arouse guilt and revulsion that override his tendency to get tumescent. He could be with someone he cares about, and dislike having his private feelings connected to a public event involving other people. He could also be a social scientist using the occasion to study his fellow men. Female observers, fewer though they may be at such an event, free of estrus and not necessarily bound by prurience, are unlikely to get aroused. Some of them, however, might identify closely enough with the female participant to start to get tumescent.

We may note in passing that human males are often relentless about their prurient eroticizing of any female activity that can be construed as sexual. Women, of course, know this. That's one of the reasons why women who have been sexual targets for men are reluctant to take their cases to bosses or to courts. They know that many men will respond—to a woman's mere talking about what happened—with covert heavy-breathing curiosity.

If we now make a last reverse time-switch back to erectines participating and observing, we have an imaginary scene, imaginary in the sense that it would not have happened. Erectines were hardly sophisticated enough for sexual rites, let alone pornographic performances or scientific studies. Their ability to understand the basic sexuality of others was nonetheless similar to that of sapiens. Unlike upright or modern apes, they could project themselves into others and identify with them with regard to basic sexuality. Going one step further, they could realize that others of their kind could make similar projections and identifications. That implies that erectines shared the sapiens desire for sexual privacy, for insuring that observers did not become unwished-for parties to their own sex acts. Which is why the scene of erectine observers of participants' sexual acts is an imaginary scene.

For a male to hear a siren sing is for him to think he has received a signal from an attractive receptive female, only to find himself rudely awakened from his projectionist fantasy, or to find he is unable to reach the source of the message or unable to confirm its nature. Would erectine males, during their long stay on earth, have continued to hear the sirens' song, the song too faint to be heard by upright-ape males, the song suggested in the erectine projectionist fantasies above? Continued to hear it? They would have orchestrated it.

Imaginative enough to envision sirens and to seek sexual privacy, erectines would have adopted social conventions that reduced the likelihood of mistaken perceptions about sexual invitations. Social customs about who-wore-what-when were not likely if erectines wore nothing that was particularly concealing, as may often have been the case. That suggests they adopted behavioral customs to dispel the sirens' song.

The sexual minds of erectine males would view a female who moved her legs apart—somewhat revealing her normally obscured genitals—as making a gesture much like a female ape's presenting her puffed pink privates to a male, a gesture that was an invitation to copulate. As soon as that kind of movement became a gesture with that meaning, erectine males would have associated the mere sight of female genitals with the promise of gratification at the pleasure site. In immediate response to that, the females developed the behavioral custom of keeping their legs together in the social presence of males.

OUTSIDE OF SEXUAL EDEN: IN SIGHT OF MAGIC AND RELIGION, MURDER AND MORALITY

We should not leave erectines without speculating further on the less practical side of their lives. Erectines users of abstract concepts and symbols did not live by bread and sexual prurience alone. If chimps charge wildly down hills in response to lightening and thunder, erectines had much in the way of activity that was of high import, though hardly practical. Their newfound ability to conceive of things that did not exist in nature led to more than ideas of axes, pelts, and, for the males, always-receptive females. If they could project their symmetry and power into an axe, they could also send their vitality into the sun, storms, other animals, and countless objects. Those things would then have seemed as if they, too, were alive, powerful, possessed of wills of their own. Such objects then had living minds or spirits within. They became agents who wanted what they did not have, and who had the power to get it.

Suppose some of these erectine predecessors of ours encountered a weird object, a misshapen, half-burned tree trunk with reaching armlike branches, or a mysterious massive monolith with eyelike marks that followed them as they hastened by. After many such encounters, they would have used their bodies' "language" of rhythmic gesture to dance around the object, at times in timid fear and submission, at other times in exaltation and exuberance. Perhaps for erectus, but surely for the erectine's sapiens successor, such a dance and its accompanying vocalizations may have been one of the sources from which they developed ritual celebration, magic, and religion.

Both seeming-spirits in the community and community spirit may have generated the erectine beginnings of what we sapiens know as morality. It was suggested in chapter 8 that it was an upright ape, an afarensis Lucy, but not her reluctant boyfriends, who ate the fruit of a certain tree in the garden of Eden. The fruit of that unmentioned-in-the-bible tree gave only freedom to be or not to be engaged in sexual activity. The famous biblical tree, whose fruit gave knowledge of good and evil, had its branches beyond the reach of any of the upright apes, but not beyond the grasp of the erectines.

Think of what might be conceived as those murders most foul, murders of innocent infants, committed by a mother-daughter pair of chimps described in chapter 4. Those murders were followed by deeds still more monstrous-seeming—the deliberate ingestion of the flesh of the babes. Those deeds were not merely unpunished by the monsters' community, but unresponded to, and hardly even noticed. Judging from her brain size, Lucy and her community of upright apes would have behaved no differently. I think apes, both modern apes and ancient upright apes, have no taste of the fruit of the tree of knowledge of good and evil.

For erectines, the situation was decidedly otherwise. It was an erectine Eve and her consort who at least tasted the bittersweet fruit of the tree of knowledge of good and evil. Worse than bitter it was for those who were the object of the pointed finger and the projectile. More exciting than sweet it was to be among the pointers and pitchers. An erectine community did not have many community values, but those upheld by the unsophisticated creatures were held as firmly as they held their hand-axes. Those among them who publicly stole, murdered, and ate babies were not regarded with indifference. Those accused—and accusation may have been enough for conviction—of such deeds were soon crushed beneath a hail of hand-axes. Or perhaps deeds like that bespoke such wild power that the offenders ceased to be people and instead became unnatural agents that required fearful obeisance and careful avoidance. On the other hand, the private enraged killing of a spouse or infant provoked no more than a brief ripple of uneasiness above the flat indifference surface of the community.

In the lives of chimps, we saw two other kinds of killings. The killing of prey animals and the killing of chimps from another "tribe," both of which I suggested might be sport of a sort. I'm sure killings of both those kinds were engaged in by erectines on a far larger scale than that of the chimps. So, then, what would have been the moral classification of these two types of killings? Would the group, which somehow makes these moral judgements with quick gesture and utterance, have approved the killing of prey? Is it possible to think not? Would the group have approved the killing of outsiders, of strangers who don't belong, of others who held desirable lands? Here, things are less clear. If the strangers were not known neighbors, if they were not females, then it was good to kill them. If they were neighbors, especially female neighbors, then it might not be good, for erectine male-dominated communities may have begun to trade females to cement relations with neighboring groups.

There is a second tree mentioned in the Bible, the tree of life, where the word "life" must mean unending life, immortal life. At first thought, it is odd that the Bible does not record any interest (serpent-suggested or otherwise) by Adam or Eve in that tree. Not so odd would be erectine disinterest in immortal life. Immortality is far too abstract a concept to come to life in the mind of an erectine. They could not even conceive of the tree of such life, let alone wish to pluck and consume its fruit.

Then why were early sapiens, like Eve and Adam, disinterested in seeking that fruit to free themselves of the dread burden of necessary death? As I hope to show in the next part of this book, second thought suggests a good reason. Like the erectines, the early sapiens simply did not know that death was inevitable. They did not need the fruit to give them the endless— endless, apart from strife, starvation, accident, illness, and the presumed action of malign nonhuman agents—life they assumed they already had.

Like all the abstract ideas that we entertain, the concept that our lives were of limited duration did not come wrapped in a gene. It had to be learned. Decidedly, it was not knowledge that anyone sought. To this day it is less knowledge in our possession than it is knowledge that possesses and haunts us.

NOTES

1. Donald Johanson and Blake Edgar, *From Lucy to Language* (New York: Simon & Schuster Editions, 1996), pp. 28–30.

2. Ibid., p. 57.

3. Terrence W. Deacon, *The Symbolic Species* (New York: Norton, 1997), p. 345.

4. Goran Buranhult, ed., *The First Humans: Human Origins and History to 10,000 BC* (New York: HarperCollins Publishers, 1993), p. 64.

5. Richard Leakey and Roger Lewin, *Origins Reconsidered* (New York: Doubleday, 1992).

6. Matt Ridley, *The Origins of Virtue* (New York: Viking Penguin, 1997), p. 199.

7. James Shreeve, *The Neandertal Enigma* (New York: William Morrow and Company, 1995), p. 148.

8. Merlin Donald, *Origins of the Modern Mind* (Cambridge, Mass.: Harvard University Press, 1991).

9. Jean Aitchison, *The Seeds of Speech* (New York: Cambridge University Press, 1996), p. 39.

10. William Noble and Iain Davidson, *Human Evolution, Language, and Mind* (New York: Cambridge University Press, 1996), p. 15.

11. Derek Bickerton, *Language and Species* (Chicago: University Chicago Press, 1990).

12. Daniel L. Pals, *Seven Theories of Religion* (New York: Oxford University Press, 1996), p. 218.

13. William H. Calvin, *The Ascent of Mind* (New York: Bantam Books, 1991), p. 181.

14. Shreeve, *The Neandertal Enigma*, p. 163.

PART FOUR

EARLIER CREATURES OF OUR OWN KIND

INTRODUCTION

EARLIER CREATURES OF OUR OWN kind? Wait a doggone minute. Wait a *while*! Do I not realize just where, in this account, we *are*? Do I not at all comprehend *who* it is that we are about to look at? A prosaic "Earlier creatures of our own kind" is supposed to introduce us to *us*?

Surely I jest. This is *it*, the sideshows are behind us, we are entering the singular sensational big time. We can't just tiptoe into the tent; we've arrived at the gul-danged big top. If nothing else, there ought to be a dramatic drumroll!

Surfaced have we, up on the high ground above the ancient muck, crawl, and four-legged walk of conceptually empty, dimensionally flat animal lowlife. Sprung free have we, from the hairy bent-and-upright apes, apes who could render no more than a few feeble gestures to represent a mental concept or two. At long last have we arisen above the earthbound erectines, those slightly human creatures who, though they lived in many places where we do, could hardly have erected a tent, let alone an edifying edifice. Verily are we, our very own selves, emerging right here and now from time's funneling tunnel. Inexorably dawning in everyone's sight is—let the years before our days be a trampoline for our triumph—no less than us, Homo sapiens, formerly known as mankind, humankind itself!

You know it had to happen; it was in the cards, the tarot, the zodiac—we have come to Homo sapiens, the guys and gals who could really *think* (after all, isn't it thinking that we do, occasionally do, with concepts?). Onstage now, ready to perform, are the ones who could not only think, but could even—yes sir and madam, I don't mean maybe—could even think about thinking. Here we are, creatures who put everyone, simply everyone else, in our long egocentric history to absolute *shame* by finally doing proper justice to egocentricity. How could we do such a thing? By thinking about *ourselves*, that's how we do it, do it without the least effort, do it without raising a drop of sweat. The ones have arrived—the only ones, by golly, gosh, and gum—who could not only think about everything *in* the world, but the neotenous grown children who could think about everything they knew nothing about, the ones who could think about everything *out* of the world!

That was the drum roll, now to earlier creatures of our own kind.

Where did we come from? What were the first people like? We humans, always and everywhere, have asked such questions. And everywhere and always we've received answers. Answers that are astonishing in their diversity. Here are just a few of them, with the first from the King James Version of the Bible:

> And God said, let us make man in our image, after our likeness: and let them have dominion over the fish of the sea, and over the fowl of the air, and over the cattle, and over all the earth, and over every creeping thing that creepeth upon the earth.
>
> So God created man in his own image, in the image of God created he him; male and female created he them.

For big time difference from the Bible, just listen to this: "A much more bizarre myth of human origins is reported from Borneo. . . . A sword handle, long ago, mated with a spindle, giving rise to offspring without arms or legs, which produced more and more human forms until they engendered not only mankind, but finally all the gods."[1] Please, don't roll your eyes or snicker softly. If someday someone patronizes you for your beliefs, you may find a clear conscience helpful for the taking of umbrage. Besides, what happens in myths is "selected less for psychological goodness of fit or narrative entailment than for formal properties. . . . Orally transmitted tales may be selected for episodic structure because of the aid this provides to memory."[2] So there might not be anything very significant about odd details in mythlike stories.

An Australian version is refreshing for its modesty, for it does not specifically mention humankind: "Once the earth was completely dark and silent. Inside a deep cave . . . slept a beautiful woman, the Sun. The Great Father Spirit gently woke her and told her to . . . stir the universe into life. . . . The

Sun Mother then went on a long journey . . . and wherever her gentle rays touched the earth grasses, shrubs and trees grew. . . . In each of the deep caverns in the earth, the Sun found living creatures which like herself had been slumbering for untold ages. She stirred the insects into life . . . then she woke the snakes, lizards, and many other reptiles. . . . Behind the snakes mighty rivers flowed, teeming with all kinds of fish and water life. Then she called for the animals, the marsupials and many other creatures to awake and make their homes on the earth."[3]

The following is a Chinese creation story:

> In the beginning was chaos, from which light became the sky and darkness formed the earth. Yang and yin are contained in light and darkness, and everything is made of these principles.
>
> When yang and yin became one and the five elements were separated, humankind was born. As the first man watched the patterns of the sun, moon, and stars, a gold being came down and stood before him. The Gold One taught the man—now named the Old Yellow One—how to stay alive and how to read the sky.[4]

Greek mythology has the gods creating five "races" of humans in succession, with our four predecessors one by one pulled offstage. "Hesiod says that a golden race of humans was created in the time of Kronos's reign. This race was later hidden away in the earth to become beneficent spirits. Then a Silver Age race of somewhat foolish and irreverent humans was created by the Olympians. They angered Zeus and were hidden away in the earth as underworld spirits. Then Zeus made Bronze Age people who were powerful, warlike, and self-destructive. These men were also buried in the earth, in Hades itself. To replace them, Zeus made a race of heroes, whom we know from the stories of the blind Homer. These heroes passed like the other races before them, but they live forever in a far-off place called the Blessed Isles. "The race that lives here now, says Hesiod, is that of Iron. What a terrible fate it is to be of this race—to toil, suffer, and die."[5]

There is also a Greek legend wherein the Titan, Prometheus (Shelley's account of whom we encountered in chapter 7), created humankind out of clay and water, and gave us fire stolen from the gods.

The Greek Oedipus myth gained new explanations both with Sigmund Freud and Claude Levi-Strauss. In the latter's structural anthropology, the Oedipus myth illustrated a conflict between the belief that mankind sprung from the earth itself and the knowledge that humans are born of the union of man and woman.

The Icelandic (Norse) version of creation (of everything) goes as follows:

Once there were two places, one in the south that was all fire and light and one in the north that was icy and dark. The first was called Muspell and the second Niflheim. The two atmospheres met in an emptiness between them called Ginnungagap. There the hot and the cold mixed and caused moisture to form and life to begin, first as the evil frost giant Ymir.

Ymir lay down in Ginnungagap and gave birth to a man and a woman from his armpits; one of his legs mated with the other to make a son. Thus began the family of frost ogres. Some of the melting ice became the cow giant, Auohumla, whose teats flowed with rivers of milk to feed the giant and his family.

As for the cow, she fed on the ice blocks around her. As she licked the ice, a man gradually appeared from it. He was Buri the Strong; he had a son called Bor who married Bestla, a daughter of one of the frost ogres. Bor and Bestla produced the great god Odin and the gods Vili and Ve. These gods killed Ymir, and from the blood that resulted, all the frost ogres were destroyed in the flood. One giant, Bergelmir, escaped with his wife and family.

The three gods took Ymir's body to the center of Ginnungagap and turned his body into the earth and his blood into the seas. His bones became the mountains and his teeth and jaws became rocks, stones, and pebbles. The gods turned his skull into the sky, held up at each of the four corners by a dwarf. . . .

The three gods made man and woman out of two fallen trees, an ash and an elm. Odin breathed life into the new pair. Vili's gift to them was intelligence, and Ve's gifts were sight and hearing. The first man was named Ask; the first woman was Embla. The stronghold, Midgard, became their home, so they were protected from the cruel giants outside.[6]

In a way, the account that follows of where we came from and what early people were like does not differ from the above attempts to answer these questions. Like them, it is a combination of what is widely thought to be true, and of what, in the writer's view, might also be true. Unlike most current examinations of our evolution, this one's main interest is in the development of our subjective selves.

We'll look at four giant steps we took in striding from nonexistence to the present. First, we'll examine the transition from erectines to sapiens: how we may have come into existence by evolving from our predecessors.

Our second subject is early language, magic, and religion, new things in the world. The last two of these were also, in a sense, out of this world. Third, we will look at what may have been our last big biological step—one or more mutations that created a language instinct—and its consequences: the perhaps independent eruptions of complete spoken language in many different places. Our fourth and last subject is how sapiens societies developed the quest for more, which, not too long ago, became the quest for endlessly more. Here the social sequence from gathering-and-hunting to herding-and-farming to science-and-industry will be compressed until the three lie like sardines in a tiny package.

NOTES

1. Susanne K. Langer, *Mind: An Essay on Human Feeling*, vol. 3 (Baltimore, Md.: Johns Hopkins University Press, 1982), p. 18.

2. Mark A. Schneider, *Culture and Enchantment* (Chicago, Ill.: University of Chicago Press, 1993), p. 107.

3. Goran Burenhult, ed., *The First Humans: The Illustrated History of Humankind* (New York: HarperCollins Publishers, 1993), p. 148.

4. David Adams Leeming, *Encyclopedia of Creation Myths* (New York: ABC-CLIO, Inc., 1994).

5. Ibid.

6. Ibid.

10

THE TRANSITION FROM ERECTINES TO SAPIENS

THE FOSSIL RECORD SHOWS A transition period after erectus and before sapiens. "Traditionally, *Homo erectus* is credited as being the prehistoric pioneer, a creature that left Africa about 1 million years ago. . . . But recent evidence indicates that emigrant *erectus*, or perhaps its ancestor *Homo ergaster*, made a much earlier departure from our continental cradle."[1] Recent redating of *ergaster/erectus* fossils put them at 1.6 to 1.8 million years ago.

Erectines were followed by what are called "archaic sapiens," creatures with head shape mixed between erectus' and ours, for the period between 300,000 and 100,000 years ago. These supposed archaic sapiens were not created whole from their predecessor erectines in one mutational big bang but rather by a succession of smaller mutations with different ones in different places. After these in-between creatures, the record shows two variations of Homo sapiens. One was Homo sapiens neanderthalensis (neanderthals for not-very-short), a heavy-boned, strong-muscled, short-necked, big-toothed, big-nosed, large-brained human, who lived during the time between 200,000 and 30,000 years ago.

The other variation is the one we're all most interested in, Homo sapiens sapiens, ourselves. We are thought by many paleoanthropologists, but not by some molecular biologists (or molecular anthropologists, as Donald Johanson sometimes calls them), to have originated roughly 100,000 to 50,000 years ago. Molecular biology was a relative newcomer to the study of human origins. That may be why it "had a slow and difficult birth. Most paleontologists ignored it, although those who did not were forced to concede the logic of its position."[2]

Currently, there are still two strongly conflicting main theories about how H. sapiens sapiens—sapiens for short, "saps" in still shorter jargon slang—came into this world. The newer theory (whose proponents are mainly molecular biologists) suggests we appeared first in one place, Africa, and then spread around much of the world, like erectines did before us. That theory further suggests we everywhere replaced whomever we encountered, erectines, archaic sapiens, or even neanderthals. Some scientists think our basic superiority somehow overwhelmed our predecessors. Others suggest "no special story about cultural and technical superiority . . . need be invoked to account for"[3] their demise; they were done in by diseases that we carried but were immune to. The out-of-Africa theory also "strongly implied that the two human types, residents [predecessor archaics or neanderthals] and [newcomer sapiens] *arrivistes*, did not breed, at least not in a way that produced fertile offspring."[4] Donald Johanson and Blake Edgar tell us (in 1996) that the out-of-Africa theory is still being evolved by molecular biologists.[5] The first such work examined the gene diversity of 147 people, and discovered a founding African mother between 140,000 and 290,000 years ago. A subsequent study, using 189 people, published similar results. Later studies suggested a founding father in Africa.

The other and older theory says we evolved somewhat independently in several places. In each place, we continued the initial interbreeding with whomever was already there, perhaps erectines, archaic sapiens, and (some experts think) neanderthals. There was also, over the many millennia, the interregional intermixing that unites us into a single kind of creature. Eventually all those earlier species disappeared. Those who study fossils espouse this theory because they find, in Asia and elsewhere, resemblances in bone structure between local archaic and local sapiens fossils that suggest local evolution.

AN AFRICAN EVE MAY HAVE BEEN MOTHER TO US ALL

The strongest evidence for the latest theory—out-of-Africa—comes from the molecular analysis of what is called mitochondrial DNA (mtDNA), which is found within the nonnuclear portion of our cells. Samples of mtDNA were taken from cells of present day people around the world. The nature of variations between individuals' mtDNA is claimed to show that our purely female line—that narrow line of mothers and mother's mothers that excluded fathers in every generation—goes back to a presumed founding female. That first sapiens woman, who mothered that long line of mothers, lived in Africa between 300,000 and 150,000 years ago, as Michael Brown tells us in *The Search for Eve*.[6] The title of Brown's book indicates how irresistible was the temptation to call her "Eve."

Despite the singularity suggested by her name, there were many other females and mothers of Eve's generation who were like her except for her mutated mtDNA. Among them there might even have been a few who had the same genetic mutation that the mitochondrial Eve had. For each of these few other females, the descending chain of mothers was at some future point broken. However, it was all the *other* DNA—that is, the nuclear DNA—in the cells of Eve and the other mothers, and in those of all the fathers, that was the major contributor to the genes of present day people. So this Eve is the founder only of our mtDNA, and not of the DNA in the nucleus of each of our cells that is the main source of who we are genetically.

Just what is meant by this purely female line of ancestors can be seen more clearly with an example. If the time between generations averages 25 years, then it takes 8,000 generations to stretch back to 200,000 years ago, the approximate time when Eve lived. An example of just three generations, however, should suffice to illustrate the meaning of an "all-female" line. Over three generations, a person has two parents, four grandparents, and eight great-grandparents. Here is a table showing the eight lines that result over three generations.

Line	Parent	Grandparent	Great-Grandparent
1	mother	mother's mother	mother's mother's mother
2	mother	mother's mother	mother's mother's father
3	mother	mother's father	mother's father's mother
4	mother	mother's father	mother's father's father
5	father	father's mother	father's mother's mother
6	father	father's mother	father's mother's father
7	father	father's father	father's father's mother
8	father	father's father	father's father's father

HOW WE GOT TO BE HUMAN

Now, every human has a mother—it was W. C. Fields who, suggesting otherwise, said that he was the result of someone having played a dirty trick on his aunt—and that includes every mother. The first line above is the line of all mothers, the only line of the eight that has females in all three generations. All the others have one or more males in the line. Likewise, of the many lines for each of us that go back 8,000 generations to 200,000 years ago, only one is all female.

In sharp contrast with nuclear DNA, which is received by an individual from the nucleus of the father's sperm cell and from that of the mother's egg cell, we get mtDNA only from our mothers. That's because the sperm that enters an egg is stripped of its mitochondrial DNA. That means mtDNA is passed down the all-female line from the mtDNA Eve, and not passed down any of the other lines, all of which include fathers who break the chain of transfer from Eve. Those who first proposed the out-of-Africa theory compared the mtDNA of about a hundred and fifty current individuals from different parts of the world. They found evidence for an African origin of all people. They also found evidence, from the apparent average rate of mtDNA mutation, for an origin time of, very roughly, 200,000 years ago. However, their work has been questioned by some, including other molecular biologists. If their time estimate, which is disputed, proves to be wrong, and the time is closer to one or one-and-a-half million years, then it could be that they detected the worldwide spread, not of H. sapiens sapiens, but of our predecessor, H. erectus. In that case they would have "discovered" what has been a widely known and accepted fact.

The out-of-Africa theory has another possible flaw: Its insistence that sapiens (after the initial speciation) had little or no intermixing with predecessor species who lived at the same time. We will see below that, despite this insistence, it is indeed possible that there was major interbreeding, even if the originators of the theory interpreted their mtDNA evidence correctly. The basic idea that I'm suggesting is not complex: There might have been massive intermixing, none of which would have shown up in the mtDNA if sapiens communities did not allow predecessor *females* or their children (with their predecessor mtDNA that might be passed on) in the group. In contrast, predecessor *males* and their children—even if they lived in the sapiens group—would pass on their nuclear DNA without leaving any telltale predecessor mtDNA mark.

To elaborate on this, we must speculate about one of the customs of early sapiens. In the following, to avoid cluttering the argument, I'll assume all those displaced predecessor creatures were "archaics." The custom at issue is that wherein almost all newly mature young of one gender stay in their native community, while those of the other gender almost all leave and join another community. This custom is found in ape and monkey communities where the group is larger than a family. It is also found in human "primitive" or "traditional" communities. With chimps, it is the females that switch to new groups. With olive baboons, the males join other groups. Among most current traditional small human communities, commonly the males stay home and the females go elsewhere.

If it was the young *females* among archaic sapiens who stayed where they were, and the *males* who transferred, then intermixing would be *undetectable* by the evidence for an mtDNA Eve. To see this, assume those archaic predecessors of sapiens disappeared over, say, a 10,000 year—400 generation—interval. During those 400 generations, archaic males who inseminated stay-at-home female sapiens would thereby father females (and males) who had sapiens mtDNA from their mothers. Some of those females would be links in the long lines of mothers in the sapiens communities. None of them, by themselves, would break those long lines. Despite tens or hundreds of intermixing archaic males, these lines of mothers all started way back with the mtDNA Eve many generations earlier, and all continued into our time.

Sapiens males who ventured out to father the children of female *archaics* would contribute to an intermixed population, with *archaic* mtDNA from those females, in the archaic communities. (That intermixed population could, in later generations, contribute some of the previously mentioned males who fathered children who had mtDNA from their sapiens mothers.) If those archaic communities died out, all of this intermixing would be invisible in the mtDNA of current people. However, it certainly would contribute to the current nuclear DNA that contains almost all of our genetic material.

If the cultural custom was for the young males to stay with their birth community and the young females to go elsewhere, the effect on the genetic transmission of mtDNA effect would be significantly different. The archaic females who joined sapiens communities and copulated with sapiens males would have had daughters with archaic mtDNA (from their mothers). Some of those daughters would have initiated lines of mothers with archaic mtDNA reaching to the present. If the molecular biologists' samples from current people included one or more of those lines, their evidence would not have pointed to a single common African origin. Therefore, confirmed evidence for an mtDNA sapiens African Eve would suggest there was little or no inter-mixing *only* if that interbreeding was the result of young adult females—rather than males—leaving their birth community to migrate to another.

If there came into being independent evidence that significant inter-mixing did occur, a validated out-of-Africa theory could be turned upside down. It would suggest that, among the young in most of the earliest human communities, it was in significant part the males, not the females, who migrated from their birth group. This in turn might suggest that male dom-inance, which usually goes with female transfer, was not the widespread case for those earliest generations when sapiens lived among archaics.

One potential source of evidence on these issues is the male counterpart of the mtDNA that is passed down an all-mothers line. This counterpart is the Y chromosome genetic material that determines that the union of sperm and egg will result in a male. This Y chromosome contains what we could call "yDNA" that is passed down an all-fathers line from the past to the present. Perhaps future work on large samples of yDNA will reveal a yDNA Adam whose geographic location and time of origin may not be the same as that of the mtDNA Eve. This in turn would imply, if nothing else, intermixing of sapiens and their predecessors. However, initial work of this kind in 1995, based on a small sample of yDNA, from less than 40 males, suggests that an initial sapiens male—a yDNA Adam—lived in Africa about 270,000 years ago.

The molecular biologists who claim no intermixing don't tell us how it might have come to be that interbreeding of sapiens and archaics did not occur to any significant extent. I can think of four reasons: war, disease, superior skills, and sexual distaste. I will ignore another possible reason—misfit of mating parts. Demographer Ezra Zubrow provides yet another explanation.[7] If the sapiens mortality rate was slightly lower than that of the others, those others would have disappeared relatively rapidly, and thus have left few mixed progeny.

I'll first explore these four reasons (that suggest no interbreeding) with the intent of knocking down that argument. In other words, I'll make a case for considerable intermixing of sapiens and archaics. Then, I'll switch sides and show that interbreeding was highly unlikely.

My reason for walking both sides of the street is a bit roundabout. All such arguments, whatever veneer of veracity they possess, are not to be strongly relied on. At this point, we are not dealing with doing "straight" sci-ence here, of either the hard physical kind or the softer social sort. We are dealing with something in some ways more important to most of us: our search for significance and meaning. The question about whether or not sapiens interbred with predecessor creatures is of significance to many of us (although it hardly compares with other questions whose firm answers are unburdened by evidence, such as the existence of an afterlife). To such high-import questions, we all seek to supply ourselves with—evidence, schmevi-dence—answers. We do that because doing so provides some of the important meaning we want in our lives.

Even archaeologists (though not as blatant about it as I am here) are like the rest of us in this regard. They "try to suck every drop of meaning possible out of [evidence from the past]."[8] For example, regarding ancient pieces of worked stone from the Neanderthal past, one prominent theory sees them as evidence that tribes had very little interaction. Another theory sees evidence that different kinds of worked stone means they were made during different seasons of the year. Yet another theory sees such stones as showing an increased complexity over time. A last theory finds them to be merely the junk that remains when the original stones have been used over and over again. So I'm trying to find meaningful answers—to a question science cannot at present clearly answer—by providing some "yes" answers and some "no" answers, from which we can each take our pick.

Let's start the argument *for* interbreeding with war, or rather, with deadly fighting. Such fighting between incoming sapiens and local archaics may well have occurred on numerous occasions in our early millennia. That is hardly to suggest that anything like holocaust or extermination of prede-cessor creatures was at the center of our earliest days on earth. A few decades ago, many young people protested politically with the slogan "Make love, not war." To suggest that the earliest of our kind fervently acted out the opposite, "make war, not love," is to stretch inference too far. We know that communities of our closest cousins, the chimpanzees, at times fight with and kill members of other chimp groups. They don't do it much of the time, however, nor do they do it everywhere. We know all too well that we humans make war. We make both war and love, plenty of the former but a lot more of the latter. If that is part of the pattern of what humans in recorded history and modern chimps do, it's reasonable to think the earliest humans and the archaics may have done the same, even with creatures who looked and acted unlike themselves. Even leaving love apart, the concepts "kill the males, copulate with the females" may have come early to human minds, and resulted in much intermixing.

HOW WE GOT TO BE HUMAN

As for disease, we know that it has always had a free ride as part of the baggage with which we have traveled. Humans from Europe, for example, brought diseases with them to the new lands in America. We also know that, disastrous as that was, disease did not begin to wipe out the native North, Central, and South American population. So it seems unlikely that sapiens from Africa arrived everywhere else with diseases so virulent that existing populations died suddenly enough to preclude many generations of intermixing.

Superiority to erectine, archaic sapiens, neanderthal, and to whatever other creature lives on our little planet is, of course, one of our main claims to fame. We'd almost sooner part with our credit cards than part with that claim. Perhaps those archaic inferior breeds beneath the sun suffered in our presence, and retreated to poorer lands to insure our absence. Nonetheless, over hundreds, thousands, perhaps tens of thousands of years of contact, many of those sapiens and archaics must have gotten to know each other quite well. "Knowing," as we know, covers a multitude of sins, including the biblical knowing that often results in offspring. Besides, the lack of handsomer features, higher skills and headier thoughts need not have inhibited either hard-headed sexual hunger or haphazard hanky-panky.

Sexual distaste of sapiens for archaics in those early days of ours is something we know even less about, if that's possible, than we know of their wars, diseases, and superior skills. We do know that, by our standards, erectines and archaics were mostly chinless, thick-skulled, brow-ridged, strong-armed, and, perhaps, untalkative creatures. Was that reason enough for comely sapiens lasses everywhere to always run from those burly boys? The (less pertinent) related question, did sapiens males have sex with archaic females, is likely to have an affirmative answer, suspecting as most of us do, that human males, sapiens or not, are usually highly prurient. But we are less interested here in those unions than we are in those of sapiens females with archaic males.

The archaic males' presumed physical and mental deficiencies might not have been, by themselves, a deciding negative factor. More important may have been how well the two communities knew each other, and how often a sapiens community had a shortage of incoming young males from nearby other sapiens communities. It is not unlikely that, over many generations, the sapiens got to know and understand the archaics. In fact, we may have known them well from our very beginning. Nor is it unlikely that, at times in sapiens communities, the supply of incoming sapiens males was too small to match the homebody females. In those circumstances, strong-backed "backward" archaic males who may not have talked much might have looked, both to sapiens females and to their communities, a lot better than no males at all.

We can conjecture that what was operating was not the impossible dream of the brute, Caliban, about the heavenly Ariel. Think rather in terms of the more pedestrian fiction wherein the socially elite Lady Jane accepts the (one might think) untouchable classless Tarzan as her mate. Perhaps she simply fell for his classy chassis.

Were their story a real one, it might have been more likely that Jane and her people had a hidden hope that out-breeding would reinvigorate her overly inbred aristocratic genetic base. If it did, that would have compensated her for a lot of discouragingly brief and tedious "me Tarzan, you Jane" conversations.

Now let's cross the street to look at the argument that suggest interbreeding between sapiens and their predecessors was *not* likely. The main argument is based on a look at our relatively recent history. Wherever there is evidence from our historical past of encounters between radically different human societies, the more powerful group increased their number, the less powerful society was reduced to negligible numbers. In *The Third Chimpanzee*, Jared Diamond present a thorough analysis of the places and ways whereby the less powerful societies in our past were decimated after contact with a more powerful opponent.[9] Among the more recent of these encounters was that, a couple of hundred years ago, whereby incoming Europeans made disappear the bulk of a society of over 100,000 people in Australia (as well as a smaller one in Tasmania) that had been living there for some 50,000 years.

Slightly earlier, Europeans who came to North America did similar disservice to over a million Amerindians descended, for the most part, from a small number of Asians who crossed from Siberia to Alaska some 10,000 or more years ago. Slightly earlier still, Spaniards initiated a similar process on millions of residents whose forefathers had been living in the West Indies, Central America, and South America for five or ten thousand years.

To those who would account for these decimations by beating multicultural drums made from the skins of dead white Europeans, all I can say is "get real." Is there any reason to doubt that there have been plenty of encounters with similar results that were initiated by nonwhite non-Europeans? For that matter, look at the extinctions, extinctions not of other human groups, but of other forms of animal life, that have followed in our wake. In *The Origins of*

Virtue, Matt Ridley describes these gross extinctions that went far beyond the slaughter need to fill human bellies: They occurred 11,000 years ago in North America, 8000 years ago in South America, by Polynesians in the Pacific long before Captain Cook, less than a thousand years ago in New Zealand, and in Australia soon after humans arrived about 60,000 years ago.[10]

It is true that currently living are descendant survivors of most of the above decimated societies. It's also true that, percentage-wise, the numbers of such survivors are pathetically small. Let's look now at interbreeding itself in these cases. In Australia, the percentage of the present population that has "mixed blood" is surely tiny. In North America, that percentage, perhaps not as small, is still quite small. In the West Indies, Central America, and South America, interestingly, the percentage of "mixed" people is not so small. In the latter places, the need for "peasant" labor and the importation of slave labor may have led to intermixing, but more so between these two "loser" groups than between either of them and the controlling Europeans.

Look now at both the pro and the con arguments about mixing between early sapiens and their predecessors; then take your pick. In any case, it is not likely that there ever was heavy intermixing of early sapiens and their predecessors. In *Reinventing Darwin*, Niles Eldredge comments that even for closely related (nonhuman) species, individuals ordinarily don't "mate with one another across species boundaries." For sapiens, what may be a bit more likely is that occasional groups of young males would have found it low-risk sport to catch and sock it to even a strong-armed predecessor female.

We can only hope that scientists someday reach a resolution of the issue. In *The Neandertals*, Erik Trinkaus and Pat Shipman note (in 1994) that scientific opinion about the out-of-Africa theory may be changing: Different computer models using the same mtDNA data suggest origins elsewhere than Africa.[12] They also note that the originating divergence in mtDNA might have occurred long before modern humans appeared.

OUR KIND OF HUMANS MAY HAVE COME TO LIFE INDEPENDENTLY IN SEVERAL PLACES

The other scientific origin theory, a multiplace flowering of the sapiens species, is of interest even apart from the intermixing that is a feature of it, and apart from the partial support for it provided by what seems to be the multiple origins of archaic sapiens.

The Transition from Erectines to Sapiens

We know that successful major mutations, the ones that initiate new species, are thought to occur in small isolated populations when environments change. What we don't know—because they leave no evidence for us—is anything about the unsuccessful mutations. I'm not interested here in the grossly maladaptive or even the slightly unhelpful ones, but rather the selfsame ones that were successful in one place and time but may have failed elsewhere and earlier. The day has not yet arrived when molecular geneticists know their material well enough to assign undisputed probabilities of occurrence to mutations. So they can't help us much. Nonetheless, a specific mutation occurring only once, not in just a few lifetimes, but in relatively unlimited time, smacks of the miraculous. Real miracles may be fine for some. For me they carry an unpleasant odor of unneeded supernatural sulfur. Both common sense and Occam's Razor suggest that if something happened once there may be a fair chance that it happened—but failed to take root—several times before.

Recall the first ape that stood erect some four-and-a-half million years ago, by mutation's grace, to found the line of hominids. That unknown creature was surely an Adam or an Eve as much to be memorialized with widespread discussion and attention as the mtDNA Eve or a future yDNA Adam. Nonetheless, that hominid may well have had predecessors with the same mutation, unsung heroes of a simian sort, who came to life in the wrong time or place. (Perhaps there exists, on a planet of the apes in some parallel universe, a chimpish poet who soulfully celebrates their—alas, barren-of-descendants—deaths.)

The fact that some scientists think there is evidence suggesting, not singular, but plural successful origins of sapiens increases the likelihood that there were even more origins, forever hidden from knowledge, that failed to bear the fruit of a new species. If there were such brief barren false blossomings of our kind, how can they be accounted for?

For the ramidus species that may have been the first of the upright apes, it's easy to see that the mutation that straightened them upright might have provided more hindrance than help in an environment that still had forests and had few open plains. In those born before their time, the mutations were not adaptive. That is to say, they failed to start a new species. But for sapiens, for us, isn't it our scientific conviction (as well as our proud belief) that in all ways of importance we were a clear improvement on our predecessors? If we came to life anywhere they lived, couldn't we truthfully have told them, "Anything you can do we can do better"? And what could they have said back to us? Nothing. (Nothing figuratively; and, as they probably lacked all but the rudiments of language, almost nothing literally.)

We were the brainy ones. The tougher the environment the sooner those predecessors would be eating the dust and other waste products we're so good at dropping behind us. In no time at all we'd be eating their lunch, and their dinner and breakfast, if not them.

"Yeah, but," a skeptic might say, "is it really so certain that sapiens would always have prospered?" Sapiens, we have to presume, could both think better and speak better than erectines and archaics could. Wouldn't that necessarily result in their prospering more than the simple native folk? When times were good for all, it might indeed have resulted in fuller sapiens bellies. As we keep telling ourselves, however, humankind does not live by bread alone. After all, we have spiritual values, don't we? We sapiens have dreams and beliefs beyond our limited nickel-and-dime desire to fill our guts with the fat of the land.

What can be said about the dreams and beliefs of our primogenitors? If they were one-fleshed with us, they indeed had them, and had them galore. What they did not have in those first millennia was our understanding, limited though it is, of the nature of dreams and beliefs. The earliest of our kind did not possess the rarified abstract concepts of "truth and falsity," or of "reality and unreality." Such highbrow ideas did not sit side-by-side in their mental toolkits with concepts for fancy flintwork and fancier footwork.

Nor, for that matter, would they have had the concepts (or words) for dream and belief. In the wakeful day, one saw things, some of them scary. In the night, between sleep and sleep, one also saw gut-wrenching things, some of them really wretched. They all happened. That was simply and obviously undisputable. At times some happenings were crudely recounted to other creatures, using some of that fabled speech that we share with no other species. The things spoken of that occurred in the night were no less believable to others than were the day's happenings.

Suppose that once upon a time, young, vigorous John was starting out on the hunt when, during a sudden heavy thunder and rain storm, he slipped on a hillside boulder, bashed open his skull on a sharp rock below, and his blood and his life left him. His companion, who saw all this, sobs that evening, although she may be too inarticulate to verbalize, "Johnnie, we hardly knew you." When sleep finally overcomes her that night she has what we call a dream. She sees the clouds dump raging waters from the sky. She sees bodies fall and with their blood color the waters red. She hears the thunder speak in conceptlike images, but lacks the lingo to hear words like,

"Your blood, too, shall I drink if, like them, you defy me! Leave you now my land!" Might not she and everyone she tells hightail it for elsewhere pronto?

Perhaps after a few miles and a few days they stop. Something has distracted them. If other private nightmares or scary daydreams become public policy, however, it is not unlikely that the spiritual values that come with imagination and language are, in this case, without survival value, are not adaptive. These people drive themselves from their Eden. Their beliefs push them from one disaster to another. Their tribe does not increase; instead, it dwindles. Before long, they disappear.

There are, no doubt, more likely disaster scenarios than that sketched above. It is provided only to remind us of what we all know. Sometimes spiritual values are at cross purposes with those of the flesh. The spirit, if it gets too far out, if it fails to respect the flesh's bond with the natural world, may push itself and the flesh it inhabits to common extinction.

Human conceptual imagination and language are instruments of awesome power, power that, in the last few (the only ones from which history narrowly reads) of the hundred or two millennia that make up our sapiens lifetime, has on many occasions produced disaster, disaster for millions of our kind, disaster beyond the capability of all the millions of other creature kinds. Those last millennia suggest those peculiarly human tools, conceptual imagination and language, might often have been turned against others of our kind during the many prehistoric millennia of our stay on this planet's thin skin.

In the earliest days, when our newborn species could barely toddle without stumbling on the rough surface of the earth, time and again we may have used our spiritual values to destroy ourselves, or at least, destroy others like us. Some remarks by Shreeve in *The Neandertal Enigma* are of interest on this issue, although they are hardly evidentiary.[13] He discusses some stone tools (found near Cape Town, South Africa) that seemed to be advanced from both earlier and *later* tools found nearby. If those advanced tools declared the emergence of a new and superior kind of human, "wouldn't these people be likely to survive into the future [where later inferior tools should be absent]?" Although there's no evidence, it's conceivable that their "superior" minds also led to a reduction in their numbers.

What is unsaid above, but clear to all (save occasional doomsayers who see our species sliding swiftly down an incline to hell on earth before we are snuffed out) is that over the millennia we have helped ourselves enormously with the work of conceptual thinking, imagination, and language among other adaptations and adaptation-byproducts of our bodies, minds, and spirits. As Susanne Langer remarks in *Mind: An Essay on Human Feeling*, "The phenomenon of mind, arising in just one primate stock, is . . . a tremendous novelty in animal evolution. . . . No single principle, however great . . . is likely to explain the rise of language and thought."[14]

NOTES

1. Donald Johanson and Blake Edgar, *From Lucy to Language* (New York: Simon & Schuster Editions, 1996), p. 46.

2. Maitland A. Edey and Donald C. Johanson, *Blueprints: Solving the Mystery of Evolution* (New York: Little, Brown and Company, 1989), p. 359.

3. Terrence W. Deacon, *The Symbolic Species* (New York: Norton, 1997), p. 373.

4. James Shreeve, *The Neandertal Enigma* (New York: William Morrow and Company, 1995), p. 5.

5. Johanson and Edgar, *From Lucy to Language*, p. 41.

6. Michael H. Brown, *The Search for Eve* (New York: Harper & Rowe, 1990).

7. Shreeve, *The Neandertal Enigma*, p. 72.

8. Ibid., p. 149.

9. Jared Diamond, *The Third Chimpanzee* (New York: HarperPerennial, 1992).

10. Matt Ridley, *The Origins of Virtue* (New York: Viking Penguin, 1997), p. 211.

11. Niles Eldredge, *Reinventing Darwin* (New York: John Wiley & Sons, 1995), p. 107.

12. Erik Trinkaus and Pat Shipman, *The Neandertals* (New York: Vintage Books, 1994), pp. 395–96.

13. Shreeve, *The Neandertal Enigma*, p. 216.

14. Susanne K. Langer, *Mind: An Essay on Human Feeling*, vol. 2 (Baltimore: The Johns Hopkins University Press, 1972), p. 215.

11

EARLY LANGUAGE, MAGIC, AND RELIGION

IN AND OUT OF THIS WORLD

O wad some Pow'r the giftie gie us
To see oursels as others see us!
It wad frae mony a blunder free us,
        And foolish notion:
What airs in dress an' gait wad lea'e us,
        And ev'n devotion!
            —Robert Burns, "To A Louse"

THESE CHARMING WORDS SUGGEST NOT only that we humans are vain, but that we might be a mite short on imagination.

Regarding the latter, it is likely that the contrary is more true. We may be mighty long, often overlong, on imagination, which may well be linked to our ability to project, unknowingly, our interior subjective world upon our exterior objective one. The giving to us by Mother Nature of a large brain—within which resides an unlocatable concept-engorged mind—has allowed us all to have a "Pow'r" quite the complement to the one Burns's narrator asks for. Like it or not, we commonly see all kinds of other beings and things, real or imagined, as we sense either ourselves or some hidden aspect of ourselves to be.

The upright apes long ago took Yogi Berra's advice, "When you come to a fork in the road, take it!" and used it to dig into the hominid road to humankind. They also started up that road with projective power far greater than what our modern chimpanzees have, who react to thunder-and-lightning storms by racing wildly down a nearby hill. They project their own power onto the black and white clouds and identify with the resulting stormy "Pow'r" introjected back from those clouds.

Their successors and our predecessors, the erectines, made Acheulian stone axes with a shape that was bilaterally symmetrical, the same balance of left with right that could be seen by them in their own bodies. They could hardly have articulated their intuitions about the powerful minds within their own bodies, but they had enough brains and conceptual capability to project that intuition of power into a hand-axe with similar symmetry. As suggested in chapter 9, they also could project their unsayable sense of their group's power into the skull or skeleton of a former leader.

Early sapiens had brains and minds larger than their erectine ancestors. What did they do to give form to their inarticulate intuitions about their capabilities? They could move beyond and outside the natural world to represent the grandeur and the grotesquery that decorated their own interiors. They could make magic and religion to show the shape of that which they hardly suspected was within themselves. They could use language to articulate and elaborate those projections.

Although hardly as plentiful as religions, there are numerous theories of religion. The old common concept of religion is simply *belief in the supernatural*. In *Creation of the Sacred*, Walker Burkert's primary interest is not the factual truth or real existence of the gods. He seeks distinctive features at the level of observable behavior, and finds three principal characteristics. First, "religion deals with the nonobvious, the unseen. . . . [It] is manifest in actions and attitudes that do not fulfill immediate practical functions. . . . Second . . . religion manifests itself through interaction and communication. . . . Third . . . [is] its claim for priority and seriousness . . . 'ultimate concern.' "[1] These three features apply not only to religion, but to some of magic as well.

In *Seven Theories of Religion*, Daniel Pals says that those of E. B. Tylor and Mircea Eliade set out to "collect the widest possible range of information."[2] Whereas Sigmund Freud and Karl Marx sought "the fundamental mechanism in human being that everywhere generates religion." His remaining three, those of Emile Durkheim, E. E. Evans-Pritchard, and Clifford Geertz, seek "some form of compromise." Although all seven of these theories are naturalistic and do not seek supernatural realities, one of them, that of Eliade, is murky about the objective existence of the supernatural.

Pioneer cultural anthropologist E. B. Tylor found that "primitive" peoples must have believed that they themselves were animated by a soul or spirit. Then they reasoned that "plants and trees, the rivers, winds, and animals, even the stars and planets also were moved by souls."[3] That sounds more like projection than reasoning, doesn't it?

To say early humans made magic and religion (with the help of language), however, is to apply modern categories for paranormal or supernatural experiences that initially may not have fit within concepts of magic or religion, concepts that had not yet been created. Furthermore, there is reason to think that even in our thoroughly modern Western world, such weird and not easily categorized experiences may still come to many of us.

Consider again a certain terror that comes in the night, that David Hufford tells us about (without making any judgement about it being real in a supernatural sense).[4] That terrible experience is known to scientists as hypnogogic hallucinations with sleep paralysis during sleep-onset Rapid-Eye-Movement brain activity. It was known centuries ago as neither more nor less than "nightmare." It is known now in Newfoundland as experiencing "the Old Hag." On one or more occasions, fifteen percent or more of people in this country suddenly experience the old Hag in the room where they have been sleeping, usually on their backs. They feel they cannot (or will not) move; they soon experience a crushing weight on their chests that pushes them down; they feel a fierce oppressive terror; they believe they are not asleep but awake; they often see or sense the presence of an unnatural being or thing that approaches and presses them down. More often than not, those who have been ridden by the Old Hag do not associate the experience with either magical or religious beliefs.

I will note here that it is never a mere impersonal object that oppresses the victim. Rather, it is something that is mentally like, but perhaps physically unlike, a real living person. Thus, the experience is consistent with our common human tendency, in situations that are both strange and of high import, to see ourselves as engaged with (humanlike) agents that are also natural or unnatural objects or forces.

Apart from some things like the Old Hag, most unnatural or paranormal phenomena can be fitted—without a shoehorn—into the categories of religion and magic. It may seem strange from our current Western perspective to couple magic with religion. In recent millennia, religion carries a thick mantle of respect and respectability. That mantle is largely lacking from the in some ways marginally acceptable activities that pass for modern magic.

By religion, I mean nonsecular high-import group-inspired evidence-lacking beliefs and rituals that often offer hope for betterment. (I don't mean religion in the sense of occasional epiphanies and other seeming direct connections with something supernatural.) By current Western magic, I mean not the activities of professional illusionists and tricksters, but rather the everyday work of many fortune-tellers, astrologers, psychics, mentalists, and others. Exclude those who are trying to do no more than earn a nice living without too much work. That leaves those who (in at least part of their minds) believe some of their doings to be powerfully effective and more than merely natural help for or harm to individuals.

Durkheim viewed magic as different than religion in that magic is a "private matter. . . . The magician . . . heals my sickness . . . or puts a spell on your enemy. . . . Religious rituals and beliefs come into play whenever group concerns are foremost in the mind."[5] This difference arises because religion, despite the magic that is sometimes happily wed to it, comes from the communal subminds in our heads. Western religions concentrate on the group's leaders, who sometimes have acquired supernatural status. Eastern religions' emphasis is on the group itself, where each individual's consciousness is a drop that is best dissolved in a cosmic group bucket, the cosmos itself, that is full of such stuff. Whereas magic, even with the religion that is sometimes still bonded to it, does not have much of a communal component. Rather, it comes primarily from the narrowly self-serving subminds that are our primary subjective endowments from nature.

Respect and respectability are among the main threads in the social fabric called morality that together we have woven. There is much that has been said and can be said about morality. Although my plan is to say as little as I can about it, I'll say something in response to the views of Robert Wright in *The Moral Animal*. "The closest thing to a generic Darwinian view of how moral codes arise is this: people tend to pass the sorts of moral judgements that help move their genes into the next generation (or, at least, the kinds of judgements that would have furthered that cause in the environment of our evolution). Thus, a moral code is an informal compromise among competing spheres of genetic self-interest, each acting to mold the code to its own ends, using any levers at its disposal."[6]

In reaction, I would ask, why not a genetic predecessor-animal-acquired *communal* instinct in our human forebears, who themselves had many *adaption-by-product* concepts? Most of the people in a group felt it right to retain or adopt those beliefs shared by almost all members of their circle, even though it meant dropping their own other ideas. Those shared beliefs were regarded as the morally right ones. Whatever its other values, this construction at least carries no suggestion of Darwinian blessings placed on moral codes.

The "higher" religions, which in recent millennia claim and proclaim morality at their core, can hardly fail to command esteem from most of us. Whereas magic often evokes from us a social sneer for the thinly veiled narrow self-serving that is at its center. Religion and morality, of whatever texture or shape, whatever else they claim, claim the high ground from which they can look down on "flawed" individuals, especially those who seek too large a share of life's booty. Magic, often considered amoral or immoral, is often used against or for individual agents, be they real or supernatural. Magic—such as lucky lottery numbers or voodoo pins in dolls—is also a recourse of individuals against a society or religion that is experienced as overweening or over-constricting.

Although magic and religion have diverged over the millennia that stand between us and early sapiens, their births from the minds of humankind may have been simultaneous. Perhaps religion came first, only to be followed immediately by its less fraternal twin, magic. Serving as midwife to the arrival of the mentation behind these less than immaculate concepts must have been the means—initially limited language—to make expanded meaning in the world.

EARLY LANGUAGE: THE MEANS BY WHICH WE FIRST EXPANDED MEANING

Early sapiens, what did they leave us that shows what they were like? Evidence dug from the rocky earth of bygone days, days some tens of millennia after our earliest days, shows many new things: The widespread use of fire; vastly improved stone tools and weapons; throwing boards, needles, darts; cut, carved, and drilled antlers, ivory, and bone; the beginnings of body adornment and jewelry; and cave paintings that may have been part of prayer to and respect for past and future prey.

The rapid spread of knowledge about these things required language to communicate. Initially, however, before word from mouth to ear made so much possible, was language mainly a means of communication? "Communication per se does not normally require a large brain. If communication is at issue it must be a matter of what is communicated rather than how communication is effected. . . . Human language first evolved in response to selection pressures for cognitive capacity. . . . Cognitive systems require large brains, and language as a cognitive system would have begun as a system from an enlarged brain."[7]

Consider the world of the earliest communities of humans, creatures who, all unknowing, could project their feelings of desire and fear into objects of the world outside them. They were creatures whose conceptual imagination also helped them create the tools, weapons, clothing, shelters, and sharing society that enabled them to prosper. Language may have begun—perhaps among upright apes but more likely with erectines—with cries, shouts, grunts, groans, and ultimately, wordlike or sentencelike utterance serving as communication related to the practical activities of living.

Language advanced us beyond that beginning, however, not with steps of communicative signals on the sticky strands of the growing worldwide web of culture, but down a different pathway. It advanced rather as an unintended part of the means of strengthening that social web by expressing fellowship with other humans. In *The Prehistory of the Mind*, Steven Mithen suggests that early human language was a social language.[8] He also says that early human language had an "extensive lexicon and grammatical complexity," with only "snippets [of it about] the nonsocial world, such as about animal behavior and toolmaking." He further suggests that social language would have moved to general-purpose language between 150,000 and 50,000 years ago.

Some may here suggest that the social use of language for the expression of the likes of fellowship could not have had anything but a low priority for creatures struggling to carve a larger niche for themselves between predators and prey. Doubtless, early sapiens had little rational intent of social congregating for purposes of glad-handing or backslapping conviviality. Nor would they have formed herdlike huddled masses to ease their anxieties whenever they came to new lands.

Consider the lions of the field, how they grow; they toil not, neither do they spin. However, at times they do find themselves decidedly red in tooth and claw. The lions do not live in large communities, but in groups of at most a few. In contrast, their prey, such as the graceful gazelle or the weird-looking wildebeest, live as parts of large herds. Prey species that have survived inherited a Darwinian maxim, "If we hang together as a group, we can hang looser as individuals." On the other hand, the unsocial lions and their like sing, "Give me land, lots of land," because that's what works for them to find enough prey.

Sapiens sapiens, the last word in the hominid line, like their cousin chimps and bonobos, are communal in nature. It goes with the absence of predatory sharp teeth and claws, although it hardly inhibits occasional bloodletting. To be communal is to seek and achieve closeness to a group of others, familiar others, of your kind. As indicated in chapter 4, apes have advanced from mostly indifferent-to-individuals community to friendly-with-individuals society. So have we, more so than they. I am not talking about communism and socialism—heaven forfend—though there may be a connection. Note however, that Matt Ridley, in *The Origins of Virtue*, suggests otherwise.[9] He presents evidence that our social nature may connect with private ownership of property, not with communism or socialism.

Sapiens, and erectines before them, could not have prospered, let alone spread over our globe, were they not communal creatures. A few sapiens, even with axes in hand, do not make much of a threat to a carnivore, a scavenger pack, or even a healthy wildebeest. And axes would hardly be in every hunting hand were sapiens not social enough to learn from and teach one another. With the coming of early language (the only partial language that erectines and even early sapiens may have had), the power of a group of like creatures was greatly enlarged. Language as they knew it could not have come into existence were our predecessors not social and communal creatures. Each family isolated by its own feeble protolanguage makes for neither large vocabularies nor full bellies.

That may be all well and good, but it doesn't mean the early sapiens knew how important community was to them. Did they know that even rudimentary language was a tool that vastly increased the group's power? "Community" is a rather abstract concept, hardly needed when what you want is to be with Billy Bob, Ginny Rae, and the others. Likewise with language, early speakers no doubt spoke without even knowing they were speaking language. They were so ignorant that they would not have known enough to marvel (as did the know-nothing, in an old joke, marvel that even little children in France or Germany knew how to speak French or German).

The earliest humans did not have minds like ours when it comes to being filled with many of our current concepts. Nonetheless, judging from skull size and shape, they did have brains genetically like ours. That suggests minds like ours when it comes to intuition. Fuzzy inarticulate intuition can make giant leaps in unpredictable directions by discerning or projecting the shape of large significant impressions. The inexpressible, almost nonconscious, intuition that they were little or nothing by their individual selves, that they owed everything to their community, may have been within their ken although not within their articulation. That utterly vague apprehension led them to seek communal solidarity on occasions of mysterious high import.

Language, even crude protolanguage, added a new dimension to their means of satisfying their human need for fearful or enraptured communion with their fellows. For animals lower than apes, the comfort and safety of communion with their fellows takes only a physical form, cheek-by-jowl closeness with the mass of the herd. For apes it is still mostly physical, but it includes the touch, embrace, and proximity of friends and family, those individuals they are emotionally close to, those with whom they can identify.

For the earliest humans, such communion was something more, decidedly more. It was something new to the old earth: shared ritual activity, a group whose members moved alike in actions such as the ceremonial dance around or near a fetish object felt to emanate or retain mysterious power. The fetish object, perhaps the skull of the community's one-time leader, would have been felt to contain, somehow, something like the strength of the group. To express their feelings about the absent power of that leader, about the remaining diminished power of the group, they learned to join together in rhythmic movement around the skull. First with slow timid fear or dread due to the absent power, ultimately with passionate exultation as their shared movements and gestures embodied their emboldening feeling that the mysterious power had returned to them. Such return of power was mostly magical gain for individuals; the rhythmic ritual group activity that marked its disappearance and its reappearance was more like religion.

It is by using their bodies that lower animals exhibit, quite literally, their power to make happen what they want to happen. Apes start to move beyond the literal with symbolizing gesture. To that limited gestural exhibition, often of power, humans added another level by using their entire bodies in gestures and movements that symbolize their feelings about power. When several people use such symbolic movements and gestures together, we might call it the body language of groups. A simpler name for it is dance. By moving expressively around an icon such as their former leader's skull, those early humans enacted the beginnings of dramatic ritual dance.

The high level of feeling during such activity resulted in utterance accompanying the dance. At a later time, revocalization of such utterance, by one person in the presence of others, brought some aspect of the ritual dance around the fetish to each mind. Such later utterances were expressions of communal feeling, valuable ones that helped maintain the cohesiveness of the community. They brought different images and ideas into the different minds hearing them, so there was less communication of information than there was communion with kindred creatures. After a while, such utterances, during or after the dance, became more precise about what external ritual things and processes they denoted. Along the way, ritual expression advanced communion beyond animal togetherness. It provided—for the first time—a way of being together with one's own group that was not just physical but mental as well.

Once it came into our lives, early speech, both fecund and seminal, often had its way with itself and gave birth to many marvels. Ritual communion was one of them. Communication, the intended transmission of information, was another. Yet another marvel was that language was a means of influencing other people, getting them to do what the speaker wanted done, getting it to happen without physical force. Language and its marvelous progeny provided much of the means by which early sapiens developed religion, magic, and the knowledge of the objective world that would someday grow into science.

Susanne Langer, the American philosopher, wrote, "Language makes every speaker a thinker. Words designating things carve out our objects. Its acquisition by an infant makes him a fully human being. It is language that orders and classifies our world. Languages make not only objects, but acts, and acts point to agents who are permanent entities. Nouns make things, verbs connect the things and refer them to the outer world, creating the entire new dimension of truth and falsehood."[10]

Wouldn't it be wonderful if spoken language left residues of itself? Wouldn't it be marvelous, even if speech left no more than droppings, coprolites of words whose meaning the listener did not fully digest? Wouldn't it be astonishing if the spoken word did not simply evaporate, but left crystallized condensations for us to find near the dry bones and stones from the past? No such luck, verbal language leaves no fossilized bits of sound, neither records nor recordings nor natural-born RAMs or CD-ROMs.

Without words from the past, even the evidence we have of activities of far-from-early humans—such as cave paintings and elaborate burials—allows us only to guess at the meaning of such activities. Cave drawings, those wonderful residues of ancient life that we have discovered, provide no clear evidence of just what they meant to their makers. Were the drawings a way of praying to their prey? We certainly don't know, nor is it likely that even their makers knew what they meant in any explicit sense. Were they the work of shamans, each in drug-assisted trance, who were overwhelmed with mental images, images which they copied onto the stone walls? In *The Roots of Thinking*, Maxine Sheets-Johnstone closes her discussion of the basic import of the drawings on the surfaces of paleolithic caves as follows: "At minimum, by penetrating insides and transforming their surfaces, our ancestors began exploring the mystery of enclosure, of contained space, of insideness, and brought the wonder of pictorial line, a wonder tied fundamentally not to a geometric eye but to a corporeal I, to an original prominence within the mainstream of hominid thinking and hominid evolution."[11] In *Origins Reconsidered*, Richard Leakey offers the opinion that many cave drawings are "images plucked from a mind in the state of hallucination, a sure sign of shamanistic art."[12]

Pictures, ancient or modern, even wonderful ones, although they may have a lot in the way of import and significance for us, carry no clear meaning. With no decipherable Rosetta Stones engraved with their ancient words, it is impossible to confirm what can at best be little more than enlightened speculations on meaning in the mental life of early sapiens.

MAGIC: THE NATURAL WAY TO GET DONE WHAT MAY NOT BE EASY OR POSSIBLE TO DO

> Glendower: I can call spirits from the vasty deep.
> Hotspur: Why, so can I, or so can any man;
> But will they come when you do call for them?
> Shakespeare's *King Henry IV, Part I*

We can inform our speculations about magic somewhat by doing some backward extrapolation from what is now known about recent primitive cultures throughout the world. One thing that is common to primitive societies studied by anthropologists is the belief in and use of magic. In *Leaps of Faith*, Nicholas Humphrey points out that, although tribal peoples are often steeped in magical thinking, they are also as knowledgeable as Western peoples about the nature of the ordinary objective world around them.[13]

The structure of magical belief shared by all recent primitive societies includes two components. First, humans live in a world where they are surrounded by powerful nonhuman agents. Second, ritual gesture and utterance can be used to invoke and influence such an agent.

Early Language, Magic, and Religion

The first of these results from what Langer has called the natural mode of thinking about events in the world.[14] In this mode of thought—not only our original one, but one that sometimes still holds us in its grasp—events in the world outside ourselves are not the result of lifeless objects acting on one another by indifferent cause and effect. Rather, events result from acts of living agents, agents motivated to achieve their own egocentric ends. Such agents need not be embodied in flesh, nor need they be mobile. They can be fetish objects like the bones of the dead, and natural objects like the sun and the storm into which were projected our own sense of power. Homo erectus, a creature with imagination but without more than rudimentary language, might have begun to populate the world with such imagined motivated agents.

Apes are narrow behaviorists when it comes to psyching out their world. They regard other creatures as bodies engaging in behavior, but not as minds engaged in feeling and thinking. Early sapiens, who projected living agents into immobile objects such as spooky stone monoliths or eerie burned-out tree trunks, were more than mere behaviorists. Those silent, still, magical objects were not engaged in any direct discernable behavior at all. To see such objects as powerful and as capable of influencing ordinary creatures is—implicitly—to comprehend that other minds exist and are powerful.

Projected motivated agents that have minds are only a part of the structure of magical belief. Another part is the human ability to influence such agents, primarily with language, although the use of gesture for the same purpose may have provided erectus with rudimentary magic. There is a real question here. Why should early humans have thought that spoken words—physically, no more than mere puffs of vibrated air—had the power to make nonhuman agents do what humans wanted?

Some people have answered this question by suggesting early humans, as well as modern primitives, were childish emotional folk with little understanding of inanimate cause-and-effect activity. This seems a poor description for creatures that were uniquely effective in the natural world, far more so than even their highly successful erectine predecessors. Others have suggested that the shaman-magician in both early and recent primitive societies was just a con-artist who used tricks to fool his credulous audience. This answer seems more bunk than debunking. Abe Lincoln's edict about not fooling all the people all the time holds here. Generation after generation of shamans could hardly have deliberately and successfully fooled generation after generation of audiences. The shaman was, if anything, a far firmer believer in magic than most other community members.

HOW WE GOT TO BE HUMAN

A professional believer, however, must learn to handle his occasional or frequent absence of inspiration when, for all that, the show must still go on. So he learns to insure inspiration, perhaps with drugs, and he learns ways to insure performance, perhaps by developing what we might call tricks and techniques not unlike those of our own day's stage magicians. These performance aids make the shaman neither a con-artist nor a mere entertainer. Rather, they make him a person who knows that he cannot rely on inspiration when he wishes to dramatize his beliefs.

A somewhat more detailed explanation for magic goes as follows. First, magic is not basically emotional, it is the expression of concepts. As such, it is symbolic and intellectual in nature. Next, magic amounts to an invisible doing. It is, in structure, remarkably like what goes on in our minds almost all the time, namely, mental activity. Mental activity—for the most part the internal unvoiced talking and imaging to ourselves that we're always doing—is unlike any and all other activity in the world. It can cause things to happen without seeming to use any physical force. Sometimes this mental activity results in the action of the silent thinker herself, when she comes to the realization that she wants to do something. At other times, the voiceless internal-talking results in the thinker convincingly vocalizing to others that such-and-such—the likes of bowing and curtsying in fear and trembling to local supernatural agents may have been high on the list—must be done.

Given the effectiveness of verbal mental activity in getting ourselves or others to do what is desired, it really is less than surprising that minds—minds, be they primitive or modern, minds that are considerably compartmentalized in their nature—minds would regard verbal magic as real and potentially effective in having its way with the world.

Lastly, magic is sometimes the reification of meaning, the reductive identification of powerful concepts and words with the real things and events represented by the words. Magic is commonly done in primitive societies by using a particular sequence of spoken words to cause events to happen to objects in the external world. To reify is to make real, that is, to misidentify the meaningful verbal symbol as the thing it symbolizes. This confusing reification is especially strong for names, particularly names of living natural or supernatural agents. In some societies this has led to the practice of keeping personal names secret so they can never be used to inflict harm on the named individual. In other societies, such reification of words has led to people being enjoined, sometimes under penalty of death, never to speak the name of their god. More broadly, it has contributed to the widespread belief in magic.

Note that a shaman's gestures and utterances often do not work magic directly. Rather, they are the beginning of an acting out of the desired doing, to be picked up and carried to completion by the (presumably as suggestible as most of us) nonhuman agent.

Such magical inducings at times may fail to influence the spiritlike agent who is invoked. For believers, when magic fails, it is either because the invocation was not correctly performed or because some more powerful agent interfered. So belief in magic does not require that magical spells reliably accomplish what is desired of them. Nor is magic easily abandoned, even when its use does not bring desired results. Why take the risk? Why abandon so powerful an instrument when its failure can be attributed to the shortcomings of its only human practitioner or, even better, to the intervention of a more powerful agent?

A recent nonhuman science fiction character frequently exclaims: "May the force be with you!" No, it wasn't Frodo, the hobbit in *The Lord of the Rings*, nor was it Odo, the shape-changer in *Star Trek: Deep Space Nine*. Yes, indeed, it was Obi Won Kenobi, the magus in *Star Wars*. His exhortation contains the following as subtext: First, "Recognize that the force just might not be with you this time." Second, "Never doubt that the force is real, that it exists!"

So magic existed in the past, when "mumbo-jumbo" could still "voodoo" us. In the present, even we science-worshipers in the West (who don't much like the word *magic*) often find that things work by the power of (magical) suggestion. Exactly how does it work now, how does it work for us? For that matter, how can it work now, if indeed it can, when we are such sophisticated creatures, creatures perhaps eager to experience an occasional cheap thrill with stage or movie magic, but basically creatures at home in the real, if too often mundane, world?

In *Stolen Lightning: The Social Theory of Magic*, Daniel O'Keefe provides a magisterial set of ideas.[15] Here is his two-step overall utterly condensed summary of how magic works:

1. Deviation via temporary *relaxation of the normal social frame* of orientation, and
2. overvaluation and patterning of the resulting experiences by *social agreement*.

Let's look at how these two steps replace conventional social more-or-less objective reality with the more subjective social reality of magic.

HOW WE GOT TO BE HUMAN

In *Culture and Enchantment*, Mark Schneider notes that when we encounter some weird event that makes no sense, we try to give it sense "by integrating it into a broader structure of phenomena that makes it intelligible."[16] Let's posit sudden bright lights in the sky that bring high anxiety. Having seen such lights, we could decide that we saw, not something like a helicopter, but a flying saucer. It is ironic that if we so succeed in making sense of what was initially in the realm of magical enchantment, we disenchant the event by bringing it into a social realm of the known, even if that realm is a loony magical one without reasonable basis in the natural world.

We normally consider ourselves as oriented to what we regard as objective reality, a reality where trees that fall in the forest—heard or unheard—are really there, a reality wherein our interaction with other creatures is equally solid. Furthermore, we regard that orientation as not requiring any special effort to maintain. We understand that some people sometimes have orientations that deviate from this norm; they are either crazy due to defective minds, or they are driven toward craziness by extraordinary stress. Thus, for example, a person stressed by a sensory-deprivation tank may hallucinate after a while. Of course, the peculiar states we enter nightly during deep sleep are not thought of as hallucinations, however, but rather as ordinary commonplace events that we call dreams.

There is evidence that it is extremely easy for us to depart from our presumed objective orientation. Deprived-of-sensory-input states less intense than sleep or those encountered while floating in dark silent tanks, states such as boredom, dullness due to a monotonous highway, or self-induced trance can also shake our attachment to reality. Likewise, intense mental states—perhaps the result of participating in emotional and active crowds or of listening to strongly charismatic speakers—can induce us to leave our realistic orientation.

So what happens when, however briefly, we leave reality? What happens when we get weird for a while? Often, that question simply cannot be answered. What happens is often not something that can be put into the words of normal discourse. For all we know, much of what happens may be quite different for each of us, although some of it (such as night terror experiences like "the Old Hag") may be common for many of us. These days, we tie all such happenings together by calling them altered states of consciousness.

After an episode of such oddness, however, each of those who has one usually tries or is encouraged to figure out what happened. The experiencer commonly figures it out by talking about it with an expert, talking with someone who really knows about that kind of stuff, such as a guru, a priest, a psychotherapist, a group-activity leader, or even a talk show host. The expert is in effect the magician. With the help of the expert, the experiencer decides both that something significant happened and that what happened was thus-and-such. Thus-and-such could be, for examples: Experiencing the mind floating out of the body and looking down upon it; a man seen climbing a rope he just threw into the sky; an extraterrestrial observed descending from a sky saucer; oneself joined at the heart at-joyous-long-last with every other creature in the universe; or (more down to earth) feeling oneself in a stunningly superb state of salutary health as a result of having ingested a certain substance.

This chapter is not the place, nor for that matter is this the book (O'Keefe's *Stolen Lightning* is one such book), to look into the countless flowers that have bloomed with modern magic. We can note, however, that "medical magic is one of the larger provinces of magic."[17] Medical magic deals with subjective experienced *illness* more than it does with objective measured *disease*. This illness-disease distinction is examined in some detail by Berry Beyerstein in "Why Bogus Therapies Seem to Work."[18]

We can also look at what might be conceived to be—or perhaps to become soon—as much a modern major manifestation of health magic as, for example, astrology, biorhythmics, energetic acupuncture, exorcism, fortune-telling, high-colonic enemas that flush out colonies of enigmatic lowlife, homeopathic dilutions, hypnotism that helps health, macrobiotic life extensions, qi gong, soothsaying, transcendental meditation, voodoo, or yoga. We also have what often works to a surprising extent for most of us, the magical medication that is not supposed to be medicine: placebos.

As perhaps one of such means of modern magic, let me suggest okra—yes, okra! Surely you've heard of it; if you haven't, you will now. Okra, in its natural form, is a tall plant with sticky green pods. It can grow almost anywhere, as we'll see in a moment. Okra could also be regarded, as we might not have learned recently, as a miracle plant.

If you're interested in what could be called a marvelous substance that might help you to maximize your entire self, one that may do wonders, wonders of course for your body, but also for your mind, and especially for your spirit, take okra. I mean take it into you, ingest it, chew it up, swallow it down.

Take any form of okra, but if you want the purest, the most concentrated, the very best, take deep-packed chopped okra. Normally, I advise against buying your okra chopped. It's a matter of the negatively charged ions; they leak out when okra is chopped. That's why you should buy it whole and organic, then chop for yourself only the amount you intend to ingest.

It could be that if you take okra, you'll soon want more. Be cautious, though, and take only eensy bits to begin with. You won't know how to describe its effects, so I'll tell you. It'll make your skin tingle, your mind sparkle, and your spirit sprout wings.

Be careful, however. If you take too much okra, especially if it's deep-packed chopped okra, not only will your skin turn green and your mind just up and leave your body, but your spirit will depart for some parallel universe. You are not ready for that now and you may never be ready, so err on the side of caution. On the other hand, you needn't go along with the homeopathic approach to okra. For the homeopaths, the stuff may seem so powerful that they'd sell water distilled from the urine of someone who once took okra.

Especially if you want an ageless mind within a timeless body, if you want perfect healing to give you quantum health, if you want to avoid the way of the wizened, if you want your spirit to do a turn with the magician Merlin, take a dab of deep-packed chopped okra. However, I should be candid at this point and caution you. Some doctors, perhaps even Deepak Chopra,[19] may not agree with me about deep-packed chopped okra's benefits.

While I'm leveling about Deepak Chopra, I'll also level with you about how they deep-pack the chopped okra. Which is to say, although I may have tried to find out how they do it, the fact of the matter is, I don't know diddly-squat about that. In fact, sometimes people tell me I've got the word wrong: They don't *deep-pack* the okra, they *de-pack* it, that is, they unpack it before chopping.

Of course, if you are a beginner, any kind of okra will work wonders for you. In fact, you needn't even buy it, since there's a way to get some free.

"Win Free Okra!" You might have seen the program by that name on the tee'n'vee. It's true enough, just like that gal says, you can win free okra. It seems to me, however, that the price is not right. In fact, it's high, way too high. You can't just let the ceremonial mistress—They call her Okra—hypnotize you and make a fool of you. You've gotta get up there and make a fool of yourself. But even that ain't enough. Only the biggest fool, the Fool of the Day, gets to win free okra.

So I say, buy your own okra right at the supermarket. Be sure it's fresh, be sure it's organic, be really sure it's grown with real manure (myself, I think the last is what it's all about), and be sure it's not chopped, so none of them negative ions leaks out.

Once you get into it, why not grow your own okra, your own body, mind, and soul food? Grow it the American way. All you need is a few bright lights, some hydroponic tanks, and a spare bedroom or a large closet. The gov'mint can't do diddly about it either. It's perfectly legal, at least it is right now. So get some and grow it—before the FDA gets downwind of it.

It won't surprise you to hear that okra has sprouted other wings in addition to its basic magical ones. There are two semireligious organizations, whose names sound somewhat alike, although they work opposite sides of the street. The *Okraholics*, they take all the okra they can get. But they can't get enough. That's why they're always working the airports and the malls, selling mantras with their greenish hands sticking out from those okra-green robes. Among the many, including me, who don't share the view that you can't take too much okra are the widely known members of *Okraholics Anonymous*, with their own heaven-sent seven-step program for those who fly too high with okra. But I can't agree with them either; they want you to kick the stuff altogether.

I say, don't walk down either side of the street. Don't traffic with either. Be a moderate. Walk down the middle of the road, just an inch or so off the ground.

Magical okra's transfiguration into religion has brought us many new faiths, of which I'll mention only the two largest. The first, as new as it claims to be old, is known as *AMORO*, the Ancient Mystical Order of Roseate Okristions. Its members know that—just like the green glasses they keep putting on and taking off—the greenish okra-laced wafers they take will turn their world rosy. The second, *The Okristic Institute*, stands by their highly controversial claim: He assumed his burden after his mother fed okra to little Jesus.

Enough, I hope, of magical deep-packed chopped okra, but I should not leave the subject without a disclaimer about my own—presumed, perhaps, by you—financial and economic connection with the stuff. I hereby solemnly swear that I have made no such attachments. In fact, I take okra in about the same way that president George Bush took broccoli: seldom. However, I cannot extend any promises to the future. Once you take okra, there's no telling where it may take you.

HIGHER RELIGION: IN PART, A WAY OUT WHEN HUMANS DISCOVERED FRIGHTFUL MORTALITY

For the earliest humans, even tiresome old age, not to speak of utterly ugly mortality, did not make the list of known causes of death. To die of old age is to die of physical weakness, to wear out one or more parts of the body. That is something early sapiens did not know much about. Once a few of them got into their forties and experienced old age, just under the surface of the notion of becoming increasingly weak and worn out lay the concept of mortality, that is, the concept of inescapable death.

The knowledge of that concept was something no human sought. Who would want to make that most awesome and awful of all human discoveries? Far worse than any plague, which at least one could seek to escape, was the death from which none could run. Who would not turn his head to avoid eye contact with the Grim Reaper who cuts down all humans?

For tens of millennia, the early humans were indifferent to the fruit of the biblical tree of Life, indifferent because it didn't occur to them that they were mortal. They were innocent. So no snitching serpent would have suggested the crime of snatching the fruit of the tree of knowledge of immortal life, the fruit that allowed escape from inevitable death. The possible consequences of the eventual violation of this innocence we'll see below. Initially for them, death meant unnatural death, the only kind that could be conceived in our early days, the kind that was commonly the result of the magical act of a supernatural or human agent. Consider this passage from Shakespeare's *Henry IV, Part I*:

> But thought's the slave of life, and life time's fool;
> And time, that takes survey of all the world,
> Must have a stop. O! I could prophesy,
> But that the earthy and cold hand of death
> Lies on my tongue.

Somehow, slowly, it must have been borne in on early sapiens that, for each of them, time would come to a stop. The earthy cold hand of death would squeeze that truth from their throats to their tongues, the truth that each, like Socrates, was slave to the syllogism of mortality.

As we still do today, early humans fought fiercely to deny themselves that knowledge of unavoidable death. For long eons they found all cessation of human life explainable not as inescapable death, nor even as accidental death, but rather as death inflicted by an agent, human or supernatural, and usually inflicted with the aid of magic. What could have persuaded them to change their views and admit that humans were mortal?

Perhaps it was in part the clear evidence that there was regular repeated annual death of all that each year grew green. Perhaps, as they increasingly mastered their environment, there were old folks around, people who had clearly declined from the peak of their strength and power, people who might be conceived of as on the road to further decline. Perhaps that led to no less than the truly horrifying notion that death was central to what we now call the human condition.

One thing that makes little sense to say about mortality is that it came to us as a Darwinian adaptation. As Stephen Jay Gould says, "The human brain must be bursting with [adaptation by-products] that are essential to human nature and vital to our self-understanding but that arose as nonadaptations, and are therefore outside the compass of evolutionary psychology, or any other ultra-Darwinian theory. The brain did not enlarge by natural selection so that we would be able to read or write. Even such an eminently functional and universal institution as religion arose largely as a [nonadaptive adaptation by-product] if we accept Freud's old and sensible argument that humans invented religious belief largely to accommodate the most terrifying fact that our large brains forced us to acknowledge: the inevitability of personal mortality. We can scarcely argue that the brain got large so that we would know we must die!"[20]

Once they got beyond the initially stupefying knowledge, humans started to finesse the issue of mortality. They sought means of seeming to accept it while still denying it.

With the objective part of their minds they managed to accept mortality, perhaps as an extension of the decline in physical strength from a youthful peak. With another mind-set, a subjective one, they found a way of denying its finality. By which I mean they found comfort in the idea of another life, a new life after death.

The same earth that, with winter, spoke in harsh tones of the mortality that none wished to hear, that earth whispered joyfully with the spring: "Hosanna, there's new life, come from the old!" Also helpful for this split-mind explanation was the firm intuition that there was something within the human body, something different from the rest of that body. That something—in my view, a something based on our apprehension that we are conscious, that we experience our lives, that we cannot localize our experiencing selves within our bodies—could be called spirit or soul. In turn, that intuition could be expanded to suggest that the inhabiting something—be its name ghost, soul, or spirit—something survived the death of the flesh.

HOW WE GOT TO BE HUMAN

However grudgingly or obliquely the idea of mortality was accepted, it started—or, more likely, speeded—large changes in how early sapiens' spiritual life was shaped. Meditate about it. The local low-level supernatural agents—initially, the only such agents—now could not extend your life indefinitely. Perhaps they could no longer take your life. Aren't your views about such agents apt to change? After all, agents with such limited power were hardly the powerful creatures they once seemed to be. (Note that even a deity so recent and so widely thought to be *the* singular supreme being as Jehovah, YHWH to some, was regarded 1,800 years ago by Gnostics as the Demiurge who stood at the very bottom of a totem pole of higher and higher supernatural beings, at the top of which was the Godhead.)

About early low-level deities, an answer gradually fallen into place here and there, ultimately almost everywhere. There was more to the supernatural world than the local fetishes, spirits of dead ancestors, rain gods, and their like. Supernatural agents with greater power were high up somewhere to control the greater complexities of life, even though such agents could be expected to be more remote and more difficult to influence. It was those higher, more distant and more powerful, deities who could give and take the lives of humans. Regrettably but understandably, human influence on such higher gods was small. Magicians and religious leaders increasingly shifted from attempts to directly influence local supernatural agents to bent-knee sacrifices that would please remote agents who might, in turn, respond positively to human pleas, who might grant a new life elsewhere. In *The Transcendental Temptation*, Paul Kurtz writes, "the quest for God has its strongest impetus in our fear of death and the unknown, and our desire to transcend it. Belief in God is rooted in the longing that God will save us from extinction and provide us with everlasting life."[21]

Save love, nothing comes free. Gods who were more powerful but less responsive, gods who were more removed but perhaps more indifferent, such gods were costly to their creators. The part of the price that goes by the name of sacrifice turned out, after a while, to be surprisingly inexpensive. Initially, sacrifice was of local humans, cast out by the community in troubled times to appease the gods. Soon, however, alternatives came to mind. Outsider humans would serve as well as local outcasts. Likewise, why not animal sacrifices? Also, the invisible high gods, spiritual creatures wholly or in part, would want to consume at most the aroma of the flesh thrown on the fire, leaving the cooked meat for mere mortals to ingest. As a more sophisticated Marie Antoinette might have said about our ancestors: "Their Gods let them sacrifice their cake and eat it, too."

Another part of the price of high gods cuts closer to the heart, as is suggested by Gerard Manley Hopkins' words, from "Thou Art Indeed Just, Lord":

> Thou art indeed just, Lord, if I contend
> With thee; but, sir, so what I plead is just.
> Why do sinners' ways prosper? and why must
> Disappointment all I endeavor end?
>
> Wert thou my enemy, O thou my friend,
> How wouldst thou worse, I wonder, than thou dost
> Defeat, thwart me?

NOTES

1. Walter Burkert, *Creation of the Sacred: Tracks of Biology in Early Religion* (Cambridge, Mass.: Harvard University Press, 1996), p. 5-7.

2. Daniel L. Pals, *Seven Theories of Religion* (New York: Oxford University Press, 1996), p. 275.

3. Ibid., p. 25.

4. David J. Hufford, *The Terror That Comes in the Night* (University of Pennsylvania Press, 1982).

5. Pals, *Seven Theories of Religion*, p. 100.

6. Robert Wright, *The Moral Animal* (New York: Pantheon Books, 1994), p. 146.

7. Harry J. Jerison, "Evolutionary Biology of Intelligence: The Nature of the Problem," in *Intelligence and Evolutionary Biology*, ed. Harry J. Jerison and Irene Jerison (Springer-Verlag, 1988), p. 8.

8. Steven Mithen, *The Prehistory of the Mind* (London: Thames and Hudson, 1996), p. 185.

9. Matt Ridle, *The Origins of Virtue* (New York: Viking Penguin, 1997), p. 227.

10. Susanne K. Langer, *Mind: An Essay on Human Feeling*, vol. 2 (Baltimore, Md.: Johns Hopkins University Press, 1972), p. 320.

11. Maxine Sheets-Johnstone, *The Roots of Thinking* (Temple University Press, 1990), p. 233.

12. Richard Leakey and Roger Lewin, *Origins Reconsidered* (New York: Doubleday, 1992), p. 329.

13. Nicholas Humphrey, *Leaps of Faith* (New York: HarperCollins Publishers, 1996), p. 56.

14. Susanne K. Langer, *Mind: An Essay on Human Feeling*, vol. 3 (Baltimore, Md.: The Johns Hopkins University Press, 1982), p. 5.

15. Daniel Lawrence O'Keefe, *Stolen Lightning* (New York: Vintage Books, 1983), p. 96.

16. Mark A. Schneider, *Culture and Enchantment* (Chicago, Ill.: University of Chicago Press, 1993), p. 44.

17. O'Keefe, *Stolen Lightning*, p. 2.

18. Barry L. Beyerstein, "Why Bogus Therapies Seem to Work," *Skeptical Inquirer*, September/October 1997.

19. Deepak Chopra, *Ageless Body, Timeless Mind* (New York: Harmony Books, 1993); *Quantum Healing* (New York: Bantam Books, 1989); *Perfect Health* (New York: Harmony Books, 1990); *The Way of the Wizard* (New York: Harmony Books, 1995); *The Return of Merlin* (New York: Harmony Books, 1995); *The Seven Spiritual Laws of Success* (Amber-Allen Publishing 1994).

20. Stephen Jay Gould, "Evolution: The Pleasures of Pluralism," *The New York Review of Books*, 26 June 1997, p. 52.

21. Paul Kurtz, *The Transcendental Temptation* (Amherst, N.Y.: Prometheus Books, 1991), p. 402.

12

THE LAST BIG BIOLOGICAL STEP, COMPLETE LANGUAGE

HERE ARE A FEW SWEET BITTER words, about words, from "On No Work of Words" by Dylan Thomas:

> On no work of words now for three lean months in the bloody
> Belly of the rich year and the big purse of my body
> I bitterly take to task my poverty and craft:
>
> To take to give is all, return what is hungrily given
> Puffing the pounds of manna up through the dew to heaven,
> The lovely gift of the gab bangs back on a blind shaft.*

In this chapter, I will suggest that the original gift of the gab—complete spoken language—came to humankind from the brains, minds, and mouths of *children*. Further, it was a *cultural* gift, one that came generations after a preceding nonadaptive *genetic* gift of potential language. Likewise, it came decades before there were enough complete-language speakers to make it a genetically *adaptive* gift for new humans who had it.

We'll look at this idea that the last big biological step for sapiens was a mutation or a number of them, that added to the brain, not complete language itself, but the potential for such complete language. Furthermore, a

---

*From "On No Work of Words," *The Collected Poems of Dylan Thomas* (New York: New Directions Publishing Corp., 1957), p. 104. Reprinted by permission of the publisher.

small community of young children—rather than older ones or adults—brought that potential into existence as complete spoken language, where it soon replaced the crude ambiguous protolanguage that preceded it. We'll also entertain the possibility that we humans acquired complete language long after we first came to exist as Homo sapiens sapiens. Likewise, we'll look at the likelihood that the language mutation was not *initially* adaptive, and did not become adaptive until, perhaps generations later, complete language was created. Before we get to all that, we should look at language itself.

There are countless questions that can be asked about complete spoken language. To consider some of them, let's first sort them into the five basic questions that the gift of the gab has given us; for English-speakers, they are what, who, where, when, why.

What? What is this thing called language? Who? Who, ignoring supernatural possibilities, brought it into being? Where? Where on Earth did it come into existence? When? When could we have done such a thing? Why? Why, from a Darwinian viewpoint, did whoever, wherever, whenever, do whatever language is?

WHAT IS THIS THING CALLED LANGUAGE? SOME INDIRECT ANSWERS ARE INTRIGUING

What is language? Whatever it is, it is the same thing whether used by the educated, the uneducated, or the ignorant, whether used by sophisticated societies or primitive hunter-gatherer groups. It is worth noting that increases in language, other than specialists' jargon, commonly bubble up from the bottom with vigorous vulgar expression and usage. This is in contrast with wealth additions, which may sometimes trickle down from the overflowing accumulations of the rich and powerful. Language is also the selfsame thing, allowing the same broad range of expression, whether it is spoken and heard by those with normal vocal and hearing ability, or it is signed and seen by those using a complete sign language.

In the opinion of experts, it is a great improvement over what preceded it, which is called protolanguage. When protolanguage is born again in modern times, it is called pidgin language, or pidgin sign language. These incomplete languages are primitive in their structure, and limited in what they can convey without ambiguity. In particular, they are more shy of syntax than bald men are of head hair; they both may have a bit, but not at the center, where it's important. Apart from their presumed long-ago use, protolanguages are used by children between ages of about one to two years. Also, users of protolanguage are those few people whose youthful years did not provide them with others who used suitable language in their presence. One such person, described by Derek Bickerton in *Language and Species*, was a young woman called "Genie."[1] Isolated as a young child, Genie was unable to learn complete language when she finally got an opportunity to do so at the age of thirteen.

Protolanguages called pidgins are also used by groups of ordinary adults who must cooperate daily for many years with other adult groups who speak one or more different tongues. There are also some scientists who believe that attempts to endow apes with nonverbal language have enabled the latter to learn a crude form of protolanguage. Protolanguage, as suggested in chapter 9, was the only lingo available to species before our own. Even our cousin neanderthals may not have had complete language. Perhaps more speculative is the likelihood that prototalk was the only tongue available to us sapiens sapiens during the first few tens of thousands of years of our tenure on earth.

The question "What is language?" can also be an inquiry about whether language is the result of a specific genetic-originated language "organ" in the brain, or whether it is just one of the many things we are able to learn and use with the general nonspecific capabilities of the human brain and mind. Many experts in this area now agree with Noam Chomsky's theory that we have a specific nonconscious capability in our brains that is shaped to provide the general structure of language. In Steven Pinker's words, language is not a "cultural artifact that we learn the way we learn to tell time. . . . Instead, it is a distinct piece of the biological makeup of our brains."[2] Likewise, Daniel Dennett writes about the claim that we have an innate language organ: "We can now be sure that the truth lies much closer to Chomsky's end of the table than to that of his opponents."[3]

Before Chomsky's arrival on the language scene, linguistics itself was "distinctly behaviorist, psychology without the mind."[4] After Chomsky, psychologists saw linguistics as part of the new field of cognitive psychology, "oriented around the systems of knowledge behind human behavior." (Unfortunately, cognitive psychology ignores the mind as a subjective entity that is based on consciousness.)

The case for such an innate potential competence for language is indirect. No one has demonstrated it to be a specific organization of a part of the human brain. Thus, it is not surprising that it is also a theory that some, such as Terrence Deacon, do not agree with. In *The Symbolic Species: The Coevolution of Language and the Brain*, Deacon denies that we have a specific brain organ for complete language. Instead, he hypothesizes that human symbol-using capability and language coevolved.[5]

Strong evidence for specific genetic language capability is seen in the effortless, speedy, spontaneous, and unthinking acquisition of language by two-to-four-year-old children. These same children find it difficult to learn other things that are far simpler for older children and adults to learn than is the task of acquiring another language. Another kind of evidence for what used to be called "nativist" linguistic capability is found in the rate at which we can identify phonetic segments of sound compared to the speed with which we can identify nonspeech sounds, the former from fifteen to twenty-five per second, the latter at no more than seven to nine per second. This higher rate of identification suggests we have some special equipment in our heads that enables it. In *Uniquely Human*, Philip Lieberman says that this ability to acquire language phonetic segments more rapidly than non-phonetic segments is a stronger claim for a genetic speech organ than the claim that such an organ is required to provide us with the basis of syntax.[6]

The human larynx, tongue, and lips are also the results of genetic changes. They are needed to produce language as we speak it, but are hardly sufficient for that task. In addition to these physiological adaptations, humans are almost the only creatures for whom control of their utterances is an option. Chimpanzees, with an effort, can limit the intensity of their food "barks" so that instead of strongly uttering them, they are sometimes able merely to mutter them. We saw that the chimp called Figan was once given some bananas just after the top males had left the scene. Fearing to alert them to his good luck, Figan almost gagged with the effort to turn down the volume of his involuntary utterances.

The "what" of language can also be meant to inquire as to just what language is to us. Is it a means of communication with others? Is it a system for representing things to ourselves? One recent book on language and the evolution of the human mind (John McCrone's *The Ape That Spoke*) suggests that language as such is the center of our human nature, and that until we had language we lived solely in the present, like other animals.[7] Can language be that central to us? I find a more useful view of language in Merlin Donald's *Origins of the Modern Mind*. He suggests language is not so much a thing in itself as a part of a long development of abstract thought and symbols.[8]

The Last Big Biological Step, Complete Language

It would be pretty hard to find an argument against language as communication, as a way of trying to get others to know what we have on our minds, and as a way for us to understand what others are blabbing about. Likewise, language is clearly a system of representing things to ourselves—we are often engaged in mental monologues, even when we don't find them engaging. However, language is not good at handling spacial information, nor has it much competence at conveying knowledge about emotions or sensations. Jean Aitchison, in *The Seeds of Speech*, records how difficult it is to convey with words something fairly simple to do, like the procedure for making a knot.[9] Aitchison also notes that there is no language that attempts to describe pain, except "bad" language (which, as we all know, conveys very little information). Also, communication with language is sometimes only rhetoric, part of a means to influence others, where the language users often have axes to be sharpened, and occasionally, oxen to be gored.

As for languageless humans living, like other animals, solely in the present, and language thus being central to our human nature, what is to be said? Language largely represents concepts; it also denotes perceived objects and processes. With relative ease, it allows us to communicate concepts to others and to represent concepts to ourselves. When language was nascent, without concepts it would have been a lot less than it now is. Then, were language and conceptual ability Siamese twins, joined at the brain, when they were born? That is clearly not the case, we have only to look at the deaf without sign language, and at Brother John, who was without language during his frequent fits—both were discussed in chapter 9—to see that humans without language have conceptual abilities that let them live human lives in which they conceive of many of the things the rest of us think about.

Language is also a basis for *meaning*. In *Culture and Enchantment*, Mark Schneider tells us that there are four different types of meaning: symptomatic, experiential, conventional, and significance. Of these, normally, only "conventional meanings . . . are possessed by language [and also] by nonverbal gestures."[10]

Another way of conceiving the "what" of language is as a secondary input and output system. As input, it augments our natural-born senses that receive information about what is outside us. As output, language emulates the muscles which allow us to embrace, confront, or escape elements of the world around us. This secondary input-output system, however, functions exclusively with the parts of the outside world that consist of creatures of our own kind. Neither sticks and stones, nor plants, nor nonhuman animals (when they are not under our thumbs), have so much as a word to say to us. Likewise, all animals (except those that are our possessions in zoos, labs, and homes) are indifferent to our attempts to muscle them with the gift of the gab. We have no one to talk with but ourselves.

Even supernatural entities, such as they are or aren't, seem at best ambiguous on this score. In the West, many of us humans still credit the word of God, and words themselves to God. In fact, Genesis tells us that God did not merely use words, he used them to make the world, he spoke everything into being. However, in recent millennia, he is also reputed to have the capacity directly both to comprehend us and to direct us, spirit to spirit, with no need of mundane intervening linguistic media.

WHO BROUGHT US THE GIFT OF THE GAB? THERE'S REASON TO BELIEVE IT WAS CHILDREN

Who first spoke complete language? If its arrival was fairly abrupt, rather than a gradual development, there had to be a beginning. So it seems fair to ask who were the first who spoke, spoke not in simple-minded, mainly nouns-and-verbs protolanguage, but in complete language? The actual individuals who mastered the deed will forever be unknown soldiers in the struggle (that each of us presumably makes) to get a Darwinian edge over other "critters."

Though they are lost to us as individuals, generically it is likely that the first complete-language speakers were of the sapiens sapiens subspecies. Even our sapiens neanderthalensis cousins are thought not to have had the larynx to produce more than a small subset of the sounds we of the sapiens sapiens subspecies can make. Also, neanderthals were both stronger and better adapted to the cold environment where we mainly confronted them. So complete language might have provided us a likely edge and ledge on which we could stand taller than could neanderthals.

In his book about our language instinct, Steven Pinker tells us about the development of sign language in Nicaragua.[11] Prior to 1979, when the first schools for deaf children were created, those children were largely isolated from one another and had no sign language, not even a pidgin. Those who ran these schools were normal users of spoken language, so "naturally" they tried to teach their young deaf students to read lips and speak "normally." Those efforts failed. In the course of playing together and sharing bus rides to and from school, however, the deaf children, no longer isolated from their own kind, quickly created a proto-sign language. The older children, those over ten or so, continued to use only this protolingo as they grew up and became adults. However, the youngest ones (who came to the school at about the age of four), and their equally young successors over the next few years, made something more from this crude proto-sign language. They made a complete (sign) language. They did it without the aid of the older deaf children, without the help of deaf adults, and without the help of the speaking and hearing community. In fact, the deaf older children and adults were unable to learn the new complete lingo of the very young.

Terrence Deacon has beliefs about language that are at odds with the history of these originally languageless deaf Nicaraguan young children.[12] He tells us that language *evolved* over time. That means that language did not merely *change*, it developed. He further suggests that language evolved (with no specific genetic language organ) in such a way as to make it *easier* for young children to acquire it. Yet successive groups of deaf Nicaraguan young children (with no past spoken or sign language to build on) made first a protolanguage, and then a complete (sign) language.

The same creation of a complete language from a protolanguage is described by linguist Derek Bickerton's book, which discusses pidgin and creole languages that have been created in places like Hawaii, Jamaica, and Haiti, where adult and near-adult slaves and wage slaves were brought together from several different language-using societies.[13] In each case, a new protolanguage, called a pidgin, was developed by adults speaking different tongues who had to work with each other. Their young children, in turn, expanded the limited pidgin to a new complete language, called a creole, that the parents never learned to use.

Jean Aitchison claims that pidgins are unlike the earliest language or languages because they are based on one or more existing languages.[14] This claim may be true about spoken pidgins, but only trivially so. Its significance is removed by the achievement of Nicaraguan deaf young children: creating a pidgin sign language without exposure to a prior language.

Recent research suggests that all young children may learn new languages more quickly than adults do, and the reason for this is that they use a different area of the brain than the adults use. A 1997 article in *Nature* reports that MRI imaging tests found that children process all their languages in one small part of the brain, while adults who learn a new language are forced to create a new storage area.

HOW WE GOT TO BE HUMAN

John McCrone's book, *The Myth of Irrationality*, suggests a novel reason—a reason other than a genetic language instinct—for the fact that young children learn language quickly and easily whereas older people do not.[15] During the first few years of life, the process of myelinization (the sheathing of neural pathways with an insulating protein called myelin) has not started, thus the "wiring of [a young child's] brain is still plastic." Whatever the validity of this theory, it would suggest that young children are very good at *learning* whatever passes for language among those close to the child, but it would not explain the *creation* of a (complete language) creole by the children of pidgin-speakers.

So, then, what was the likely scenario when a few of our long-ago ancestors created the first complete language? There must have been a particular society of protolanguage users wherein complete language first emerged. Let's assume that a single mutation occurred—although it may well have taken several to converge—and was manifested in an infant who had all of the language instinct. Yet that child had no one to speak with who spoke other than the prevailing protolanguage. Thus, this child, during the significant years between, say, two and ten, added nothing, or at most some slight improvements, to the protolanguage of the community.

This suggests that, in this scenario, the language mutation was not *initially* adaptive in the normal Darwinian sense of contributing significantly to the survival of the individuals who had it. However, the genetic language *capability* would not have been strongly maladaptive to its bearer (if it were, it would not have survived); it might have been of minor negative or positive adaptive value to its unknowing holder. Of course, if that bearer was barren of issue, then the gift of gab would have been lost forever, or until it popped up somewhere again. Even if that tongue-tied gift-holder had several children, and even if they all inherited the language mutation, chances are that a spacing of several years between each of the children that survived life's early adversities would have kept language from flowering in that first family with the gift for (but not of) complete gab.

It might have taken generations or centuries before there came together a critical mass of young children—all with the language instinct—who had the opportunity to spend enough time together in whatever passed for playgrounds to create the first language. Their parents would no doubt have regarded their children's talk as valueless gobbledygook, child's play. Only when that fabled few grew up and began to demonstrate to themselves the power of a complete language, only when some of *their* children, those with the mutation, in turn began to speak, only when such creatures began to make a brave new world that all (even those without the new talent) could see, only then would complete language have taken sturdy root in the human stock. Then, new humans with the language instinct would have found that innate capability to be *adaptive*, as it enabled them to learn complete language from others.

Though there must have been one group of children who created the first language, does that mean that all subsequent languages grew from that first tongue? Walter Burkert (in *Creation of the Sacred: Tracks of Biology in Early Religions*) thinks that was the case. Language "has never—in tens of thousands of years—been reinvented."[16]

The facts of the matter may never be known. Nonetheless, I think it's likely that, long after that ancient first flowering of language, some later genes for the language instinct went dormant for decades before promoting the growth of new languages that were *independent* of predecessor tongues. We'll see how that might have happened when we look at the "where" and the "when" of complete language.

WHERE ON EARTH DID COMPLETE LANGUAGE COME INTO EXISTENCE? IN MANY PLACES, PROBABLY

Where did complete language first come to be? There is next to nothing on which to base an informed opinion of the "where" of the original language creation. In principle, some might think it possible, or at least conceivable, to seek commonality and difference in existing languages, so as to sort them into language families. Then, more work, inevitably tinged with more speculation, could result in combining and separating the families into superfamilies. The locations where the component languages are spoken provide only an increasingly broad territory. From what seem to be concepts *common* to members of each superfamily, it might be possible to get clues to less broad geographic territories suitable to each superfamily. You know, like the concept of "reindeer" just might suggest a cold climate. Then, perhaps, families of superfamilies might be found eventually to converge on a "first tongue," conceivably with clues to some first location.

In fact, there have been attempts by hardy explorers of our linguistic past to arrive at a handful of giant language superclusters. These efforts have not resulted in any suggestion of the location of the first speakers of a complete language. One thing that suggests that such explorations may be more factitious that factual is a belief shared by many linguists that a language cannot be traced back more than roughly ten thousand years. Over that interval, all threads of the original fabric are thought to have been replaced or so twisted as to defy recognition.

In a second method of locating language's birthsite, a strong connection between language and humankind's more lasting cultural developments is suggested. European archeological digs and explorations have yielded the seemingly sudden appearance of new things in the form of cave paintings, stone figurines, bladed tools, and other human-made artifacts, all dating back about thirty-nine thousand years. The argument is that the sudden appearance of new kinds of things must be tied to the sudden appearance of new mental capabilities, and that the largest, if not the only, thing fitting the latter description is complete language. James Shreeve, in *The Neandertal Enigma*, notes that many influential scientists who consider the profusion of new kinds of objects found in the record from the Upper Paleolithic believe them the result of new fully human language skills.[17]

There is some face value to this idea of a connection between the spoken word and the hand-crafted artifact or artwork. But the argument has a couple of weak points. First, similar artifacts, made tens of thousands of years earlier, may have since decayed into dust or may remain unfound to this day. More important, for all the doubtless glories of language, it is largely of less than primary value in the creation of things that now go by the name of art. Those who make moving images of things, real or imagined, that are of high import to them don't seem to need a lot of language as part of the doing. The conductor of an orchestra, for example, uses a few words and a lot of gestures to convey his concepts to the musicians. Thus it may be that our ignorance—of the nature of improvements in the human mind over our long past—allows us to tie together what is only remotely connected.

There is another somewhat thin bridge of reason that can connect language and location on earth. However, as it does not connect the very first language with a place, it will not be described here, but below, under the "when" of language. It will suggest the possibility of a connection between subsequent rebirths of language and a couple of places; the time is some fifty thousand years ago, and the places are Australia and New Guinea.

WHEN WE MIGHT HAVE BROUGHT LANGUAGE INTO BEING

When did we first make complete language? The most obvious choice for "when" was when we ourselves, Homo sapiens sapiens, came to live on earth some 140,000 to 290,000 years ago. It has a nice ring to it. Like Athena who sprang whole from the brow of Zeus—herself complete, with no need for improvements or revisions—we came to life complete with all our present genetic capabilities, including complete language. The idea is in a way like one we cherished before archeology showed us small-brained upright apes as the "missing links" between apes and humans. Back then, we expected to find a predecessor that, although it stooped like an ape, had a large brain, "our" kind of brain. So now we are inclined to think that the African "Eve," who presumably was the partial mother (partial because she was the first with our kind of mtDNA and not with all our DNA) to present humankind, but not only our physiology and large brain, had also had within her brain all the genetic capability we have in our brains. Thus, the language instinct was inherent to humankind after the birth of the Eve of Africa.

The simplicity of the idea, plus the trick of language that suggests two creatures with the same name, ancient mtDNA Eve and biblical Eve, are similarly complete humans, tends to seduce us into accepting it. But seduction is presumably more vice than virtue. So (even though all such looks are more speculation than we'd like) let's look closely at the time when the language instinct untied our tongues.

Pinker suggests that somewhat scientific speculations (that attempt to lump clusters of supergroups of language groups) propose that there are six superlumps of languages.[18] One of the latter includes all tongues in Eurasia, northern Africa, and all of the Americas. Two more superclumps are found in sub-Saharan Africa. A fourth is in Southeast Asia and the Indian and Pacific Ocean lands. The fifth and sixth are New Guinean and Australian languages.

These six superlumps, of course, may not be real. They may also be not only real but divisions of an all-Earth lump that started somewhere and ended everywhere. In any case, they are really worth considering, especially since I can suggest the means whereby language could start up afresh, independent of then-existing tongues, in one or more places.

One such means for language to start anew, elsewhere than its first locale, is obvious, though how unlikely we can't say. It says that the language mutation came to human life on earth more than once. Its appearances subsequent to its first were in lands where the first such mutation and its embodiment as spoken language had not yet spread.

There is another way that multiple independent flowerings of language could have come to exist. Remember the stories of Gypsies who stole babies? (As their name suggests, the Gypsies were originally thought to have come from Egypt. Gerry Altmann, in *The Ascent of Babel*, tells us that analysis of Romany, the Gypsie language, has since shown India to be its source.[19])

What if (long before the Romany-speaking Gypsies) some ancient people—themselves protolanguage speakers without the language inheritance—had stolen a baby from people in a society that had both the language instinct and an actual complete language? What if the thieves (both unspeakable and unable to speak as we do) had hightailed it for far away with their stolen babe—where neither the language instinct nor its realization as complete language existed? All unknowing, they might have recreated the same circumstance wherein the first language came into existence. With the same good fortune as the original group, they, too, within decades or more, would have initiated a new complete language produced by children.

Nor do we need a melodramatic story about stolen babies to instantiate this sequence of events. Move the venue to Australia, to an Australia populated, not with Gypsies, but with Homo sapiens sapiens who lacked the language mutation (and, of course, lacked complete language). This time, the baby comes to them with its parents and others in one or more boats that crossed the sixty-or-more-mile watery gap between New Guinea and Australia. The scenario now calls for the aboriginal Australians—who themselves came by boat earlier—not to steal the baby, but to save it. Save the baby, that is, after the adults and older children among the new arrivals are killed by those same old-timers. That sounds more like us than does baby-stealing, doesn't it? As does another way of getting the new genetic material: don't take one or more babies; rather, take some females and get the babies next year.

Finally now, after all this meandering, we can wonder when this event that initiated new language—getting (languageless) babies with the innate language capability—took place. Australia is thought to have been first populated about fifty thousand years ago. If those first immigrants had complete language, that language would, in some likelihood, resemble the tongues then spoken in nearby New Guinea. Then the later immigrants would have used a language related to that of the earlier immigrants. Thus, the arrival of the baby, linguistically speaking, would have been a nonevent. But if the superclump of

Australian languages is unrelated to that of New Guinean tongues, the infant might have brought both the language mutation and the possibility of complete language to an Australia mired in the twisted tongues of protolanguage. It is evident that a similar argument can be made about recreation of a new first language in New Guinea and other locations at other times in our prehistory, long after our arrival as a new kind of creature in Africa.

The final point is that humans in Australia would not have been possessors of the language instinct when they first arrived fifty or more thousand years ago. (It's worth noting that "new dating evidence . . . suggest that humans first arrived [in Australia] 60,000 years ago."[20]) Whatever the arrival date, there may be a long gap of ninety thousand or more years between African Eve and our putative Australian baby with the language instinct. If that interval is unreasonably long for Eve's descendants to have radiated— radiated with complete language—to Australia (which I think is the case), then complete language did not come to humankind until tens of thousands of years after Eve came to life in Africa.

WHY, FROM A DARWINIAN VIEWPOINT, WE MAY HAVE CHANCED ON LANGUAGE

For the function of language, here is a recent suggestion (from Susan Blackmore's *The Meme Machine*): First, "Imitation evolved, something like two and a half or three million years ago."[21] Second, imitation brought to life the "meme," which is any element of culture that is passed on by imitation. Third, "the function of language is to spread memes." I interpret this as suggesting that when brains and minds of social creatures got large enough to develop culture, understanding cultural aspects of the social world became important (in addition to understanding aspects of the natural world). And language helped with understanding the cultural social world. I must confess that, for me, adding the meme to this less-than-exciting explanation adds nothing and subtracts clarity.

In my view, it is not memes but more meaning that is the heart of complete language. How do I know this? I could have learned it from books, but I may have learned it some years ago from a practitioner of some distinctive language, a Los Angeles "valley girl." I was walking along a busy San Fernando Valley street and talking to myself about language, talking out loud. Human memory is recreative, as you know, so here's how it might have gone:

So, I still have to ask: Why did we humans make complete language?

"Like, you know, so we could answer questions. Like, like that one, guy."

Hold on! Where did you come from, young lady? Do you alsways sneak up on people like that? And is that answer anything more than stupid or obvious? What is it supposed to mean?

"There, see, like you demo'd it."

Demonstrated what? Can't you at least speak more clearly, if not more grammatically?

"Like, talk's gotta mean stuff."

Stuff? Are you trying to be profound as well as obvious?

"No way, dude. Like, there's like no meaning to speak of without lingo."

That's what linguists and philosophers say: it's mostly language that makes meaning; you think I didn't know that? What's that got to do with why we made language?

"Chill, dude. Cool down and like I'll walk ya through it."

Is this conversation really worth continuing, young woman? Is it really?

"Check this out. Like, once talking started, everyone dug it. Like, it meant something."

You're being deliberately maddening, aren't you? Isn't that so?

"Your major dude, Darwin, that's like what it's all about."

Wait a minute—are you trying to say that complete language has adaptive value, survival value for us, because we make more meaning with it?

"Crazy."

So if the question "Why did we make complete language?" has a Darwinian answer, it's that meaning has survival value by improving communication and other conceptual interaction, and we can make meaning mainly with language?

"Crazy! You got it. I'm outa here."

Wait, what about deaf people without sign language? Or even apes? Are you trying to say there is no meaning, no significance, no import in their lives?

"Meaning, no way. Well, not a lot anyway. Significance and import, like way, okay. Later, dude."

Wait a minute, isn't it the same with music and art—significance and import, but not specific meaning? I knew all this before this conversation started, didn't I? Where'd she go? Who was that girl?

NOTES

1. Derek Bickerton, *Language and Species* (Chicago, Ill.: University of Chicago Press, 1990), p. 115.

2. Steven Pinker, *The Language Instinct* (New York: William Morrow and Company, 1994), p. 18.

3. Daniel C. Dennett, *Darwin's Dangerous Idea* (New York: Simon & Schuster, 1996), p. 388.

4. Randy Allen Harris, *The Linguistic Wars* (New York: Oxford University Press, 1993), p. 55.

5. Terrence W. Deacon, *The Symbolic Species: The Coevolution of Language and the Brain* (New York: Norton, 1997), p. 328.

6. Philip Lieberman, *Uniquely Human: The Evolution of Speech, Thought, and Selfless Behavior* (Cambridge, Mass.: Harvard University Press, 1991), p. 37.

7. John McCrone, *The Ape That Spoke: Language and the Evolution of the Human Mind* (New York: William Morrow and Company, 1991), p. 120.

8. Merlin Donald, *Origins of the Modern Mind: Three Stages in the Evolution of Culture and Cognition* (Cambridge, Mass.: Harvard University Press, 1991), pp. 215–16.

9. Jean Aitchison, *The Seeds of Speech* (New York: Cambridge University Press, 1996), p. 18.

10. Mark A. Schneider, *Culture and Enchantment* (Chicago, Ill.: University of Chicago Press, 1993), pp. 42–43.

11. Pinker, *The Language Instinct*, p. 36.

12. Deacon, *The Symbolic Species*, p. 102.

13. Bickerton, *Language and Species*, p. 118.

14. Aitchison, *The Seeds of Speech*, p. 11.

15. John McCrone, *The Myth of Irrationality* (New York: Carroll & Graf Publishers, 1993), p. 117.

16. Walter Burkert, *Creation of the Sacred* (Cambridge, Mass.: Harvard University Press, 1996), p. 18.

17. James Shreeve, *The Neandertal Enigma* (New York: William Morrow and Company, 1995), p. 272.

18. Pinker, *The Language Instinct*, p. 254.

19. Gerry T. M. Altmann, *The Ascent of Babel* (New York: Oxford University Press, 1997), p. 228.

20. Donald Johanson and Blake Edgar, *From Lucy to Language* (New York: Simon & Schuster Editions, 1996), p. 49.

21. Susan Blackmore, *The Meme Machine* (New York: Oxford University Press, 1999), pp. 93, 107.

13

SAPIENS SOCIETIES DEVELOP THE QUEST FOR MORE

THE LONG AGES OF GATHERING, SCAVENGING, AND HUNTING, WHEN OUR NUMBERS GREW SLOWLY

WOLF PACKS HAVE INDIVIDUAL LEADERS, as do chimp societies. For communal species, leaders serve a useful function (but not one with origins in the adaptive genetic endowment of the individual leaders). Leaders provide structure and they reduce contention. Animal leaders are always more fearless than their followers. Does this mean that leaders achieve power because they have less fear? Surprisingly, the answer to that question is "no." It is the other way around. After a creature becomes leader, it becomes freer from fear. This is because more freedom from fear comes from the higher self-dosage of rewarding internally manufactured drugs, of hormones like testosterone. "But surely in normal males the amount of testosterone increases at the outset of an aggressive episode?" asks Marvin Harris in *Our Kind*.[1] His answer is, "Not so! Typically, the amount of testosterone is highest at the end of an aggressive encounter. . . . Monkeys that fight to form dominance hierarchies have higher testosterone levels after they become successful, not before."

After leadership is achieved, frequent successful exercise of power makes these attitude-changing substances continue to be generated in the leader's body. As Robert Wright remarks in *The Moral Animal*, "Yes, there is a correlation between serotonin (a hormone, like all neurotransmitters) and social status. . . . Serotonin level, though a 'biological' thing, is largely a product of the social environment. . . . It's nature's way of equipping [creatures] for leadership once they've gotten there."[2] Both individuals and group may be better served by a relatively infrequent change in leadership; hormones are the source of much of that stability.

There is also a price that is usually paid by leaders. The self-assertive urge that is present in all of us egocentric creatures is somewhat suppressed by the social mindsets of followers. Leaders are freer to assert themselves. This frequently puts high ranked animals into stressful situations. Both the internal stress and the external risk can contribute to damage to the leaders' bodies. They may live a briefer life than do their followers.

Mainly, what animal followers do in the presence of leaders—especially when the latter are aggressive—is grovel, that is, make themselves small. Sometimes they make themselves kissy-faces, at other times uptight anxiety dictates a brief loosening of their rear-end sphincters. Followers, as well as leaders, intuitively understand power. As well they might, for power is both coin and big bucks in the realm of social egocentricity. If you've got it, flaunt it, use it to get whatever you want, and to keep the next creature from getting what you don't want it to have. If you don't have it, curry favor and get near the one who has it; you may feel a bit of reflected power and you will feel safer.

Dogs and wolves crawl on their bellies to get close to the leader. When two canines fight, the loser acts with similar obsequious behavior toward the winner. This sucking-up to power is done partly out of fear and partly out of desire for the comfort of closeness to a superior. We saw in an earlier chapter that after an aggressive encounter between two male chimpanzees, their behavior is more complex than that of lower animals. The loser wants to get close to the winner. Sometimes he tries to avoid playing by the chimp rules. Those rules call for the proper procedure of showing: "Thou art great, great a lot. And I am not, not-not-not." The loser would like to receive the winner's embrace but he is reluctant to pay his dues by making himself small while the winner puffs himself up into as high and mighty a posture as he can reach. The winner demands his price, "No obeisance, no absolution," and he usually gets his way.

Regarding the earliest humans, there is no real evidence about leaders and followers. There is some indirect evidence, from not-very-early twenty- or thirty-thousand-year-old burials showing valuables buried with the body. These things may be perks or payola showing that the sapiens creature who died was a leader or perhaps a member of a leader's family. Evidence present or absent, there were always leaders and followers among the social creatures called sapiens. As with lower animals, the leaders were doubtless more courageous, thanks in part to their bodies' pharmacopeia.

The earliest sapiens were gatherers, scavengers, and hunters of food. With imagination, conceptual ability, and language (probably limited language, but still superior to that of their predecessors), they began to do what we still find difficult—to plan ahead. With greater ability to project themselves into and to identify with others with whom they were close, they could share what they had. The leaders led the search for carrion and prey, and they led the struggle against predators. Both when times were good and when they were bad, they probably got more than an equal portion of grub. It is likely that most members of a sapiens group led but a brief life of twenty to thirty or rarely forty years. A leader's life may have been even briefer. In their conception, death came mainly from malign magic. In our view, they died on the hunt for prey; they died on the run from predators; they died from hunger and cold; they died from infection and plague. Only after they finally wised-up to its existence would they have shared our belief that they also died from the common cold fate of mortality.

So far we have sorted the earliest people into two categories: nonleaders and leaders. There developed a third category, one unknown to our predecessor animals—that of shaman, wizard, magician. Although most individuals had some skill with dealing with the supernatural that bubbled into existence with early sapiens or their erectine predecessors, sooner or later a separate class boiled over out of the common pot. At that point most major magic fell into the hands of specialist shamans. Perils that could not be seen by the leaders were confronted by these magicians. They bought off or fought against the supernatural sources of hard times and illness. That means they prescribed savory sacrifice to satisfy unseen agents. They made magical spells with purging herbs to expel other agents from the body in order to regain health. When magic didn't work, which was the case far more often than not, the magicians seldom lost their reputations, nor did magic lose its glory. It was far easier to believe the failures were the result of new supernatural interventions or of bad attitudes on the part of the ill or the hungry. Besides, who would have wanted to tangle with a powerful magician, even when his magic did not seem to work?

This new class, that of magician, was devoted to the area of subjective knowledge as well as knowledge of the objective variety. That subjective/

objective division is a modern one and unlikely to have been within the ken of the early workers of magic. However, in the pursuit of their arcane and occult arts, they developed the beginnings of semiscientific understanding and manipulation of the physical world they lived in.

In this—to say the least—compressed account of the development of sapiens societies, we have arrived at the end of our first and longest stage, that of hunter-gatherers, a stage which lasted something like 100,000 years, a stage in which a few societies continue to the present day.

TWELVE MILLENNIA OF HERDING AND FARMING, OF MORE FOOD AND MORE PEOPLE

As we all know, some sapiens societies shifted to a second stage, the agrarian stage, wherein group members made a living by herding and farming. This second stage started about 12,000 years ago with what is called the neolithic revolution. Were sapiens pushed by the needs of too many often-empty bellies into this new way of making a living? Or were we pulled by the insight of geniuses into this once nonexistent niche? It was both, for geniuses don't invent useless square wheels. In any case, agriculture and the domestication of prey animals were not the work of a bunch of Johnny Appleseeds who brought to fruit a sudden one generation transition. It must have taken hundreds of years to take local root and thousands more to spread. This nongenetic "mutation" in the way we lived did not start and spread from any single location. It was a multiplace and not-far-apart-in-time flowering, initiated in what is now Iraq, Egypt, India, China, and Central America.

If, in our first stage (as hunter-gatherers), we had a motto, or perhaps a banner proclaiming our desires, it might have read, "We want enough!" For a hundred millennia, all we wanted was enough food to stay alive in bad times as well as good. As we know, there are still a few societies among us who live by this rule. Most of the time they live well enough to devote a good part of their lives to social interaction rather than the search for food. Sometimes they may not have enough to eat; when that continues, their numbers soon tumble.

Those who joined the neolithic revolution that became our second social stage may have done so when their minds' eyes read a new banner, more suitable for revolutionaries than its predecessor. It proclaimed, "We want more!" We want more than above-starvation wages with occasional layoffs from the dinner table. We want more than catch-as-catch-can foraging and an occasional lucky catch of fish or flesh. We want our bread daily. We want the makings for next winter's bread stashed in the silo.

In reality, of course, there was no revolution, nor were there revolutionaries. Nonetheless, these fancied revolutionaries had their demands fulfilled. They got results because they were party to another struggle, one hidden in the minds of humans. This struggle was between the subjective and the objective sides of their minds. Objective knowledge was beginning to take territory from the subjective world.

In subjective life, it was pretty much "anything goes"; anything widely believed became obvious accepted fact. In objective life, impressions of things and events needed not only the wide confirmation of others, but confirmation from the out-there world before they qualified as truth. Among the processes observed, confirmed, and expanded on were the ways the fecund earth often brought food sources into being without request or demand from the shaman providers of magic. Wild grains, once haphazardly harvested by sapiens gatherers, were here and there (initially by accident and subsequently by design) saved and planted each year to yield greater harvests. Primeval Luther Burbanks repeatedly selected varieties that produced still better yields. A more or less parallel development of animal domestication and selection provided handy supplies of meat on the hoof to go with the bread from the ground grain.

The presumed revolutionaries got "more" indeed, they got more than they asked for. They got the attention of everyone who wasn't directly producing the newly invented surplus, all those who saw more stuff and thought themselves more entitled to the product than were the producers. These keen observers, both inside and outside the community, were real or potential leaders, the alpha, beta, and gamma individuals, most likely males. Like the chimpanzee leaders we saw in chapter 4, who found new uses for their power when scientist-provided large quantities of bananas appeared like manna from the sky, human leaders quickly realized what to do with surpluses. As anthropologist Ernest Gellner tells us, the situation changed "drastically with the emergence of agriculture . . . [which] gave birth to extensive storage and wealth and thereby turned power into an unavoidable aspect of social life."[3]

About that time in our history (when stored resources came into being), Steven Mithen (in *The Prehistory of the Mind*) has some interesting views. Stored resources came when the ability was developed to integrate "natural history intelligence" with "social intelligence."[4] What developed then was "the propensity to use animals and plants as the medium for acquiring social prestige and power." Yet, I must ask, if apes can use sudden overabundance of food as an instrument of social power, why would humans doing something similar have needed any *new* ability to integrate different kinds of intelligence?

For the powerful, surpluses are obviously heaven-sent to command and control. Whatever occurred in the way of strife between hunter-gatherer groups was amateur sport compared to the professional strife that now went on. To cope with potential raiders from outside, defenders on the inside were needed. To organize those defenders the leaders themselves had to expand: What was needed was a ruling class. That ruling class had little trouble realizing it had needs of its own, needs for as much of the surplus as it could get by virtue—not the best choice of word—of what could be called taxation and confiscation, or the protection racket.

A surplus of grain also provided the opportunity for expansion of and specialization in the production and trade of other things besides food, things that also were good. The appetite for things that we call "goods" came easily. Goods were not only good to have as weapons, tools, clothes, and shelter, they were also good to gain higher status. They were good to make evident, to oneself and to others, one's long-suspected superiority. To make goods available, new classes of people, artisans and traders, even a few artists, learned new work and emerged to join the existing food producers, political leaders, and "spiritual" leaders.

The three older classes had much to learn as well. The food producers learned that more is often less, less than enough to live on with anything that approached comfort. They learned that, with the exception of a rare peasants' revolt, they had no alternative to a lifetime of long days of hard labor for adults and children alike. They learned that they had forgotten how to live the hunter-gatherer life. They learned there was no place or way to live other than where they were with what they had.

The political-military class learned to defend their lands and workers. It was either that or allow attackers to learn to defend what they had won. The powerful also learned that it is often better to attack than to defend, even if success imposed a new obligation on the nobility to defend an enlarged territory. With little effort, they also learned what we have seen that even chimpanzees can figure out: It's great sport sometimes to beat the bejesus out of weaker others. Those holders of physical power also learned to live a symbiotic life: sharing some power with the spiritual leaders. Together, they could persuade the peasants to accept the community values and beliefs that made coercion only rarely necessary.

Apart from their uneasy compromise with the politicos, the spiritual leaders learned to add pomp and circumstance to their interactions with the supernatural powers. They learned to hold regular feast days to honor their supernaturals with the aroma of food that was simultaneously available for redistribution as fresh victuals to the masses. They also learned to expand knowledge, both imaginary and real knowledge, categories that were still not clearly distinct from one another. Alchemy and astrology are the names of two of the areas of expansion where the search for powerful magic led to real knowledge that was the foundation for future sciences of chemistry and astronomy.

A FEW THOUSAND YEARS OF SCIENCE AND INDUSTRY, FAST GROWTH OF GOODS, AND OF OUR NUMBERS

Our third and most recent stage of social evolution started with some mathematics and science at the dawn of civilization in ancient Middle Eastern pre-Assyrian pre-Babylonian Sumer. Its rapid acceleration in recent centuries is called the industrial revolution, its current phase is known as the information revolution. As we are ourselves viewing it from within, we have difficulty comprehending its nature and its metamorphosis. Perhaps our descendants will call it simply marvelous. Perhaps metastasis will someday prove to be the hardly better but conceivable more accurate word with which to describe this revolution's seemingly endless and increasingly rapid growth and change.

We saw that our modest motto during our long first stage as hunter-gatherers was, "We want enough, enough food to be free of hunger!" That first goal was replaced by a more revolutionary one: "We want more, more than enough food, and we want other goods, too!" That idea represents our many-millennia second stage as masters of domesticated grains and animals. As we saw, we did get more goods, even though it was a small powerful "we" at the top that got most of them. In addition to goods, we also got some "bads," such as despotisms largely unencumbered by enlightenment.

Always bent on self-improvement, after a dozen millennia of farming to produce more than we once had as hunter-gatherers, we developed another new idea. This third stage idea is so new, only a few thousand years now, that it is less than clear how best to characterize it. Perhaps it can be stated as:

HOW WE GOT TO BE HUMAN

"We want more and more, we want endlessly more!" Agriculture was only one step in the right direction. Yes, we want more steps, but we now want more than more steps, we want endless steps. Astonishingly enough, by inventing science and technology, and then mass production and marketing, we got more and more steps; we got so many closely spaced steps that they amounted to an inclined plane.

Is this inclined plane rolling our continuously reinvented wheels of progress up to the really good life? Are we on a smoothed stairway to heaven on earth? It's too early to say; our merely hundreds of years on the slope are far too few to tell us much.

It may not be too early to note that there is a price we pay for the endless flow of goodies that our society produces. One large part of that price is the sheer quantity of the increase in our numbers; we are beginning to see it as an increase the good earth that nurtures us can hardly cope with. In "Sixteen Myths About Population Growth," William Ryerson provides tabular data on the Annual Crude Rate of Natural Increase per 100: For the years from 1650 to 1900, an average rate of less than 0.5; from 1900 to 1950, less than 1.0; from 1950 to 1990, a little less that 2.0.[5] Ryerson notes that the very high rates of growth of the world's population results primarily from declining mortality levels. This suggests medical science and technology as the major forces behind our overall rapid increase.

There are now places on this planet, Edey and Johanson tell us, where a large fraction of the population "chronically cannot get enough to eat, . . . [and others where] people are burning up their country's last pieces of wood to cook their food."[6] Another part of the price is that the more we make, the more we must consume. Otherwise, we fall into the pit of despair and depression. To keep stimulating our occasionally saturated hunger for goods, we have developed, directly and indirectly, whole new huge industries. Advertising is the direct appetite-stimulating medicine. The indirect means for stimulus are print and electronic media stories, the jam-smeared bread that helps the medicine go down.

Yet another part of the price is that we have learned to expect the future to be better than the present. That future betterment takes the shape—we hardly know there exist other shapes—of more things for us and for our children to consume. When demand and supply get out of sync with each other, we have recessions, sometimes deep and lasting ones that make us fearful that the growth of consumables will not soon resume. Then we get nervous and start looking for someone, someone else, to sacrifice. The "we" who largely control our societies get sore heads and sour bowels about sharing our wealth with those who have less, even though that per-head wealth is grossly more than that of others among our fellow humans who still live closer to the bone, either at home as unemployed or unemployable, or elsewhere as small agriculturalists or hunter-gatherers.

If time must have a stop for each of us, it's not out of line to suggest that someday growth of wealth will slow to a stop for all of us, taken together, in Western society. Our third stage of development can come to—I know, it's hard to believe—an end. What will we do when the meat of progress gives out? Do what will we when science and technology run out of seminal new ideas? Who does whom when the golden goose of more goodies gets old and no longer lays fertile, fresh eggs for us to incubate?

Consumption, also the wasting disease, was once a name for what we now call tuberculosis. Will our modern, ever-growing growth and consumption prove to be a cancerous condition? Will our accelerating numbers finally exceed Mother Earth's means of feeding them, as Malthus once predicted? Will time tell our posterity that our earthly heritage from our animal and ape ancestors has been wasted away? Will our bottom line, the one just above our signature, be little more than "bigger brains and minds make bigger appetites?"

"It ain't necessarily so." For example, Jared Diamond's experience over the years as a consultant to the Indonesian government was contrary to his expectation that preservation of species and habitat would not be pursued seriously by a third-world country. He found that in the Indonesian New Guinea province a nondemocratic government with pervasive corruption had the beginnings of a nature reserve system comprising twenty percent of the province's area. "All . . . measures were adapted not out of idealism, but out of cold-blooded, correct perception of Indonesia's natural self-interest."[7]

Certainly our future disappearance (nuclear or otherwise), although a scary scenario, is scarcely a realistic one. Short of a collision with a rock from space far larger than the one that may have brought about the demise of the dinosaurs, humans should continue to live on earth. Although our species will continue, that does not mean that, for each of us, our own line will continue. The odds on that have always been poor. As Niles Eldredge tells us in *Reinventing Darwin*, demographers showed long ago that "most people alive at any one time do not have living descendants, say, 500 years later."[8]

As well they should, our species' numbers will stop increasing. They should stop increasing if you consider that (as Eldredge remarks in *Life in the Balance*) the earth's population would have to decrease from its present nearly six billion to a third or a sixth of that if its median standard of living were to approach that of the average middle-class American.[9]

When both our numbers and our consumption stop increasing, the way we live will change drastically. Too many old folks will be a larger burden on too few working-age people. The discipline needed to deal with that may mean that much else of what we value in our Western culture disappears. But hey—not to worry—you and I won't be around to experience it. For that matter, neither your descendants nor mine are apt to experience that (or anything else—as we saw, for most of us, our line of descendants soon disappears).

A slightly less downbeat scenario for the future can be formed by noting that our Western culture is moving the world in the direction of less "material" interests. By that, I don't mean moving toward more "spiritual" interests, but rather toward interests in more two-dimensional virtual objects (flat images on paper and on electronic screens) instead of real three-dimensional ones. It takes a lot less in the way of fuel to make an image of a moving vehicle than it does to make the vehicle and to make it move, let alone less fuel to make a moving two-dimensional facsimile of a person that to make a real multidimensional human. This switch to a less "material" future may delay our hitting of the Malthusian wall of 10 to 14 billion (compared to about 5.7 billion now), according to Eldredge's *Dominion*,[10] but it's not apt to prevent it. For all that, delay is not to be disparaged. After all, all there is between now and an eventual dead universe is—you got it—delay.

NOTES

1. Marvin Harris, *Our Kind* (New York: Harper & Rowe, 1989), p. 265.

2. Robert Wright, *The Moral Animal* (New York: Pantheon Books, 1994), p. 243.

3. Ernest Gellner, *Plough, Sword, and Book* (London: Collins Harvill, 1988, and Chicago, Ill.: The University of Chicago Press, 1989), p. 145.

4. Steven Mithen, *The Prehistory of the Mind* (London: Thames and Hudson, 1996), p. 222.

5. William Ryerson, "Sixteen Myths About Population Growth," *PCI Member News* (summer 1997): 5.

6. Maitland A. Edey and Donald C. Johanson, *Blueprints: Solving the Mystery of Evolution* (New York: Little, Brown and Company, 1989), p. 388.

7. Jared Diamond, *The Third Chimpanzee* (New York: HarperCollins Publishers, 1992), p. 367.

8. Niles Eldredge, *Reinventing Darwin* (New York: John Wiley & Sons, 1995), p. 153.

9. Niles Eldredge, *Life in the Balance* (Princeton, N.J.: Princeton University Press, 1998), p. 184.

10. Niles Eldredge, *Dominion* (New York: Henry Holt and Company, 1995), p. 120.

PART FIVE

MIRACLES FOR MODERN HUMANS

CHILDREN, LOVE, DEATH

INTRODUCTION

THREE MAJOR CONCEPTS—ALL AT the heart of our contemporary lives as humans—warrant examination in the light of our expansion of mentality. At one point in "Sweeny Agonistes," T. S. Eliot's narrator boils down the facts of life: To be born, to have intercourse, and to die.

The first of these, coming to life by entering this world, is a natural-born miracle. Our own so-doing we don't experience as miraculous, the arrival of our children we do. Having a child may be enchanting, at least in part, because parenting feels like it gives us an almost unbelievable chance at another life for ourselves.

The second natural miracle is not simply copulation but love-making, which we often disassociate into occasional real love and more frequent making-out. Real love is something many of us with a bit of luck do experience, if sometimes only now and then. Copulation, on the other hand, is common. For most of us, love is love of more than merely the beloved's body. For some, it is love apart from love-making; it is love of those miraculous children of ours, even after they are no longer new-born marvels. For some, it is also love of those once-gigantic creatures who are now our shrunken parents.

The last miracle, death, is bitter when it happens to people we care for, and painful even when it relieves their agony. Beyond its happening to close or remote people, it has scary meaning for us as the opposite of life. Nonetheless, for ourselves death is something we don't experience at all. Experiencing things is being conscious of them: the dead don't do it. The miracle of death is that this solid flesh of ours should contain consciousness-based natural spirits that, like the stars in a time's-end firmament, simply blink out and disappear.

They may be marvels in some sense, but are the beginnings of human life, love, and life's end *miracles*? Perhaps I am being a romantic and exclaiming: Gosh, ain't these things, like, grand? Maybe so, but I'm also trying to suggest something else. Miracles are *enchanting* and *magical* in a sense other that either the supernatural one or the romantic one. As sociologist Mark Schneider points out, there is a sense of the word "enchanted" that is different than being "charmed" or "deeply delighted," a sense "where we are faced with something both real and at the same time uncanny, weird, mysterious, or awesome."[1] Schneider also describes three dimensions of our conceptual structures that influence the degree to which they seem enchanted. One dimension is *complexity*, another is *variety*, and the third is the degree of *disorganization* or unconnectedness. Conceptual structures that are highly complex, varied, and disorganized are the ones that seem enchanted. I think there is something about all three, about pedestrian-sounding "birth, copulation, and death," that is sometimes, in that sense, enchanting.

I also hope that words like *miracle, enchant,* and *magic* don't make believers in a narrowly scientific view of what constitutes explanation roil in their grave insistence that all of human life is straightforward and algorithmic. Science has done a lot to help us humans understand how we got to be us. There's no reason to expect that science can and will explain everything about us.

We'll examine these modern miracles and their close relatives in the pages that follow. In so doing, we'll see some of the ways in which our beliefs are based on the interaction of our biological inheritance with our learned concepts. We'll also see some of the ways in which enchantment is still with us.

NOTE

1. Mark A. Schneider, *Culture and Enchantment* (Chicago, Ill.: University of Chicago Press, 1993), pp. 3, 31–32.

14

PARENTS AND CHILDREN

ONE OF THE THINGS ALL of us human adults share is that we were once children. We all have a childhood, including events of high import, that we can and do look back on now that it's gone. We all know that, wonderful or terrible, we can never go back to that time. But is that childhood *entirely* gone from us? Isn't each of us, in some sense, as much the embodiment of the patterns impressed upon us in our childhood as we are the incarnation of the configuration of our genes?

OUR CHILDREN: TO US, THEY ARE ALSO NEW VERSIONS OF OURSELVES

Examine the attitudes we modern humans often have about our children. Even before they come into the world, most of us want our children to be our own in the sense of being "our blood." But what do we mean by "our blood?" It could be, of course, that we intend nothing more than the obvious meaning of our genetic heritage, our kind of people, the ones from whom we descended.

If it is our own genetic heritage that we want our children to have, it is more than odd that most of us know very little about that family tree. Beyond parents, sometimes grandparents, and a rare great-grandparent, most of us not merely don't know our genetic past but make no attempt to find out about it. Our studied disinterest in our family history makes suspect the idea that it is that heritage we want our children to embody.

HOW WE GOT TO BE HUMAN

Instead, I suspect we mean something much closer to home than our family tree of ancestors—we mean our very own selves. All of us want our children to be new versions of ourselves. Oh, sure, it's not only ourselves; we compromise. We decide that a child is a combination of oneself and one's spouse. It's a common indulgence we all allow ourselves, to parcel out an infant's features and attitudes mostly between its parents.

It's true enough that we hope that these new young versions of us will be improvements on the originals. But not improvements in the basic stuff of self. Rather, improved as they develop under our guidance, improved by the superiority of our sensitive but firm instruction, compared to the often thoughtless manhandling provided us in our own childhood.

How often do we see parents of grown children feeling the need to direct those children's activities, feeling anxiety or anger when those activities are not the recommended ones? To be sure, part of the explanation of such things is that it is easy for symbol-using creatures to regard not only their pets but also their children, and, for that matter, their spouses, as objects that are their possessions.

Regarding our young as our new selves may be related to a biological attachment found in animals. I refer to the animal mother's bond of identification with the young. The animal bond is an almost literal one, based on the fact that the young were indeed once part of the mother and came out of her body. For lower animals that bond fades with time. The mother lacks the ability to recall the past as the young grow less like their infant selves and more like other adults in the group.

For sapiens (and, I suspect, our erectine predecessors) who see their children as part of themselves, the animal bond is extended to the conceptual level, where it can be brought to mind throughout the lifetime of the parent. It may be easier for a mother, for whom the literal identity of the child with herself before its birth is present together with the more abstract conceptual identity. But for Homo sapiens fathers, who long ago learned where babies come from, the knowledge that seminal fluid provides the seed from which a child will sprout works as well. The identification at the conceptual level allows men to join women in conceiving of their young as part of themselves even before they are conceived. It lets us all continue to regard our children as part of ourselves even when they no longer are children.

To feel we live in our children, by easy extension, is to feel we live on in our children. That is a marvelous way to extend our lives, and one of only a few ways left to creatures who may once have thought that, if no malign spirit intervened, their lives need have no end. To have children is to have a magical passport to a realm we ourselves will never, in this flesh, visit: the realm we call the future. So it can be wonderful to be a parent.

That's not all it can be. To see detached parts of oneself—the only parts that with anything like certainty will live on in that fabled future—do things that are simply not right by one's deepest beliefs, that can indeed be maddening. To see them do things that endanger that same future is frightening. To see those same independent parts of ourselves, parts to which we are bound for as long as forever lasts, to see them subject to pain, pain that we can neither take from them nor diminish, to see that is agony made flesh. To see our children subject to the unripe termination of life, to see them pulled untimely from the womb of protection with which we once thought to cradle them, to experience that is to feel our own flesh torn open.

Of course, as children, once we reach a middling age in our teens where we start to think of such things, we want no truck with the notion that we are objects or possessions, let alone detached parts of those troublesome creatures who once bestrode our little lives like gods. So it can also be maddening to see what we in the West conceive to be our birthright of freedom and independence denied by those who should care for us most, our parents. Breathes there an adolescent with soul so dead, who never to herself has said, "Though once they fed me, with them more than up I am now fed"?

CHILDREN LEARN TO BECOME ADULTS LIKE THOSE AROUND THEM: VILLAGE LIFE IN INDIA

Childhood is more than the period in which we change from being dependent on parents to becoming independent of them. It is also more than the interval during which we grow physically and mentally into an adult human individual. It is the time when—despite our development into what each of us conceives of as a person unlike our parents or anyone else—with our every breath and with our daily bread, we all pick up and make our own most of the culture in which we live.

Consider, for example, what Clifford Geertz (in *Local Knowledge*) tells us about a child in an African Azande community who stubbed his toe on a tree stump and developed an infection.[1] That child knew that what happened to him was witchcraft. No one need ever have warned him about stubbing toes on stumps. Rather, he learned the common sense of how things happen in the world; he learned it from countless cases he knows of where witchcraft had resulted in harm to someone. Thus, if the child grows up to become, say, a skilled pot-maker, he will know that a pot coming cracked out of his oven is a result of witchcraft.

Well, some of us may think: Witchcraft, we know something about that. It is peculiar, we had some problems with it ourselves, in our early days in America, didn't we? There are or were lots of places, in South America, Africa, the South Seas, where people have peculiar ideas like that. Most of them have learned better by now, haven't they? With that kind of thinking, we can dismiss what seems a backwater tribal culture where children grow up to believe in witchcraft. More to the point, we can continue to think our own children grow up having learned the simple common sense of what is real and true about the world.

That may, indeed, be the case. However, it's conceivable that we are wrong in thinking so. Perhaps our commonsense knowledge is, in some ways, as weird and outlandish as we think the knowledge of presumably primitive societies to be. How could we find out? Is there a way for us to look at our culture from outside? With Robert Burns, we "would a gift the Giftie gie us, to see ourselves as others see us." Perhaps we could read or listen to individuals of other cultures who have examined our belief systems.

All right, what if we asked someone from another society, say, someone from Zande who knew Westerners? Of course, he would tell us of our odd blindness about witches among other things. Aye, there's the rub, we're hardly going to believe what some odd duck from some oddball place tells us about ourselves. In Mark Schneider's words (from *Culture and Enchantment*), "so well socialized are most of us that we come to see our own conventions as natural, and (quite unaware of alternatives) believe ourselves able, through them, to see the 'objective' shape of [things]. In this circumstance interpretations wildly at variance with our will be deemed 'strange.' "[2]

Instead of pursuing some independent and outside view that will teach us about ourselves, a more modest approach may be appropriate. Let's leave ourselves out of it, leave ourselves to heaven, as it were. Let's not look at more small backwater societies for examples of presumably primitive magical and religious beliefs. What seem like freak shows, at the circus or elsewhere, may be fun to look at. However, far from convincing us that we might also be in some ways peculiar, if not downright freaky—which is my covert and immodest hope—they indirectly affirm for us the nobility of our own form.

Instead, let's find a large modern society other than our own. Let's look not at their magic and religion (what we see as the oddness of other people's religious beliefs never seems to suggest to us that our own might have oddities), but at the ordinary social life that a child learns about as it grows up. If we find that a child learns to adopt what seem strange attitudes and beliefs, perhaps that will suggest at least the possibility that our own notions and convictions might also be, if not fully freaky, then more odd happenstance than they are eternal truth.

Let's look at what the young learn about everyday living in most of the countless villages in India (as abstracted from David Mandelbaum's *Society in India*).[3] City life, where the age-old local culture has been diluted by Western ideas, will be ignored here. An Indian village child learns that there is one aspect of life that is fundamental, one dimension that is dominant: purity and its absence. Life itself, in the world where that child lives, is mainly a struggle to achieve purity, purity that, unfortunately, is fleeting and brief at best.

The absence of purity is pollution, pollution that is almost always present. Pollution is sticky, like gum, glue, or—more vulgarly and more accurately—like stepping into animal excreta. Purity, or rather as much of it as a given person can achieve, is gained by ritual cleansing, by the combination of proper washing and ritual invocation of supernatural assistance. Pollution, as its cure suggests, is one or another kind of dirt, commonly dirt that is emitted by the body, dirt that soon soils the newly cleansed person. Pollution is also dirt from being too close to the wrong kind of person: a person who is already polluted by uncleansed bodily emissions, or a person who can be described as filthy with inherited dirty flesh.

Closely related to pollution and purity in the minds of villagers in India is hierarchical social position. In animal social interaction, the dimension of hierarchy is usually no more than physical power, power as first seen in pecking order. In our Western social structures, hierarchy's most common dimensions are power, wealth, and, perhaps, intelligence. The power dimension is usually political rather than physical. At one time, but less so lately, the wealth dimension, like some wine, had a component of value that increased with age. Intelligence, when we do value it, often has to be evidenced by prizes, Nobel or otherwise, as the basic stuff itself is too difficult for us to discern.

In Indian villages, hierarchy often has components of power and wealth. However, at its heart is something more important: purity. The highest social groups (no surprise) are the purest; moving down the ladder is to encounter the increasingly polluted. At the bottom are those so dirty that they are untouchable. Indian hierarchy is unlike our social classes where, always in principle and occasionally in practice, a person can move up. The hierarchy in India is based on the rigid refractory concept of caste, wherein one lives—as did one's ancestors and will one's descendants—with no prospect of more than minuscule change.

An infant born into an Indian village family learns after its first few years that it defiles both itself and the person that takes care of it with every production of urine or feces. Washing, clean clothing, and ritual restore it to a state of higher purity. The child learns that every meal, each ingestion of food, increases its pollution somewhat. Again, cleansing restores purity. The child also learns purity-improving habits, such as to use its right hand for ingestion, its left to cleanse after excretion. It quickly learns that all that was once a physical part of a living human body, but is now separated from it, is—yes—impure. To the list that started with urine and feces is added with time the following: spit, sweat, cut hair, nail clippings, women's menstrual products, elder brother's and father's seminal fluids (not that these male or female emissions would by discussed with a child). When the human body ceases to live, it itself is terribly defiling and dangerously impure. Likewise, birth, whereby the new creature is separated from the mother's body of which it was once part, is—equally with death—the occasion for a severe crisis of pollution.

Note that all these defiling things were physically part, as distinguished from spiritually and mentally part, of a living human. I will step out on a limb and suggest that it is in essence the spirit and the mind—and perhaps the body, only because it can't be separated—that gets polluted. Stepping further from the supportive trunk of what is in fact fact, I suggest that, for Indian villagers, it is the body itself that soils the spirit and the mind.

If the child is a girl, she soon learns that she is less valuable than a boy. Her parents will spend less to keep her fed, clothed, and healthy than they would were she a male. Their celebration at her birth was much less than it would have been for a boy. A girl needn't feel bad about it. She knows it's not personal, it's just the way things are socially in her society.

She and her male and female siblings also learn that their parents are deadly serious when they demand respect, especially respect for the father. The children grow up without praise—except for an occasional and swiftly regretted slip of the lip—from their parents. The elders are frightened that praise will feed the child's ego and make it headstrong enough to reject the obedience and obeisance that define a well-trained child. (Recall how, in chapter 4 above, the ape Evered crawled toward top-ranked Goliath—after the latter beat him up—to seek and regain comfort and protection. Likewise, Gerard Manley Hopkins' protagonist, in "Carrion Comfort," kisses the rod or hand of his God after the latter's "wring-world right foot" rocked him. Looks like parallel construction, doesn't it?)

When a girl child approaches puberty, the disciplining of her behavior lightens, as she will soon be leaving the homestead to live in the harsh house of her new husband's parents. There she will have a new and difficult task: submitting to the discipline of the hard-hearted mother of her husband. That task teaches her, if nothing else, that her childhood training was not that heavy handed after all. The in-law mother's heart will begin to soften only when the girl gives birth to a child, preferably a male.

Nor will the new husband provide comfort for, let alone defense of, his bride. In the presence of others, he must have a distant and formal relationship with his wife. In private, they may have only somewhat furtive sexual acts. Even these acts do not necessarily warm his heart toward her. Certainly he wants sexual gratification from her. He also suspects, however, that her desire and capacity for sex are greater than his, which pleases him not. In *Intimate Relations: Exploring Indian Sexuality*, Sudhir Kakar comments, "Physical love will tend to be a shame-ridden affair, a sharp stabbing of lust with little love and even less passion."[4] Kakar also says that interviews with low-caste women portrayed "sexual intercourse as a furtive act in a cramped and crowded room, lasting barely a few minutes and with a marked absence of physical or emotional caressing." Nothing in that new husband's knowledge leads him to believe that he can satisfy his wife sexually. He has learned to believe that she is easily capable of absorbing what he regards as his limited supply of precious bodily seminal fluids, and, yes, still wanting more. All this heightens his conflict about sex, for by now he believes that a rapid diminution of his life-giving fluids may weaken him drastically.

A male child is the preferred and favored sex. Thus (the logic of Indian village life makes that the right word), his father acts more sternly toward him than toward a girl child. The boy begins to be held responsible for his acts at an age of eight to ten. The higher the caste, the more strict the discipline is apt to be. Among the top category of Brahmins, the boy learns to prostrate himself in obeisance before each parent daily, as well as not to touch either of them. The low castes cannot afford the time for such training in family values, but they do their best. Thus, a boy child of cultivators learns to take orders from his father without daring to ask for clarification, let alone discussion. Fortunately for the child, his mother is more tender and less aloof than his father. Her care and authority over her son's life continue after he brings home a bride, despite her added duty of overseeing this young and doubtless unworthy stranger.

In what becomes an extended family once the children are more or less grown, the young man and his wife live with his brothers and their spouses, and of course with the parents. Each of these male children knows that, from birth, he holds an equal share, with the others and their father, of the family possessions. They also know that they still must show utter respect and obedience to their father, who controls and decides everything. To step again away from the tree-trunk of fact, this sounds like more of the Indian villager's genius for making life hellish. In any case, the boys, especially those in groups toward the bottom of the unclimbable ladder of caste, before long begin to demand their share of family wealth, and to depart to set up their own household. (Which is hardly surprising, is it?)

When grown, the young man devotes all his time and effort—beyond the considerable time and effort needed to make a living—to protecting and attempting to improve his social status. Which is to say protecting and improving the purity of himself, his family, his family line of male offspring from an ancestor, and his "jati." His jati is the roughly occupational group within which his family line marries, and which is his caste. He must be careful of every interaction with a person of a lower, and thus more polluted, jati. One wrong move, like taking water or food from the wrong person, can pollute and thus lower not just him and his family, but his entire jati. He must also fight to improve, be it ever so slightly, his jati's status in relation to higher jatis and those more or less equal to his own.

The never-ending struggle is further carried out to improve his status against other men of his family line and against other men of his family, for if he doesn't, he will suffer painful losses here, too.

He will strive to take maximum advantage of the marriages of each of his children. These marriages are his main opportunity to improve his family status by making connection to a family whose status is a bit higher than his own. The price he pays for this victory is the larger outlay of gifts required from him and his family. An alternate course is to arrange a marriage down, a marriage to a less prestigious family. In doing that, he gains prestige from the large marriage gifts he will negotiate from the other family. Arranging the marriage of a child is one of the few likely win-win situations for the parents in Indian villages. Perhaps that is why the children's marriages are the high point of a parent's life.

Parents and Children

What are we to make of the life of Indian villagers? Let's look at it in terms of the earlier mentioned three "commandments" from nature for animals to live by. The first, suggesting egocentricity, went something like, "Thou must look out for number one." The second was about communality: "Make thy life with others of thine own kind." The third recommended sexuality: "When opportune, relieve thine itching private part by rubbing it against another's." To me it appears that the second commandment, the one about what for humans is the social dimension of life, has been heavily accentuated and elaborated. The other two have been bent, as much as possible without breaking, to serve the second law. Social life is almost everything. Egocentric needs and desires have been strongly subordinated to serve social purposes. Sexual desire and gratification have been discouraged and darkened to serve the same social purposes.

Indian village life, however, is not about community values and family values in general. It is about a particular kind of community and family value: purity, and the absence of its opposite, pollution. Its egregiously grungy treatment of life has a sour and stingy flavor. After all, purity could have been conceived of as not too difficult for many to achieve, heavy pollution as only occasionally a problem. For people in this culture, there is little abundance in social life, there is mostly scarcity. For the most part, to get something is to deprive someone else. Life is usually a less than zero-sum game, one with quickly tarnished prizes for the vast majority whose winnings are not inherited.

It would not be surprising if among Indian villagers there were some English-users who read and repeatedly reread the following words we have from "Carrion Comfort" by Gerard Manley Hopkins:

> Not, I'll not, carrion comfort, Despair, not
> >   feast on thee;
> Not untwist—slack they may be—these last
> >   strands of man
> In me or, most weary, cry *I can no more*.
> >   I can;
> Can something, hope, wish day come, not choose
> >   not to be.

Of course, Indian culture in general is a towering structure that includes many aspects that even those of the West admire. Nonetheless, Indian villages somehow created the above-described narrow and cramped society of self-imprisonment. The present caste structure has existed for only the last thousand odd years. Before that, the further you go back in time, the fewer and less constricting were the bars of that prison. Thus it looks like, starting from whenever the beginning was, Indian villagers looked up toward heaven as they slowly spiraled their social life down from earth toward hell.

> Ah, but a man's reach should exceed his grasp,
> Or what's a heaven for.

These words, which we have from Robert Browning's "Andrea del Sarto," conjure an image of Adam's limp extended fingers reaching toward the limpid life in the hand of God on Michelangelo's Sistine ceiling. Indian villagers, as guilty as Adam, reached toward a heavenly goal of perfect purity. Their eyes were on the prize in an impossible dream as they built a social world that mired their minds and spirits in filth and pollution.

In my own more pedestrian view, a man's reach should equal his grasp, and heaven is for tall tales around the fire that show the reach of human imagination.

The reader may not need a keen nose to detect a whiff of ethnocentrism flavoring my comments. My response to that is this: Conceiving the world on the basis of my own culture? Luxury, indeed, would such single-mindedness be! Not only do I look at Indian village life from the point of view of my own society, I look at the life of everyone else in my society from my own limited point of view. That's not the end of it. I have a mind for this, for that, and for the other situation, and some of those mindsets are often opinionated about the others. Nor, in my view, need the reader flare his nostrils to nose out his own similarities to me.

Yet cultural relativism has its limits. As Ernest Gellner writes in "The Stakes in Anthropology," "Cultures are not cognitively equal, and the one within which alone [the study of culture] is possible cannot really be denied a special status."[5] In *The Roots of Thinking*, Maxine Sheets-Johnstone expresses strong opposition to a cultural relativism that would declare "the roots of thinking unexposable."[6] Also, cultural relativist obstacles "are all in a robust biological sense biodegradable."

Interperson differences also have their limits. E. B. Tylor (who pioneered the disciplines of ethnography, ethnology, and cultural anthropology over a century ago) was convinced that "the principle of psychic unity . . . within the human race" would be made apparent by those sciences.[7]

How united we humans are, as individuals or as cultures, remains unclear. In these pages, I've offered my opinion that we all share a biological human nature that makes us self-centered, social, and sex-driven. We all engage our subjective minds and objective brains in frequent conceptual activity when we're not in sound sleep. All of us sometimes put our behavior where our beliefs are, even when those beliefs conflict with one another.

I'm sure what's common among us hardly begins, let alone ends, with these items. Note one other thing we Westerners share with Indian villagers: An aspiration for a life that is more spiritual, one that seems freer from the earthbound body. What is this shining halo of spirituality that we'd all like to wear like a crown? Mircea Eliade (who spent some of his formative years in India's countryside absorbing the heritage of folk religion) believed religion and its spirituality should be explained on its own terms, and not reductively.[8]

In my own view, the fact that religious experiences are real, subjectively real, does not imply that they don't have components. One component of spirituality is its esteemed selflessness. We'd all like to have a self that is not selfish, a self that seems selfless. Why such aspirations? After all, don't most of us know, or at least intuit, the truth that our most basic nature is narrowly egocentric? Yes, but we also know that, although we aren't that way often enough, we can be generous and loving, which is to say, selfless or spiritual.

Finally, I'll repeat the question raised before this inquiry into Indian village life. Isn't it possible that, all-unknowing, we, too, could be raising our children to have some peculiar if not downright freaky ways?

NOTES

1. Clifford Geertz, *Local Knowledge* (New York: Basic Books, 1983), p. 78.

2. Mark A. Schneider, *Culture and Enchantment* (Chicago, Ill.: University of Chicago Press, 1993), p. 143.

3. David G. Mandelbaum, *Society in India* (Berkeley, Calif.: University of California Press, 1970).

4. Sudhir Kakar, *Intimate Relations: Exploring Indian Sexuality* (Chicago, Ill.: University of Chicago Press, 1989), pp. 18–19.

5. Ernest Gellner, "The Stakes in Anthropology," *The American Scholar* (winter 1988): 29.

6. Maxine Sheets-Johnstone, *The Roots of Thinking* (Philadelphia, Pa.: Temple University Press, 1990), p. 8.

7. Daniel L. Pals, *Seven Theories of Religion* (New York: Oxford University Press, 1996), p. 20.

8. Ibid.

15

BUSH-LEAGUE SEXUALITY AND, AT TIMES, MAJOR-LEAGUE LOVE

WITH ITCHY ESTRUS GONE, HUMANS STILL SEEK THE RUB OF LOVE

AS ELSEWHERE IN THIS BOOK, we will look at the subject from the viewpoint of my central theme: How we got to be us, us as subjective minds and spirits with objective bodies and brains. Perforce, this strategy will require here that I shine light on subjectively underexposed private parts.

The first thing we must do is to rearm ourselves with some key words about sex and their meaning. Apart from its physical components, female heat (estrus) is said by ethologists to consist of "receptivity," "proceptivity," and "attractivity" for a female animal. The first and second terms suggest a passive and an active interest in copulating—I've included the second in a broader subjective definition of the first: receptivity as the female's desire for sexual interaction. The third term, "attractivity," is not really part of the *female's* sexuality. Her receptivity, as evidenced by her sexual swellings and aromas, is what makes her attractive to the *male*. So the attractivity of in-heat females is part of the sexuality of the male animal. In fact, it is the essence of his often monomaniacal "rut," which asks one question about a female, "Is she hot or is she not?"

One other word we'll need again is "prurience," which refers to lustful desires and/or lustful concepts. Lower animals have little or nothing in the way of concepts, so their prurience is limited to desire for copulation. Human prurience, as most of us know, is loaded with lustful concepts as well as desires. One of the fascinating things about apes is that they are like us in having some concepts, in addition to desires, about sexual activity.

Now we can turn to books by scientists about human sexuality, which are full of objective information about the Darwinian adaptive value of sex and about the way it is practiced. In *Anatomy of Love* for example, Helen Fisher provides us with an odd bit about human practice: married women copulate one to seven times per week.[1] As this number hardly matches up to captive bonobos' performance of several times, even dozens, per day, it suggests to me that, in this "sport," we humans are in a bush league.

About the subjective aspects of our sexuality, however, such books are often not helpful. Consider the first sentence of the first chapter of Jared Diamond's *Why Is Sex Fun*: "If your dog had your brain and could speak, and if you asked it what it thought of your sex life, . . . its response . . . would be something like this:"[2] The answering words that are put in the dog's mouth convey only the objective notion that it is peculiar that human sexual activity frequently does not satisfy its (presumed) purpose of producing babies. The words convey nothing at all of what a thinking dog might think of the (subjective) experienced aspects of human sexuality. As Jack Gibbs notes in *Control: Sociology's Central Notion* about the work product of sociobiologists, they "write as though impregnation is the ultimate human concern, although all the while denying the relevance of consciousness."[3]

Much more significant in science writing about sex is the confusing use of words like "receptive." Diamond, for example, follows scientific custom by referring to female human sexuality as based on "unceasing receptivity" (p. 67) or "constant receptivity" (p. 79). Such usage is immediately puzzling— surely scientists don't believe that women are unceasingly or constantly desirous of sexual interaction. Diamond, himself, provides evidence of that when he writes, "it would be hard for a woman convincingly to fake sexual receptivity if she felt turned off" (p. 76). Here, it is clear that he has switched to the common subjective meaning of receptivity.

What, then, is the presumed objective scientific meaning of "receptivity?" In *The Evolution of Human Sexuality*, Donald Symons explains that receptivity consists of "the female's reactions that are necessary and sufficient for fertile copulation with a potent male."[4] But do women have such "reactions" unceasingly or constantly? Scientists know as well as you or I do that women are not receptive in that sense. Symons confirms this by quoting F. A. Beach, another scientist struggling with the word: "human females definitely are not continuously sexually 'receptive.' "[5]

Bush League Sexuality and, at Times, Major League Love

So, is the matter cleared up? Unfortunately, it is not. Instead, the nature of women's sexuality gets even more muddled. Listen to Beach as he continues his explanation: Women are not continuously sexually receptive; rather, "they are continuously 'copulable.' " This odd idea of how to characterize women's sexuality is something that Symons buys into: "Human females are, in Beach's expression, constantly copulable" (p. 122).

So science writing has led us from female animals, who desire sexual interaction when they are in estrus, to estrus-free female humans who—to define "copulable"—can be pressed to have intercourse whether or not they desire it. This decidedly strange view of what should be said about women's sexuality is confirmed by Diamond when he writes about "the peculiar physiology of the human female" (p. 72). What on earth can he mean by "peculiar physiology?"

Put plainly, for scientists to refer to a woman's "unceasing receptivity," "constant copulability," or "peculiar physiology" is to say that her primary sexual part is a hole, the kind of a hole that is fillable by a powerful desiring man's firmed-up sexual organ.

Look further at this clear implication that an important thing about a woman's sexuality is simply that her sex organ is an orifice. Note that a woman has—as does a man—a couple of other orifices, the mouth and the anus, which are sometimes used by a man for sexual connection. Yet scientists who write about human sexuality don't refer to the cake-hole and the butt-hole as having "peculiar physiology," do they? All the science jargon quoted above refers to the hole that is the vagina, which we know is the place that human males commonly desire as the object of copulation.

Thus, besides the banal fact that the vagina can be viewed as an orifice, such jargon expressions covertly suggest something no doubt unintended but known to virtually everyone, including scientists: Men have frequent prurient interest in that organ (and in the woman whose organ it is) as a sex *object*. Further, men often have that interest while surmising that the *subjective* wishes of the woman may be either congruent with those male desires or they may be unimportant. Science jargon usage of "receptivity" and "copulability" makes a veiled connection between (1) the ordinary meaning of "receptive": desire to engage in sexual activity, and (2) the fact that a woman's sexual organ can be regarded as a hole.

Given that such attitudes by men toward women exist, I am obliged to ask about another hypothesis held in Western society: Are the peculiarities of male sexuality nothing other than the result of human males sometimes being "Demonic Males?" In their book with that name, Richard Wrangham and Dale Peterson suggest that males of the human, gorilla, chimpanzee, orangutan species, but not bonobos, are demonic.[6] Part of this genetic-rooted demonism consists of males using their power to sexually domineer and tyrannize females. Noted in Timothy Taylor's *The Prehistory of Sex* is an expansion of this concept of demonism: "When men rape women, according to sociobiology, they could be pursuing a strategy—albeit a violent, criminal one—of genetic maximization."[7] However, the fact that most men do not rape implies that this expansion—which suggests such rapacious male sexuality is a Darwinian adaptation—is simply wrong.

The *demonic male* theory of male human sexuality fails in another way. Consider something else that most of us know: Men regularly, reliably, and relentlessly eroticize women. That kind of activity—which sometimes includes the firming-up (for sex with women who are simply not present in the flesh) of the appropriate part of their bodies—is just not explainable under the demonic male theory.

So I must take a closer look at a different hypothesis about male sexuality, one that I introduced in earlier chapters. Before I spell it out in greater detail, I must note that, about the idea that men's sexuality differs from that of women, some of us are undercommitted and others are overcommitted. In the first group is Murray Davis, who writes in *Smut: Erotic Reality/Obscene Ideology*, "I hope to show that sex . . . is related to gender . . . only incidentally. . . . I have emphasized the major aspects of sexual experience about which men and women . . . share the same perspective."[8] In the second group are those who are overly committed to doctrines about the demonic aspects of male sexuality. For them, I have a cautionary note: The existence of ancient genetic roots in life forms does not imply those forms must bear foul-smelling flowers, let alone putrid fruit.

The idea at issue is this: Human males have retained genetic animal rut, rut that endlessly lusts for females who are sexually receptive. Without further genetic sexual changes, men have transformed that rut—by potency of large brains with concept-engorged minds—into prurience, prurience with endless lustful desires and interminable lustful ideas.

In saying that men have retained rut, I am being somewhat reductive, as men are hardly as monomaniacal in their pursuit of what seem to be receptive females as are lower animals. In the same sense, my suggestion that humans have retained an adaptive communal or social instinct is somewhat reductive. In both cases, something ancient is retained, and is also modified in its expression in the human species.

Bush League Sexuality and, at Times, Major League Love

In *Why Is Sex Fun*, Diamond also writes: "[We humans are] bizarre in our nearly continuous practice of sex, a behavior that is a direct consequence of our concealed ovulations" (p. 64). These are strange words. Why should human concealed ovulation (which is part of the abandonment of female estrus, and thus of periodic female sexual hunger) result in the nearly continuous practice of sex?

Another writer notes a connection between the absence of female heat and the presumed continuous sexual availability of those females. Daniel Rancour-Laferriere, in *Signs of the Flesh*, discusses the consequences of a prior hominid species shifting to an erect stance and also no longer having a female estrous ovulating signal. "The female hominid *seemed* to be always signalling sexual availability. . . . A similar problem crops up in the frequent but mistaken assertion that present-day female hominids are continuously sexually receptive."[9] Connecting the absence of heat in females with the continuous practice of sex or with continuous female sexual availability makes no sense, not unless my thesis—about human males having a conceptually prurient expansion of animal rut—applies. It is the males who make the mistaken assumptions about hot females.

Note the similarity of (1) these mistaken assumptions about hot females to (2) the curiously inappropriate use by scientists of expressions like "unceasing receptivity" and "peculiar physiology" about the nature of the human female. Note again another odd usage by scientists: The suggestion that the "attractivity" that results from estrus is somehow a property of the female rather than the male animal. All three of these strange usages can be explained by observing that all of us in the West, including scientists, live in a culture that is still somewhat patriarchal. Thus some old odd male notions slip unseen into the mindsets of all of us. Here, some that originated in male minds are projected by several writers into females.

Here's a truth that's almost a truism: For both men and women, great and basic is the pleasure of sexual activity, the rub of physical love.

Enjoying sex, however, is one thing. Seeking such satisfaction is quite another. With estrus gone, the task of initiating the search for sexual interaction had fallen from the females. Per my hypothesis, the only genetic force remaining to push our ancestors toward frequent sexual activity was rut, the disposition that had long ago developed in animal males to search for what attracted them—females that were receptive to sexual activity—to search for them one after another. Unlike animal males, our erect prehuman and human male ancestors had the much more difficult task of dealing with estrus-free females, females whose sexual receptivity was no longer easy to psych out. Only because those males had expanded mentation—minds that could use prurient concepts to think about sex—could and did they succeed with that task.

Although we can guess (as I have in previous chapters), we can't really know any details of how early men or their predecessors handled this problem. If we turn to modern men, there is much that we do know.

Modern men look at an unfamiliar woman for many reasons. High on that list is the prurient desire to assess what was once clearly determined by estrus, her sexual receptivity. "Hot: Is she apt to be—or not to be?" That is the question with which men are—one way or another—often engaged when their eyes encounter an unfamiliar healthy young woman.

Getting a sound answer to that question is certainly not easy for men. Apart from getting an answer, however, many men are often uncertain about exactly what the question is. When possessed by sexual desire, they are unclear on the nature of a woman's sexuality, not to speak of being muddle-headed about their own.

I suspect that few of the males among you would dispute that men are commonly uncommonly interested in searching for a woman who is, or might soon be, sexually receptive to them. More often unclear to men is the idea that a woman *chooses* to be or not to be sexually receptive to a given man on a given occasion. Men who have difficulty (with this idea that the woman has an *option* about her receptivity) sometimes get overly opinionated and build theories about a woman's sexuality. Some pick up on the idea of year-round female receptivity. They use it to suggest a woman is indiscriminately hot to trot toward bedding a man, any one of many men who will satisfy her desire for sexual rapture. That allows a briefer description of the woman with one of our four-letter words, "slut." Others fabricate a different kind of theory; they go along with Jean-Paul Sartre in his famous statement, "sex is a hole." Such a description suggests something indifferent and mechanical about a woman's sexual availability. Sexually, she's merely a machine with a slot, a "slot machine" into which a man can slip his penis when he seeks for her to put out a sexual jackpot. Such ideas sometimes lead only to idle speculation between males:

"That broad over there, one with the big behind, wouldn't surprise me if she's a slut."

"Good to know. But the other one, behind her, she looks like she's out of it. A machine, y'know? Bet her slot's up for grabs."

Bush League Sexuality and, at Times, Major League Love

Much of such sexual theorizing about women is no more than male wishful thinking, somewhat like the presumed wishes of those in hell for ice water. Some of it, however, is neither male, nor wishful, nor thinking. For example, in *Anatomy of Love* Helen Fisher writes, "With the loss of estrus, a female [upright ape] was continually available sexually. . . . [Upright ape] Lucy could finally begin to *choose* her lovers."[10]

There are times, however, when muddled male thinking—about women as sex objects—is not idle. When such thinking gets social support, it can result in demonic behavior, as we shall soon enough see.

To expand (less ideologically, I hope) on my own ideas about randy men and their uncertainty about female receptivity, consider, for example, the anxious protagonist in T. S. Eliot's "The Love Song of J. Alfred Prufrock," who, at one point, wonders: Should he have screwed up his courage and somehow just asked this unfathomable woman the question, the big question? After all, she might have curtly responded: "To ask me a question like that, whatever were you thinking?"

It just might be that the pressing question J. Alfred fears voicing is some modern variation on a man's eager and ancient inquiry, "Shall we play the two-backed beast?"

In contrast with men, women may look to see if a male is attractive, but they don't often concern themselves with the man's potential interest in sexual activity. They take it for granted. Well, if we concede that, for reasons with genetic roots, men are often on the lookout for sexually hot objects they might wish to possess, and women usually are not, does that imply men are more promiscuous than women?

It implies that men are genetically more inclined than women to seek and evaluate potential sexual opportunities. It follows then that—all other things being equal—more window-shopping leads to more acquisitions of what often seems like goods, sexual goods.

However, to put it ridiculously mildly, other things have not been equal between men and women. First, as anthropologist Marvin Harris says in *Our Kind*, "It is the female's body and not the male's that endures the risks and costs of pregnancy, birth, and lactation."[11] Second, as far as we know—and we know little about the first half of our time on earth and not a lot more about the rest before history proper left us writings—most human societies have been patriarchal. Men have been able to shape our cultures in countless ways that encourage male sexual enterprises—especially, those of "alpha" males—while discouraging or punishing the same for women.

HOW WE GOT TO BE HUMAN

Do you, perchance, think the reference to alpha males is unfair to our leaders? Listen, then, to Matt Ridley in *The Red Queen*, where he considers some of the rulers of our six earliest civilizations: "Babylonian king Hammurabi had . . . thousands of slave 'wives.' . . . Egyptian pharaoh Akhenaten procured 317 concubines and 'droves' of consorts. The Aztec ruler Montezuma enjoyed . . . 4,000 concubines. The Indian emperor Udayama preserved sixteen thousand consorts. . . . Chinese emperor Fei-Ti had ten thousand women in his harem. The Inca [sun-king] Atahualpa kept [thousands of] virgins on tap."[12]

To this day, in Somalia, Egypt, and nearby countries, girls are denied even the possibility of future sexual pleasure by the cruel, heartless, and demonic practices of clitoris excision and stitching genital lips together until they are partly fused. In most of the Middle East, the activities of women are carefully controlled in ways that virtually eliminate the possibility of unsanctioned sexual interaction. In *Sexual Attitudes*, Vern and Bonnie Bullough tell us that local beliefs in the Middle East dictate that women adulterers and fornicators "be locked in their houses until they die."[13]

Even in the presumably "advanced" part of our own society, most women must still contend with thinking ill of themselves—let alone contend with male accusations of "bimboism" and "sluttery"—if they engage in sexual activity too freely. Thus, in Western and in most other societies, patriarchal culture, in addition to genetic inheritance, has contributed to the fact that men are commonly more promiscuous than women.

If we look further at the issues of female attractivity and receptivity for human males, a couple of interesting things can be seen. First—with estrus and its telltale swellings and scents gone—what is regarded as attractive and potentially receptive is largely cultural. (An exception is the attractivity of youth and good health, which may well have genetic roots.) That means it varies greatly from one culture to another. Females were felt to be attractive in Japan if they exposed the backs of their necks, and in the West a few generations ago if their clothes revealed their ankles. Presently, in many Asian societies, women with small breasts—which may symbolize desirable youth and innocence to Asian men—are regarded as sexually attractive. By contrast, Western men frequently find large breasts attractive.

Second, what humans regard as attractive and what they regard as receptive are not different kinds of things. They are coupled for us, although not quite as they are with animals. For whatever is thought attractive, "more" (beyond the current threshold) of the same thing is thought to imply inappropriately high receptivity. In Western societies, women's clothes that reveal "enough" are thought to be attractive—women generally like to be found sexually attractive as much as men like to find them so. However, clothes that reveal "too much" imply the wearer to be receptive, and women dislike being regarded as receptive in a general or indiscriminate sense. Men tend to be of two minds, a prudish one and a prurient one, depending on whether or not it is "their" women who are showing too much flesh. The line between enough and too much is a moving one that varies both with community and with time.

A couple generations ago, many young American males eagerly sought the *National Geographic* magazine. It was one of the rare places where they could see photos of young females with bared breasts. Let's ask what may seem a question from a real stranger, such as a visiting saucer-pilot: "Why would men do a thing like that?" Certainly, there is an obvious answer that we earthlings all understand: It was *exciting* for young men to do so, thrilling even for adolescent boys who hadn't yet experienced a proper improper kiss.

If that saucer pilot came from a place like ours, she might respond: "That's it? So what else is new?" I hasten to say that I pose the question at a different level. I'm asking: What is the *meaning* behind the looking at those magazine photos? Apart from probing the earth's resources for our daily flat or risen bread and for a place to flop during each long night, "the imposition of meaning on life is the major end and primary condition of human existence" (as Clifford Geertz remarks in *The Interpretation of Cultures*).[14] Let me suggest that there was meaning being imposed on life by those long-ago young men, meaning behind what was then the rare excitement of looking at pictures of half-naked young women.

My hypothesis is that those (now aged or dead) men were seeking affirmative answers to the still-demanding male mammal question: Is she hot or is she not? Imbued as they were with Western cultural beliefs, it was their implicit knowledge—not really taught by anyone but nonetheless usually learned—that such sights normally could be seen only as part of an active sexual relationship with a female. The pictures had "intersubjective meaning," which expression refers "not to those beliefs . . . of which the individual members of a society are aware . . . but rather to ideas so foundational . . . that they exist and function beneath awareness" (to use Mark Schneider's words in *Culture and Enchantment*).[15] Schneider further suggests that the analysis of such hidden cultural significance is not to be viewed as part of an experimental science that looks for laws; rather, it is better seen as part of an interpretive science that seeks meaning. What the pictures meant to those grown boys, meant in some hidden but powerful sense, was that the depicted engaging females were indeed sexually receptive. Those dames were actually interested in doing it!

The photos provided opportunity for nonconscious projection by the young males of receptivity into the females depicted. This resulted in the introjection from the photos of the stimulating feeling of being in the presence—not of females with protruding parts suited for the purpose of nursing infants—rather in the presence of sexually attractive and, yes, available women, as evidenced by their exposed large-or-small heavenly all "hooters."

Note that at that time, such photos would have been of little interest to the young men who lived in the cultural community shared by the photographed women. Their learned local criteria for attractive and receptive women were different. The passage of time has resulted in the importation of Western customs into those tropical countries. Thus, young women there now often cover their breasts in public; I suspect their young men now begin to find bare bosoms to be erotic.

As the threshold between attractive and receptive in women's wear—or the absence thereof—also moves with time, young Western men today may not be as moved by such sights, which no longer have as much significance. Much more sexually explicit pictures, still or moving, are now available to move men. Although some of these pictures may pale with time, most of them are too explicit to ever bleach out.

Among these images are some called erotic and others called pornographic. The latter, lewd likenesses, show women who are engaged in direct sexual activity. The former, unchaste pictures, show enticing bodies of engaging women who, at the particular moment, are not engaged in sexual activity. None of the porno images are likely to ever become so colorless in the minds of young men that they are not soon sexually arousing.

Note, however, that around the middle of the twentieth century, American men seeking pornographic images often had to make do with crude, illicit, and illegal "dirty little comics," samples of which are provided in Adelman's book, *Tijuana Bibles*.[16] Redrawn there are the protagonists of then-popular comic strips and movies. They are all graphically engaged, when push comes to shove, in one kind or another of intercourse.

Pornography, crude or refined, is arousing because young men, or rather the sexual part of their brains and minds, commonly think—and that's close to the right word, it is in part conceptual activity or its nonconscious near-equivalent—think that a female who is sexually receptive to someone is a sexually receptive woman. A sexually receptive woman, it follows as the night soon follows the day, soon will be sexually receptive to them. This prospect for male bodies means sexual excitation; for male minds it means major sexual excitement.

This hypothesis (genetic-rooted rut activating human male conceptual prurience) also helps explain a minor mystery: Why do men respond to woman-with-woman pornography—which, after all, suggests disinterest in males as sex partners—by getting sexually excited? The sexual mind, which is as simple-minded as it is old, cannot comprehend the concept that such women would not want a male lover. It understands only two classes of females, those who are hot and those who are not.

But I misconstrue a bit when I categorize only two classes. There are always in-betweeners in the intensity of male responses. Look for a moment at the difference, in the significance to male observers, between artful photos of unclothed "sexy" women and artists' depictions of what could be the same women in similar poses. Artful paintings of attractive women, relative to photos, permit the attention of male viewers to be less monomaniacally sexual. That's because the votes of other subminds are relatively stronger. Erotic art (compared with erotic photography) is somewhat like being old (compared with being young): Both art and age lend distance; both borrow desire.

Two separate points were made about human male sexuality, with the implicit claim that they are related and have a genetic base. The first, that men look at an unfamiliar woman as a potential sexual "object," is not very controversial. The second, that the looking seeks an answer to the unstated question, "Is she apt to be receptive?" may be more arguable. To look into it more thoroughly, we move (I must alert you) to a slightly seamier and steamier side of the community of ideas.

Before we move on, allow me to note the following: If I write—or try to, here—in a light vein about human sexuality, it is because there is something more than a little absurd about how much attention is paid by men (and women, too, with breast implants now, and with waist-cinching, breath-denying corsets not long ago) to the mere signs and symbols of what can be one of the grand realizations of human life—embodied love. Such over-attention is often not sad enough for us to weep about. Sometimes, though, it seems mad enough for us to laugh about. We will see soon enough more of the not sad but downright disgusting aspects of male human sexuality. For now, it's on to the filth and smut.

We go first to a bar where there is entertainment in the form of a series of young women, wearing little or no clothing, each of whom dances alone before an audience of mostly youngish men sitting behind a counter or tables. Another disguised nonearthling, one who's trying to spy on us, might imprudently blow his cover by again posing what proves to be an earthy question: "What are these men doing?"

Certainly any fool can see what their eyes are doing—they are looking at (or, more accurately, ogling) the women. Yes, of course, the men are also drinking the beer or booze that makes the place economically viable. ("Americans now spend more money at strip clubs than at Broadway, Off-Broadway, regional and nonprofit theaters, opera, jazz, and classical music performances combined."[17]) Yet many of them are covertly doing something else, something undercover. Beneath their underwear under the table or counter, as they watch the bare young chicks, the incoming tides of the blood are quietly raising their expectations. The concept, "It's morning in America, where anything can happen," comes readily to the randy crowing sexual brains and minds behind the rising cockadoodles.

Surely those risings are an activity that call for some explanation, even for stuck-in-the-muck earthlings. A disapproving moralist might say that in such places the women often are available for sex in exchange for money, and the men are responding to the well-known weakness and strength of the flesh. In any case, the second part of that explanation does not explain, and the first part is weak. There are too many such bars where sex is not for sale and there are too many such men, sitting erect though they may be, who do not intend to buy sex after the entertainment. Still, a dust-dry skeptic, without moralizing might say: "With alcohol, music, and tantalizing live female flesh so close, it's only natural that the men are aroused."

To deal with that pseudorebuttal about men getting sexually aroused, we can learn more about what "natural" means by switching to a more unnatural setting. Let us remove ourselves from the alcohol, music, and nearby female flesh, and shift to a setting that is possibly more scientific, but nonetheless by no means sedate. In this arena we can see another example of the same marvel of a stiffening penis with poor prospects of gratification.

One of the presumed blessings of modern technology is a device called a penile plethysmograph. It consists of a small narrow expandable band—containing electrical transducers—that is placed around a subject's penis, plus a device connected to the band that records the expansion, presumably of both band and penis. This recent addition to the brimming cornucopia of semiscientific invention is used to measure any change in the circumference of the penis while the subject looks at still or moving pictures of unclothed individuals. This understandably controversial device is sometimes used on sex offenders to demonstrate their sexual interest in the group that includes the individuals against whom they may have committed offenses.

Let's be scientific. Let's attach ordinary heterosexual young men to such devices. Now, for stimulus, show them films of unclothed young women dancing. What is their response? Most of them register both sheepish expressions and somewhat hard yards. Without the real women the boys in the bar were ogling, the question is sharper: In such a situation, why do they do something as unlikely to be useful as engorging their Old Adams?

The answer to this puzzle has two components. First, each of these sheepish young men is of more than one mind on the question: "Am I in the presence of an attractive female who would be receptive to me?"

There is a simple-minded sexual part of the mind that has been watching the moving pictures of a dancer revealing her joyful come-and-get-it jugs-a-jiggling hallelujah-hipped build. It demonstrates its wordless belief—belief that the answer to that question is: *Hey, why else is she naked!*—by mutely commanding the genitals to rise to the occasion. There is also an objective mind that watches the pictured uninspired dancer present her bouncy hippy body. For it, the answer is: *Please! Don't be such a fool.* It insures that the rest of the body does nothing as loony as attempting to do one of the female images.

Before examining the implications of having two minds, I'll present the second component to the puzzle's answer by suggesting we might well ask how this sexual mind can be so seemingly stupid. Wouldn't this dim-witted sexual brain-corner and mindset learn—after a few such experiences viewing real or pictured bare female flesh—that the situations that it conceives are incorrect, the sights seen are not reliable indicators that an attractive sexually receptive female is present? With either the live dancers or with the plethysmograph-related pictures, there's simply no payoff. Put outright, none of them is about to put out, right?

With no real receptive female present, why do the males ready their bodies as if they sought to pump their seed-full spunk (as seminal fluid was defined a century ago)[18] into absent partners? The answer is something we all know. It's part of the reason we have laws about child abuse and child pornography: Certain kinds of learning are virtually impossible to unlearn. For beliefs of high import—be they upheld or denied in the family and the community—that are conceived as true before the individual is mature, learning is often forever, and unlearning, never. (By the way, about children and sexuality, it is known that children are sexually active in some societies. I won't elaborate: for most of us, hearing about things like that makes our social mindsets twitch and twang too much.)

For lower mammals, the male sexual mentality has a direct connection from the senses. Those senses speak of sexual opportunity in terms of female genital swellings and pheromonal aromas. If we took a young stallion who had never been near an estrous mare and presented it—by vice of the miracle of modern chemistry—with continuous fake human-made signals of this kind, the poor creature would be driven ceaselessly to seek the promised fulfillment.

For human males, the senses (tempered by other mindsets) speak their piece in more muted tones to the sexual part of the mind. The sight of unclothed healthy young adult females, in particular those with genitals revealed, is one of those instantly learned signals or symbols that suggest female sexual availability. The enormous cultural systems we humans have created, consisting of verbal, gestural, bodily, and yet other kinds of symbols also have things to say to that sexual mind.

Thus, half a century ago, a photographic image—showing a young woman with a bare pair of "bodacious tatas"—spoke volumes to most young Western men, but had nothing at all to say to young African or South American Indian male villagers. Nor can that long ago speaking be fully erased from those Western minds. It now directs what have become a few remaining weary old men. Although age does indeed both lend distance and borrow desire, these men still keep a hopeful eye on young female chests.

Lest we think that it is males of only the sapiens species who use *indirect*, as well as direct, representations of the receptivity of females, use them to "turn on" their sexual minds, consider again the behavior (described in chapter 6) of gorillas in captivity, discussed by Anthony Rose in his article "Orangutan, Science, and Collective Reality."[19] Primatologists have observed that, in certain circumstances, captive males "court" and assertively initiate copulation with *nonestrous* females. On occasion, when such a female was in close proximity, the male first engaged in a (sexual) chest-beat display and then proceeded to copulate with the female.

Bush League Sexuality and, at Times, Major League Love

Primatologists were at a loss to account for such behavior, at a loss until they realized what was going on in such "laboratory" situations. This behavior amounted to a duplication of the wild male's normal response to an *estrous* female when she came closer than a few yards to him. It is evident that these male gorillas had gone beyond the exclusive use of estrous swellings and aromas as evidence to their sexual minds that they had been presented with an opportunity to satisfy their prurient interest in females. They had *learned*, from experience in the wild, that a female's close proximity carried the same message.

It's true enough that the captive male gorillas, because of the cramped quarters that forced them and their females into "unnatural" proximity, had *misread* the message, but they had done so only because they were in a situation created by the reflective minds of humans.

Likewise the above-mentioned boys ogling the unclothed girls in the bar, likewise similar young men buying "phone sex" from similarly working women, likewise men providing strong support for our gross national product by regularly renting porno videos and buying copies of, say, *T & A Today*, they all make mistaken decodings of female messages. Likewise human males generally (far more than ape males who must have had suitable past experiences) both read and misread the symbolic sexual import of the activities they observe. Men, and to some extent, male apes, expand on genetic-originated rut that seeks receptive females. They expand with conceptual prurience that is often mistaken about the presumed messages from females. (Of course, even a blind hog gets an acorn now and again.)

Nor is what we can glean from these accounts (of human and gorilla males with sexual minds that are interested in "doing" females) solely that prurient primates can learn to abstract sexual symbols from the activities they observe. In general, larger brains with their expanded mental capabilities, especially sapiens brains and minds, have added countless kinds of learned new dimensions to the "flat" world of more literal-minded lower mammals. Likewise, using our multiple mindsets, human males and females both often read and misread the messages we find in the higher dimensions of our human world. These are messages about everything we hold significant, not least of which are messages about our own "no-man-fathomed" nature as creatures with body, mind, and spirit.

Being of two minds about something is a subject we began to examine in chapter 4. There we saw numero uno chimp Mike first benignly pat infant

Goblin and then, when he (Mike) gets highly excited a little later, indifferently grab and almost throw little Goblin like a broken branch. Recall also the ground-nesting plover in chapter 2 whose first conflicted reaction to a nearby predator was to try to protect its nestlings by staying near them, and also to try to protect itself by getting away from the predator.

Why is there, within each of us, such a thing as conflict? Why don't we animals and humans evaluate the situation, come to a decision, and then stick with that choice? One might argue that we all sometimes simultaneously desire incompatible things. But why do we do that? We could answer that by saying the desired things belong to different realms, such as wanting a smoke and also wanting to be free of wheezing coughs and the possibility of lung cancer. Still, as we are each a single being, why can't we resolve the conflict between realms?

The best explanation (as I suggested in chapter 4) is that, *mentally*, we may not be the single unitary integrated creatures that we humans like to think we are. We may experience conflict mainly because our minds consist of parts, parts with differing intentions and desires. This need not be viewed as anything radical. As Daniel Dennett reminds us in his contribution to *The Oxford Companion to the Mind*, we are "used to thinking of a person's mind as an organization of communicating subminds."[20] These parts control us sequentially, they wrest control back and forth from one another. The parts have means to influence one another, they share access to memory, perception, and activity, but such influence is limited. The intensity of the struggle between these parts is a measure of how unintegrated they are.

Significantly, there is no clear evidence in the brain or the mind—other than the subjective "I" that claims, and usually convinces us, that it is running the whole show—evidence of a separate judge or chief executive that stands above the parts and resolves conflicts. Those of us humans who are, somehow, more integrated may experience less conflict. Those of us who are considerably less of a piece are more apt to be torn—if not downright shredded—by conflict. Those of us who, for whatever reason, are off the Richter Scale of isolated mental components may be more than a bit loony. In *The Society of Mind*, Marvin Minsky makes a case for humans having not just a few subminds but a whole society of mental "agents."[21] However, members of his society are better thought of as brain structures that are both smaller and more mechanistic than the mental agents I have in mind.

I think it reasonable to believe that, for each of us, our states of mind are less than united. They form, somehow interconnected with the brains within our skulls, no fully federally United States of Mind. Might they not make something more like a Confederacy of Mind?

Want more high-minded inquiry? Sorry, I must take you back to the filtered filth and smoothed smut for a few words about do-it-yourself sex, now known as masturbation. A hundred years ago, a male inclined to that activity was called a "frigster" who would "bring off by hand" his thing or "dash his doodle," whereas a female so disposed was a "frigstress" who would "digitate" herself or "tickle her crack."[22]

When conditions are suitable, sexual minds (commonly of young people) direct human beings to take sexual gratification into their own hands. Why then can't those powerful minds simply direct the bodies they control to do whatever it takes—without physical stimulus—not only to levitate those penises and clitorises, but to achieve sexual rapture? Why can't spirited humans do it directly, by mind over matter: "Look, maw—well, no, not maw—no hands?"

They can't because human sexual minds need representations that seem, somehow, like sensory input that implies the presence of an engaging partner. That input cannot be only at the conceptual level, such as that provided by pornography or erotica. It can indeed be only at the physical level (not long ago known as devil-inspired, madness-inducing, life-debilitating self-abuse) where hands—with or without vibrators—provide sensory input that seems, to elementary sexual brains and minds, like contact with a passionate partner.

Even young Romeos and Juliets with Shakespearian imaginations cannot achieve sexual bliss with only the virtual input of pure impure fantasy. For better or for worse, Romeo's mind and spirit are forever married to his body, as Juliet's are always wed to hers.

In *The Kinsey Institute New Report On Sex*, June Reinisch notes that male and female "episodes of arousal . . . occur several times each night . . . usually during [rapid eye movement] sleep. . . . [They] may or may not involve a sexual dream."[23] These episodes are not disproving counterexamples to the thesis above. Although they are episodes of *physical* arousal, they are not always signs of *sexual* arousal. Rather, they sometimes are physiological inhibitions of inhibitions, inhibitions (during nonerotic, dreamful, REM sleep) of the wakeful state's normal inhibitions of arousal. To stoop to computer jargon, the "default" state of genitals, present in some kinds of sleep, is not inhibition but excitation. However, once the doubly disinhibited genitals are tumescent, pressure against bedclothes and bedding may wake the sexual mind and invite the experience of dreams—be they dry, moist, or wet—of more than mere firkytoodling (which is century-old slang for "indulging in sexual endearments, preliminary caresses").[24]

COPULATION IN A CULTURE THAT IS NOT SEXUALLY MALE-DOMINATED: BEFORE THEY BECOME MEN, BOYS MUST LEARN TO GIVE PLEASURE

At this point, it should prove interesting to examine the interaction of modern men and women in a culture that differs greatly from our own in its concepts of appropriate sexual behavior. Turn to sex on the Polynesian island of Mangaia, extensively described by Donald Symons.[25] In sharp contrast with that within our own culture, sexual behavior in Mangaia is dominated by female notions of what is appropriate. As part of the rites of passage into adulthood at the age of thirteen or fourteen, boys are indoctrinated with the explicit practical details of sexual technique that will enable them to be successful in their sex lives. For each of the youths, the last step of this training is one-on-one practice: intercourse with an experienced woman. From all this, what the boys learn is how to bring each future female sexual partner to orgasmic satisfaction several times in the course of a single copulation. In so doing, they also learn the necessity of delaying their own singular orgasms until it appears that the lady is ready to be sated with a last "big bang." Girls of the same age are given similar knowledge by an older woman, but there is no hands-on classroom sequel by either a man or a woman.

As in our culture, Mangaians engage first in foreplay and then in intercourse. In our own society (it is said by those with a claim to public knowledge of this kind of private information), typically a couple minutes of intercourse is preceded by a couple tens of minutes of foreplay. In Mangaian sex, the numbers are said to be opposite, a few minutes of warm-up are followed by several tens of minutes playing the interactive game of intercourse. The singular goal of their sexual foreplay is to arouse the female sufficiently to desire intercourse. The males, it is fair to assume, have no such problem. A young Mangaian male understands the seriousness of each sexual encounter. He who fails to pleasure her properly may lose much more than his present partner, he may lose the likelihood of finding another. A Juliet on

Mangaia has no interest in Romeos who flee-by-night, fly-by-the-seat-of-their-pants, or fire before they see the finally tired whites of her eyes.

Some Western men may find these Polynesian women monomorphously perverse. Some Western women may find Mangaian men better classified as awesome than as awful. Despite this big difference between Mangaia and the West, please note that other aspects of their sexuality seem similar to those in the Western world. For example, men who have been married for a while are often less interested in bringing their spouses to a number of orgasms at each copulatory occasion than they are in increasing the number of copulatory occasions with their spouses. Also, Mangaian married women find it difficult to understand their spouses' not uncommon interest in pursuing extramarital sex.

If something is suggested by this brief look at the coupling of Mangaian couples, it is the following: Men and women have different genetic sexual endowments. Men's way is to get brief sexual satisfaction frequently and—other things equal—with many women. In contrast, equal-women's way is frequently to get plural climaxes from each sexual encounter, with less interest in a plurality of encounters and partners. Women's way seems congruent with their possible retention of part of their presumably abandoned estrous heritage: The capacity to get lots of rapturous gratification on the occasions when they are in the mood for any. Men's way seems compatible with the retention of rut.

RETAIL AND WHOLESALE RAPE, BROTHELS FULL OF CHILDREN: WHY CAN'T MEN BE MORE LIKE WOMEN?

Turn now we must to the more serious problems men have with women, or, more to the point, to those that women have with men. Certainly most of these difficulties have ancient origins that lie in the likelihood that men and women are no more created equal in their sexuality than they are in their size or shape. "Why can't a woman be more like a man?" is the rhetorical question posed by Henry Higgins, the bachelor protagonist of *My Fair Lady* (the musical comedy version of George Bernard Shaw's play *Pygmalion*). Surely there are countless women who have voiced with vexation its converse.

In part, it was evolution that created us as we are; that includes our basic male-female difference in sexuality. We know the last big biological change in our sexuality occurred when our upright-ape (or, conceivably, human) ancestors abandoned female estrus. At least once, perhaps many times in many places, one or more random mutations in the genetic material resulted in an individual female without estrous aroma, swelling, and arousal. Even though nature did not regularly press that creature to find males that would seed her eggs, males found her and so did. She had young, and in time her line, which included that mutation for estrusfree females, prospered. Indeed, her line not only continued, it replaced the line of estrous females, although it may well be that this mutation, before it succeeded, occurred and failed to spread many times.

Might it be a good thing if something—be it mutation or magic wand—could remove estrus's counterpart, the root in rut of men's sex drive? Might we be better off if something could free men—and women as well—from the forever spinning sex in the male head, the endless ogling, the unwanted headstrong hitting-on, the gratuitous groping, all of which occur everywhere that men can find women? No doubt there are many men who would recoil with horror from the prospect of such ending. The recoilers are those who believe men must never abandon their coat of arms, showing man erect on a field of supine women, above the sacred words, "Vive La Difference!" They are males who, at other times, may feel somewhat like the man who—even as he pushes his broom behind the circus elephant—says "I should quit? And give up show business? Never!"

Slip your conceptual drive control from neutral to forward. When our upright-ape and erectine ancestors stood erect in the wilderness and began to see a broader horizon, the genetic rutlike roots of male prurience may well have improved the prospects for survival. Now, when we have tamed not only most of the plants and creatures we think we need, but more importantly here, when we have somewhat tamed and domesticated ourselves, do we still need our males to be so occupied with at best a partially tamed monomania about female receptivity? Now that countless other plant and animal species are disappearing as the earth's surface itself approaches exhaustion, all because the pressing by men of sex on women contributes to our carpet-bombing that earth with hundreds of millions of our own kind, do we still need that?

Male demonic behavior is not restricted to certain aspect of male sexuality. We all know that male war-making sometimes includes demonic behavior. We also know that demonic behavior is not restricted to men, as accounts of women killing their young children show us. For all that, I will here restrict myself to a look at some of the more demonic aspects of male sexuality. Do we need victorious foreign armies that can be relied on to rape the females that come into their "possession"? Do we need, right here in America, a prison system where sexual abuse is virtually a fact of life for incarcerated women? Do we need vacation tours for men in southeast Asia that highlight the unjust pleasures of brothels populated with girls who are just children of local societies? Do we need the situation wherein a sizable fraction of young children in our own country are sexually abused by some male in or near the family, the school, or the team?

For that matter, though it be less demonic, do we need professionalized prostitution with its pimps' recruitment of adolescents to serve as sex objects? Do we need the hyperactive, though hardly demonic, sex life of many young male homosexuals, men who—freed from the limitations imposed on their heterosexual peers by prospective sex partners who are females and thus not often overdriven by prurience—find themselves having sex with strangers several times on certain days and hundreds of times in a year?

To get back for a moment to Henry Higgins' question, if women were sexually like men, our heterosexual men and women, as well as our homosexual women, would find themselves struggling with the same mind-bending burden of seeking endless sexual acquisitions, a burden that is presently carried by many young male homosexuals. Whereas if men were sexually like women, heterosexual men as well as homosexual men would have love lives more like those of women—who mostly are more into caring about, and on occasion pleasuring and getting pleasured by, the embodied minds they love to be with than they are into obsessively seeking sex objects.

We may not need the horrors and absurdities listed above, but we have them. They are all related—via the patriarchal societies we have developed that have shaped them—to the inability of men to cope decently with an engorged sex drive that is unlike that of women.

Despite all that, there's another truth to be faced: Concepts that are used as baseball bats—as I have to some extent waved this one—when wielded too vigorously sometimes come around to strike the user in the back. What else about men might change were their sexuality no longer driven by such genetic-rut-rooted prurience? We really have no idea. It's hard to imagine what the world might be like, let alone to determine if it would be a better world, if men were freed of the burden of their present biological sexual nature.

HOW WE GOT TO BE HUMAN

Despite our ignorance of consequences, we can still ask a related question. Might a random genetic change bring us a world where men are no longer subject to such prurience? There are two reasons, each of them sufficient, to think the answer to that question is "no." Let's give ourselves the benefit of a simplifying assumption: Male prurience has a single genetic rutlike root that is coded by a simple gene. Thus random mutation might easily result in the existence of individual human males without that excessive sexual drive.

The first reason to disbelieve in a widespread change: Are such males likely to have more progeny than their sexually overdriven peers? That seems unlikely. Obviously, a genetic change is not adaptive if it does not spread better than the preexisting competition.

The second and possibly more important argument goes as follows. Perhaps we cannot imagine how much it is that men are worn down by the rub of the endless itching of their mental phalluses, so assume we are wrong about the first reason. Thus, men with the itch-removing mutation would have more progeny than existing men without it. Unlike past successful mutations, each of which took over a small isolated population, this one would have to take over a population unprecedented in size. However, as Niles Eldredge tells us in *Dominion*, "there is simply no way an increasingly homogeneous breeding pool of 5.7 billion people can be altered genetically in any truly significant evolutionary sense."[26] Thus, the answer is: No, mutation will not lift this burden from humankind as we know it.

It is worth noting, however, that even if such mutated men—men with muted sex drives—are not apt to displace the others, that does not mean that they could not come to exist in good-sized numbers. In fact, they may well be among us. Are there thousands of them? Millions? Like so many of the interesting inquiries we have about ourselves—starting with what is the nature and value of consciousness—questions like these don't easily lend themselves to scientific inquiry, our latter-day foundation for faith and belief.

There is one last question on this subject that pleads for voicing. If neither the good fairy nor random mutation can help men, is anything to be done? Can men somehow help ourselves? The question is a fair one, even though it cannot reliably be answered. It is fair precisely because it does not call for each man to be a Hercules swabbing clean the Aegean stables next to the depths of his own bowels.

It is fair, and it is also difficult, because each of us humans has a nature that is only in part of his or her own making. A large part of that nature is communal or social, created from the culture we live in. We are not like other creatures in this world: Poor fish who swim only in a sea, a sea that is entirely made by Mother Nature. We are rich fish who, swimming together in cultural schools, have fashioned water-filled mobile enclosures that allow us to remain in the sea even as we leave it to travel where the Grand Dame may not have expected her youngsters to be, to roll on the land, to fly in the air, and even to enter the airless void above. Within each of those communally made mobile vehicles of nurturing culture that take us far from the sea, we must still swim, swim in the oxygen-holding water that our gills need for us to live within our genetic heritage.

Our Western culture has in many ways changed—and in some ways improved—the way men think about and act out their sexual imperatives. That rapidly changing technological culture is now, for example, bringing us computer-generated virtual reality. One of the uses for that marvel that is now upon us is the birth of personal interactive pornography, whereby a young man can strap on an image-making helmet and electronic gloves. With these he can entertain himself endlessly with attractive and receptive virtual sex objects. Soon, no doubt, a further improvement will allow a randy modern-day dandy to slip the monkey-business end of a lubricated grasping plethysmograph around his soon-to-be-slick willie. The purpose? To make more real his penetration of what his sexual mind thinks is a receptive woman's orifice. Who can say whether this and other future marvels will tame or inflame masculine sexuality?

Less ludicrously, who can say whether women, in the recently (only yesterday on the grand scale) started process of liberating themselves from a once-tight eons-old patriarchal lubricous grip, may thereby help men to further loosen the ties of an outmoded animal bond? After all, why shouldn't women lend a hand? A major relaxation of nature's sexual bondage may have come four million years ago when Lucy and her fellow female upright-apes were released from the shackles of estrus. That change permitted a future emancipation of males from the ties of estrus's mating part, male rut. Such emancipation, however, has not yet been realized. Those ties that bound males during that four thousand millennia past have—for many but perhaps not all men—been stretched but not broken. So the task is a more difficult emancipation; men need to seek freedom not from those ties, but despite them. For that self-emancipation, men can use all the help they can get.

Women now are no longer the delicate homebodies we men once thought they were. No longer does it seem that their only options are to husband their energies and to energize their husbands. If nothing else, "women are unlikely to leave the workplace, even if families could afford it, stop having abortions, and devote full-time to their families."[27] In helping men to help themselves by changing their social mindsets, women may gain a lot more than what is likely to be at least a few bruises.

Having possibly said more than you care to hear about sex, dare I follow with a few pages about what may be sex's classier sibling, love? Dare I switch from sometimes unsublimated sexuality to occasional sublime love?

Dare I do, indeed. But fear not, I've not a lot to say. First, I am hardly alone in seeing love as related to sex; almost everyone does. After all, we commonly refer to copulation as "making love," don't we? The same hot rubdown—embedded, embroiled, and ensnared in a larger fabric—becomes our love-life.

Do I hear you ask: What about mother love and that of other family members? What about impersonal love, what about the love of God? I'll put them on hold for a while. The line to the last in particular—the Information Highway to Heaven—seems to me to be strung on a one-way street.

IF ANIMALS PERCEIVE OTHERS AS MERELY MINDLESS BAGS OF BEHAVIOR, THEY CAN'T EXPERIENCE LOVE

Cats and dogs and other lower mammals engage in sexual relations; are those sometime suffused with the likes of human love? Cat and dog mothers, and those of other four-legged mammals, seem to treat their young with care and concern. Is that rightly to be called mother-love? Among apes, mothers and their young maintain a close relationship not seen with cats and dogs. Are they thus by love possessed? In *When Elephants Weep*, Jeffrey Masson and Susan McCarthy seems to suggest human love's possible existence among apes, wolverines, geese, cockatoos, and elephants.[28] To imply that—without some effort to account for the vast difference between the "loving" of these animals and that of humans—is to illustrate why social scientists have been leery of examining love.

Before I try to answer these questions about loving animals, let me seek help from the bard. In "Sonnet 116," Shakespeare's narrator says:

Bush League Sexuality and, at Times, Major League Love

> Let me not to the marriage of true minds
> Admit impediments. Love is not love
> Which alters when it alteration finds.

With a prosaic alteration, I'll suggest that love of that animal kind is less than human love if it is not the true marriage of embodied minds.

Recall now that apes seem to see one another as no more than mere bodies engaged in behavior. They have no perception of the minds that activate the bodies of other creatures. Nor, for that matter, do they understand themselves to have minds or mental abilities. However, that chimpanzees may sometimes show something close to human love is suggested by Jane Goodall. Here are the words she quotes of someone who spent ten years with a female chimp named Lucy: "[Lucy's affection was] sufficiently intense and enduring that I would not hesitate to call it love. It is always there, side by side with a protectiveness and concern for [her human family] that is touching and tender to see."[29] That may be so, but nonetheless chimps show no evidence that they understand that mental activity, as such, exists.

I can't agree with Jean Aitchison, who says in *The Seeds of Speech*, "Possibly only one primate branch, the great apes, has a true theory of mind, the ability to attribute intentions to others."[30] Her words ignore the large distinction to be made between attributing intentions to *behave* in certain ways, and attributing intentions that are *mental*. After all, even the ability to recognize themselves in a mirror, as some chimps, bonobos, and orangutans learn to do, implies no more than that they perceive themselves as bodies that engage in behavior. George Gallup, who initiated studies of mirror self-recognition in apes, makes a similar distinction, "[Self-awareness is] the ability to become aware of your own existence, and mind [is] the ability to monitor your own mental states."[31] Which is not to say that apes, and some lower animals, don't care about certain others of their own kind. I'm sure they do, and such caring is a root of what we call love, but it's little more than a root.

Although it may sound flippant, I hope it's safe to say you can't marry what you don't know exists. Although many kinds of creatures care for and are attached to others, when it comes to true love, I think we can throw out the cats and dogs, and even the apes. As we humans developed the concept of love with reference to ourselves, why make its already fuzzy meaning fuzzier by applying it to animals?

What's needed for human love is not only the body that we share with animals, but a mind. We humans not only have minds, but we know that we have them, and further, we know that those we care about have minds.

I don't agree with those who think we should walk gingerly, if not pussyfoot, around the notion of claiming capabilities for humans that are beyond those of animals. We should credit ourselves for the ability to love, however briefly or infrequently, and it's not credible to so credit animals. In the same sense, as creatures who know what morality means (regardless of whether or not morality has the biological roots that I think it has in our being a communal species that has access to learned concepts), we should give ourselves discredit for the grossly vile things we do to one another. Animals from aardvarks to apes have a limited range of mentation, a range that includes at most no more than a hint of morality and of love.

I have half a mind to go back before I go forward, and take issue with my earlier mind's words about the marriage of true—or rather, embodied—minds as the measure of love. The problem is that each of us is a collection, a plurality, of mental centers that can be called mindlike. So, when it comes to love, which ones within each of us get married, and what of the others? That depends on how integrated we are, how well interconnected our multiple minds are. Perhaps love comes as often to those whose minds are less well put together, but maybe it falls asunder more quickly for them. This same lack of a more fully integrated mind may account for the brevity of so many love affairs. Before long, some of the minds within each of the participants start to protest: "Not only am I not married, I'm not sure I was at the wedding!"

THE GIVE AND TAKE OF LOVE: IT'S ALL GIVING, AND TAKING PLEASURE IN SO DOING

So what, then, is love—or, at least, love at its highest level? I'll give you my take on the give and take of love. To love someone is to take pleasure in giving what you have to that person, and also to take pleasure in getting whatever that person gives you.

To give without taking pleasure in the giving is not love, it is cold, dry duty. To take pleasure in getting what you demand or can grab, rather than in whatever is given, may be second nature to us—or rather, narrowly egocentric first nature—but it's hardly love.

Can the best of us meet these austere criteria for love perfectly? That's hardly likely. The gravity of our situation as natural embodied minds and spirits pulls us down to earth, where our basic narrow egocentricity more easily feeds itself. That's why we value love—and altruism, and also spirituality—so highly. We all yearn for some respite from the self-serving that endlessly motivates us. Can the bulk of us measure up to love even imperfectly? Sure we can. Which means we do it now and then. We can fly high for a while. Frequently, we can fly higher than a fish out of water, better than can a sky diver. That's not great, but for creatures lacking the lovely wings we'd like to think that angels have, I suppose it's not so bad.

Is love a Darwinian adaptation? Evolutionary psychologists seem to hanker for such an explanation. For example, in *The Moral Animal*, Wright says that evolutionary psychology's working thesis is that the "various 'mental organs' constituting the human mind—such as an organ inclining people to love their offspring—are species typical."[32] His justification for this working thesis is this: "Every organ inside you is testament to [natural selection's] art—your heart, your lungs, your stomach. All these are 'adaptations' . . . and all are species-typical." But isn't it far more likely that—rather than a one-by-one multitude of organ-by-organ adaptive mutations for each species—there is an adaptive mutation that changes the *pattern* of all or most of such organs? If so, for *some* organs, the change might be an adaptation byproduct that is not itself adaptive.

Wright also says that natural selection can "give males a 'love of offspring' module." These are strong words, too strong in their specificity. I prefer his more moderate words, "Love between man and woman appears to have an innate basis," which imply only a foundation. I suggest a pause in the search for genetic modules for human love. Better yet, let's switch to human love as an adaptation byproduct. A byproduct of what? To my thinking, love is a byproduct of a pile of things, not only of our sexual nature, but also of our social nature, and of our large largely unfathomed brains and minds, all of which have biological origins.

Do you think "love is an evolutionary byproduct" is hardly an explanation? I must agree. Love is an example of what Schneider suggests are concepts so complex, so varied, so loosely connected that they seem "both real and at the same time uncanny, weird, mysterious, or awesome."[33] None of us may have means sufficient to give love anything approaching a scientific explanation. Which, as we all know, opens Pandora's box to peering minds that provide explanations that cover a small backward-seeming range from edifying to ectoplasmic. (About those of my own explanations throughout

---

this book that are less than scientific, I can only hope that they are also more enlightening than they are vaporous emanations from a medium's trance.)

About love, there is another possibility. Maybe love is not much at all; maybe love is at most the "affectionate feelings"[34] with which it is identified in *What Women Want—What Men Want*, a book that is mainly about sex. Perhaps love is one of those concepts we use so easily—such as unicorns and centaurs—that do not refer to anything real. I have suggested that it is not real in the sense of a biological adaptation. Perhaps love is not real even as an adaptation byproduct. Perhaps it is more like a do-nothing, taste-free icing on the cake of human life. I don't doubt that some people live their lives without providing any love in the sense of giving freely of themselves. However, I think those who feel they experience the real thing don't have any doubt about its reality.

Love is not an impossible dream; many of us engage in it now and again. At other times we slip from the grace of love into the unexalted and all-too-common opposite, a graceless default state: We give less that we can; we want more than we get; we take pleasure in complaining.

Neither are the true marriages of embodied minds made in matchmaker heaven. Such minds can be unbelievably different from one another. Yet minds as unlike as chalk and cheese can become as plainly close as a base-ball player's cheek and chaws. When two of them share a space and time, each learns that much of the other is quite unlike itself. Fortunately, we are all capable of learning, and many of us succeed, however imperfectly, in learning about the embodied minds we have grown to love.

Sex is at the root of this kind of love between peers, perhaps because it gives us the means to both give and get rapturous gratification in the same spirit. Sometimes, though, the getting of ecstatic sexual satisfaction gives us reason to add insult to misconception when we're hardly giving anything—when we desire complete possession of the "object" of our affections—by calling our own activity "love."

Sometimes we find ourselves "in love." That state is rather different from that of "love," although it may lead to it. To be in love is to take pleasure in giving what you have, and to take pleasure in thinking your beloved can give you everything you want. To be in love is to have a marriage in your embodied mind with a magical combination of a person and a phantasy. In *Anatomy of Love*, Fisher uses the term "infatuation" with somewhat the same meaning as I use "being in love." For her, such infatuation fades with time and, at best, is replaced with what she calls "attachment," which is "perhaps . . . the most elegant of human feelings, that sense of contentment, of sharing, of oneness with another human being."[35] We cannot count the ways that love has meaning for us. Although it's not love as I have described it, such attachment may be another aspect of human love.

SOME LOVE MAY BE OTHER THAN SEXUAL—MAYBE

The love of parents for their children, as we saw earlier in chapter 14, may be in part an elaboration at the conceptual level of the old animal perception that the newborn are a somehow-detached part of the mother. It may also make use of the projection onto and identification with others that we all do so well. Is parental feeling for children more than all that, is it "real love?" It might be such love, now and then, for some of us.

Is the feeling of young children for their parents "real love?" Please, it's no secret to anyone who has spent time with an infant: The younger children are, the closer they are to being totally narrowly egocentric. Babies want their needs satisfied, period. So do children, but less and less narrowly so as they get older. Childhood is about getting love, not giving it. Some of us, as we move from childhood to adulthood and get some diluting expansion of our central egocentricity, may well end up with love, now and then, for our parents.

Self-love is another variety of love we are given to discussing. Let's ignore the oxymoronic and onanistic overtones of the term. At best, this overblown expression refers to the degree of integration between the several minds that each of us has. If these minds have a high measure of acceptance of one another, we can, when in a congratulatory mood, call it self-love.

As for impersonal love and the love of God, I have nothing authoritative to say. Nonetheless, I will say that at the center of everything we do is the possibility of past, present, or future gain, and from whom can we get more than from the Almighty? Also, love as I have described it does not require reciprocation right here and now. Some of us, for a while, can love without those we love loving us. We can take pleasure in giving what we have without getting much, if anything, to take pleasure in from those we love. So it would not be surprising if some of us, now and then, might be able to love a conceived superhuman entity, whether or not anything corresponding to it exists.

For a real authority on this subject, however, let's hear from Paul, who in 1 Corinthians 13 gives us his take:

HOW WE GOT TO BE HUMAN

> Though I speak with the tongues of men and of angels, and have not love, I am become as sounding brass, or a tinkling cymbal. And though I have the gift of prophecy, and understand all mysteries, and all knowledge; and though I have all faith, so that I could remove mountains, and have not love, I am nothing. And though I bestow all my goods to feed the poor, and though I give my body to be burned, and have not love, it profiteth me nothing. Love suffereth long, and is kind; love envieth not; love vaunteth not itself, is not puffed up, doth not behave itself unseemly, seeketh not her own, is not easily provoked, thinketh no evil; rejoiceth not in iniquity, but rejoiceth in the truth; beareth all things, believeth all things, hopeth all things, endureth all things. Love never faileth: but whether there be prophecies, they shall fail; whether there be tongues, they shall cease; whether there be knowledge, it shall vanish away. For we know in part, and we prophecy in part. But when that which is perfect is come, then that which is in part shall be done away. When I was a child, I spake as a child, I understood as a child, I thought as a child: but when I became a man, I put away childish things. For now we see through a glass, darkly; but then face to face: now I know in part; but then shall I know even as also I am known. And now abideth faith, hope, love, these three; but the greatest of these is love.

Sounds wonderful, doesn't it? It also sounds less within the reach of a human than that of what is, to me, a semigod like the mythical Prometheus or the mythologized Jesus.

NOTES

1. Helen E. Fisher, *Anatomy of Love* (New York: W. W. Norton and Company, 1992), p. 185.

2. Jared Diamond, *Why Is Sex Fun?: The Evolution of Human Sexuality* (New York: Basic Books, 1997).

3. Jack P. Gibbs, *Control: Sociology's Central Notion* (Chicago, Ill.: University of Illinois Press, 1989), p. 136.

4. Donald Symons, *The Evolution of Human Sexuality* (New York: Oxford, 1979), p. 77.

5. Ibid., p. 106.

6. Richard Wrangham and Dale Peterson, *Demonic Males* (Boston, Mass.: Houghton Mifflin, 1996).

7. Timothy Taylor, *The Prehistory of Sex* (New York: Bantam Books, 1998), p. 84.

Bush League Sexuality and, at Times, Major League Love

8. Murray S. Davis, *Smut: Erotic Reality/Obscene Ideology* (Chicago: The University of Chicago Press, 1983), p. xvii.

9. Daniel Rancour-Laferriere, *Signs of the Flesh* (Bloomington, Ind.: Indiana University Press, 1992), p. 55.

10. Fisher, *Anatomy of Love*, p. 187.

11. Marvin Harris, *Our Kind* (New York: HarperPerennial, 1990), p. 254.

12. Matt Ridley, *The Red Queen: Sex and the Evolution of Human Nature* (New York: Macmillan Publishing, 1993), p. 198.

13. Vern L. Bullough and Bonnie Bullough, *Sexual Attitudes* (Amherst, N.Y.: Prometheus Books, 1995), p. 38.

14. Clifford Geertz, *The Interpretation of Cultures* (New York: Basic Books, 1973), p. 434.

15. Mark A. Schneider, *Culture and Enchantment* (Chicago, Ill.: University of Chicago Press, 1993), pp. 19, 56.

16. Bob Adelman, *Tijuana Bibles: Art and Wit in America's Forbidden Funnies, 1930s–1950s* (New York: Simon & Schuster Editions, 1997).

17. John Townsend, *What Women Want—What Men Want* (New York: Oxford University Press, 1998), p. 20.

18. J. S. Farmer and W. E. Henley, *Slang and its Analogues* (New York: Arno Press, 1970). This book is a reprint of the original edition, published between 1890 and 1904.

19. Anthony L. Rose, "Orangutan, Science, and Collective Reality," in *The Neglected Ape*, ed. Ronald D. Nadler, et al. (New York: Plenum Press, 1996).

20. Daniel C. Dennett, *The Oxford Companion to the Mind*, ed. Richard L. Gregory (New York: Oxford University Press, 1987), p. 163.

21. Marvin Minsky, *The Society of Mind* (New York: Simon and Schuster, 1986).

22. Farmer and Henley, *Slang and Its Analogues*.

23. June M. Reinisch with Ruth Beasley, *The Kinsey Institute New Report On Sex* (New York: St. Martin's Press, 1990), p. 88.

24. Farmer and Henley, *Slang and its Analogues*.

25. Symons, *The Evolution of Human Sexuality*.

26. Niles Eldredge, *Dominion* (New York: Henry Holt and Company, 1995), p. 138.

27. Alan Wolfe, *The Human Difference: Animals, Computers, and the Necessity of Social Science* (Berkeley, Calif.: University of California Press, 1993), p. 163.

28. Jeffrey Moussaieff Masson and Susan McCarthy, *When Elephants Weep* (New York: Delacorte Press, 1995), p. 83.

29. Jane Goodall, *The Chimpanzees of Gombe* (New York: Cambridge University Press, 1986), p. 382.

30. Jean Aitchison, *The Seeds of Speech* (New York: Cambridge University Press, 1996), p. 70.

31. Sue Taylor Parker and Robert W. Mitchell, "Evolving self-awareness" *Self-Awareness in Animals and Humans*, ed. Sue Taylor Parker, Robert W. Mitchell, and Maria L. Boccia (New York: Cambridge University Press, 1994), p. 414.

32. Wright, *The Moral Animal*, cited above, pp. 26, 58, 106.

33. Schneider, *Culture and Enchantment*.

34. John Townsend, *What Women Want—What Men Want*, p. 248.

35. Fisher, *Anatomy of Love*, p. 162.

16

LIVING, LIFELESS, DEAD

ALWAYS TRYING CONCEPTS

THE MOTHER OF ALL MENTAL WARS: SUBJECTIVITY VERSUS OBJECTIVITY

ALTHOUGH THE TWO ARE INTERTWINED, subjective knowledge is quite unlike objective knowledge. Subjective knowledge starts with (conscious) *experience*, but all experience is dependent on brains, which are part of the objective world. Objective knowledge begins with *information* about the world that includes our bodies, but it doesn't become knowledge until that information is worked over by our conscious minds. The two kinds of knowledge are so different that they are often in conflict.

Consider aspects of this conflict in some of the work of Descartes, C. P. Snow, and Isaiah Berlin. "I am thinking, therefore I exist." That's what Descartes said four centuries ago, showing his belief that (subjective) experience was, in a major sense, primary, and (objective) information about his reality was secondary. Forty years ago, in "The Two Cultures and The Scientific Revolution," Snow showed us an associated conceptual struggle, a "culture war" between subjective knowledge-centered art and objective knowledge-oriented science, where those who worked in one of these two areas had little interest in or use for the other. "The Divorce Between the Sciences and the Humanities," Berlin's essay two decades ago, described a related contest: the abrasive friction between the more subjective humanities and the mostly objective sciences. A few years ago, a well-known actor, convicted drug-addict and parole-violator described his own objective-subjective conflict as follows: "It's like I've got a shotgun in my mouth [with] my finger on the trigger and I like the taste of gun metal."

These disputes are part of a war in our minds, a struggle between concepts based on subjective experience and concepts based on objective information. The subjective viewpoint often deals with *subjects*. They act for their own benefit, they are motivated agents. Dominating the objective perspective are *objects*, they follow the laws of nature. They are things driven by causality or probability. We will see that long after the long-ago beginnings of this war, art, and science became opposing leaders, one for the subjective frame of mind and the other for the objective mindset.

What is this war about? It's a war about *meaning* in our lives. Whenever we put aside our interest in the basics of survival, the search for and satisfaction of meaning in our lives has high priority. We find important meaning in the subjective, experienced aspects of our lives. Such meaning can be destroyed by the objective information that comes to us.

Before getting into this contest, I'm obliged to note that there are some who think it is a phony war. "Once we get rid of the . . . notion that there is external reality on the one hand, and internal . . . reality on the other hand, as two different kinds of reality, the mind-body 'problem' vanishes."[1] That's what Howard Hintz writes in his chapter in *Dimensions of Mind*. No doubt too much can be made of the subjective-mind versus objective-body problem, but to think it can be dispatched so blithely is to make much too little of it.

The contesting sides in our heads will be dramatized as follows. First, in the bright corner—the subjective frame of mind—we have not only ourselves but also those like us: other subjects, natural or supernatural, real or merely experienced as real. Taken all together, they are the often irrational, the moved by motives, the subjective Living! Second, in the dark corner is the indifferent foe of the Living—the objective mindset with information about uncaring billiard ball or quantum particles, sticks and stones and other things that fall on each other or on us by motiveless cause and effect or probability. Pulled together—and treated somewhat as if they were precisely what they are not, motivated agents—they are the orderly, the indifferent, the objective Nonliving!

How long has this struggle between often disorderly Life as it is experienced (which does not include nonconscious plants and animals) and mechanistic informing Nonlife been going on? In *The Prehistory of the Mind*,

Steven Mithen makes the case that the human mind's modern structure made its appearance after a "series of cultural sparks that occur at slightly different times in different parts of the world between 60,000 and 30,000 years ago."[2] He suggests that the earlier human mind had more isolated realms of intelligence, and after these "cultural sparks" a new cognitive fluidity allowed easier connection between realms. In my view (leaving apart culturally reduced viscosity between domains), the *genetic* structure of human mentation included a subjective mindset and an objective frame of mind when first our Homo sapiens feet hit the East African ground a couple hundred thousand or so years ago.

Our large brains allowed a large expansion of concepts within both the subjective and the objective frames of mind. Look first at the subjective viewpoint. Evolutionary psychologists Leda Cosmides and John Tooby, in their article in *Mapping the Mind*, note that even in young children there appears to be a domain-specific cognitive system that "allows one to represent the notion that *agents* can have *attitudes* toward *propositions*."[3] Within a larger version of such a system, we think of ourselves and other people as motivated agents. However, we did not stop with people. Here's the way Susanne Langer puts it in *Mind: An Essay on Human Feeling*: "Whoever the 'dawn men' were, the world their minds created was apparently not modeled on crudely conceived laws of matter and motion. [For them] the natural way to imagine an event [was] in the form of an act [by an agent]."[4] Also, "the first true human beings . . . were . . . the first symbol-mongers." From the beginning, we intuitively knew we were doers, motivated agents with a simple agenda: to get more of what we wanted and less of what we didn't. Symbol-mongering and projection allowed us to see our likes wherever we looked at important objects around us.

At the same time, those dawn humans lived with an objective mindset where information—that helped them to stay alive and make a living—was vital. Our ape ancestors used (objective) information, such as that telling them where and when to find which food items. Their erectine successors expanded objective thinking to let them live and increase their numbers in environments unlike that of their African origins. Human minds, in turn, further elaborated the erectines' mental structure for objective knowledge.

So it was long long ago when we started projecting, seeing in the world around us conceptual aspects of the subjective essence we intuitively understood within us. Projection allowed us to believe ourselves surrounded by guts as greedy as our own. The hills were alive, as was everything else, not with the sound of music but with powerful ancestors, spooks, sprites, demons, demigods, and other seeming animated agents. We had little in the way of our present creature comforts in those days, but we lived in a world packed with the meaningful acts of projected beings that were like ourselves. Long before we could conceive of a lone high god, we were possessed by a many-entitied version of Gerard Manley Hopkins's vision, in "God's Grandeur," that

> The world is charged with the grandeur of God.
> It will flame out, like shining from shook foil.

In those primitive days of projection, heavenly objects, odd earthbound things, bones of dead elders, all such things were charged with the significance of motivated agents inside them, charged with glory, grandeur, or gut-wrenching behavior that could "flame out."

Despite all that subjective projection of meaningful acts of motivated agents into the world, the objective Nonlife enemy's weird tactic—to absorb Life's blows by relaxed rope-a-doping without even lifting a glove, to exhibit absence of motivation—that tactic continued. That is to say, objective knowledge gradually increased, often at the cost of subjective knowledge about imagined agents. The ever-caring Living were slowly giving ground for most of our ancient days, back-pedaling from the foe's heedless activities. The enchanted circle of Life shrunk sluggishly but steadily as Nonlife knowledge advanced. (*Enchanted* phenomena, we learn from Mark Schneider's *Culture and Enchantment*, occur at those borders where objective knowledge is not successful in expanding.[5])

Though still long ago, the informing Nonlife forces then delivered against experiencing Life a massive blow, delivered a body blow called mortality. The objective knowledge was inexorably borne in on our ancestors: They would not have what once seemed the possibility of endless life— before long each of them would have death forever.

To knowledge of mortality, the response from the Life side was no concession speech, let alone funeral plans. To see why, consider the dead and their dominion. The dead played an astonishing role in the battle between the experiencing Living and the information-providing Nonliving. The dead rose from their objective role as inactive disintegrating matter to become the vanguard in subjective Life's army.

Living, Lifeless, Dead: Always Trying Concepts

Before humans entered the field of play and expanded the idea, death as a concept had no meaning beyond what it signifies to animals: at most, the cessation of a particular life. Then, as we saw above, there came a time when the objective knowledge of death-as-inevitable forced itself on us. Before long, however, those imaginative, ancient folk—whose blood still pumps through our arteries—began to wonder about the silent, still bodies of those of their fellows who ceased to act. They wondered what had happened to the once-embodied selves within the fallen flesh.

Necessary death was such a distressing idea. Perhaps the newly dead were somehow not really dead. What about the "I" within each of those now-still bodies? Where was the "I" that, while they lived, was an unlocatable seemingly nonphysical something that could even be viewed as including a "spirit?" Yes, they, the still-living spirits within the still bodies—it was such comfort to believe—they had somehow risen from the overthrown flesh. So the dead were reclaimed from the armies of the Nonliving. Projection turned the previously dead into double agents and returned them to imagined subjective being, to being Life's living agents.

Other than the gain for objective knowledge when humans realized they were mortal, and despite the regaining of lost territory when subjective knowledge came to include belief in life after death, the struggle between experience and information continued at a slow snail's pace for untold aeons.

Those slow aeons ended a dozen millennia ago, when objective knowledge delivered domesticated plants and animals. With the delivery there was an invoice specifying the price to be paid. Countless imagined agents, such as rain gods and magical animal spirits within our subjective mindsets, were sacrificed during the change.

ART AND SCIENCE PROVIDE LEADERSHIP IN THE BATTLE

Then there appeared—where for prior millennia there was little leadership—there arrived six thousand years or so ago in what is now Iraq a leader, a Nonlife leader who would eventually acquire the name "Science." As a youth, this charismatic leader offered "A Better Life for the Living" by organizing the expansion of useful, objective knowledge that resulted in cities, wheels, pottery, and smelted metal. Then, in his adulthood thousands of years later, this leader turned inside his coat so he could go—I'm obliged to say it—so he could go both ways. In his new direction, this leader became interested in objective knowledge *as such*, useful or not to subjective life. In this second role, Science gradually became a warlord who—with indifference rather than harshness—swore: Death will come to the subjective Living! This information-gathering and theory-building-and-verifying machinery was not born; rather, it was assembled. It was first put together in pre-Babylonian Sumer, built with the piecemeal creation of formalized objective knowledge, worthy of the name Science.

I hasten to interject: I am in no way nominating science for the role of Bad Guy in the affairs of humans. On the other hand, neither do I join Carl Sagan in *The Demon-haunted World: Science as a Candle in the Dark*, where he sees science as the Good Guy in a world where the Bad Guys are magiclike pseudoscience and overlydogmatic religion. For example, Sagan tells us that medical science "permits the Earth to feed about a hundred times more humans, and under conditions much less grim, than it could a few thousand years ago."[6] I must ask: If science gets credit for feeding more people, shouldn't it get discredit for overpopulating the earth? And is it clear that conditions were much more grim for our hunter-gatherer ancestors than it is for so many of the people within the billions that now populate the earth, and not the other way around? More to the point, I'm sure Sagan knew the shortcomings of science quite clearly. However, as a scientist and popularizer of science, science was for him both objective knowledge and subjective knowledge. So it is hardly surprising that, like the rest of us, he projected some of his subjective self into what was important for him in the outside objective world.

In heaping uncritical praise on science, Sagan is joined by Steven Mithen, who says that one of science's three properties is "the use of metaphor and analogy, which are no less than the 'tools of thought.'"[7] This suggests there may be little risk associated with the use of these presumed tools of thought. We know, however, that figurative language can as easily misguide us as it can guide us. We are misguided when we take as literal truth that which may be true only figuratively. This is often now obvious with magic and religions that have fallen out of favor. It is sometimes equally true in the realm of science, where striking expressions, such as "the selfish gene," "the genelike meme," and "the brain is a computer," seduce us into accepting them as simply accurate. (My own use here of "subjective Life warring with objective Nonlife" is, I hope, no more than dramatic figurative language, and not seductive misguiding.)

In its long subservient childhood, Science had been no more than the nameless search for useful knowledge about the external world, knowledge that would be useful to Life, that is. With Science taking the lead as it grew into maturity in recent centuries and turned itself freer of its childhood bonds to Life, the pace of Nonlife's conquests quickened dramatically. We will look at the effects of the recent work of Science, effects so corrosive of subjective Life's vitality as to suggest that workers in science are digging, not their own, but for many of us, our conceptual graves.

Listen, for example, to some voices in *Speaking Minds: Interviews with Twenty Eminent Cognitive Scientists*. Joseph Weizenbaum (developer of ELIZA, the pseudo-psychiatrist computer program) recalls some of the vivid rhetoric of other scientists whose defense of science includes attacks on subjective life. Thus, Marvin Minsky made the "now famous statement . . . that the brain is merely a meat machine."[8] (I find Minsky's flourish, *meat machine*, revealing: after all, meat is *dead* flesh.) Likewise, Daniel Dennett said, "If we are to make further progress in Artificial Intelligence, we have to give up our awe of living things." Also, Herbert Simon once called Weizenbaum a "carbon-based chauvinist" (perhaps for claiming superiority to silicon-based computers).

To return to the main war, the armies of the Life-force always had leaders whose task was easy because their soldiers were so well motivated. From near the beginning, the Life leaders were the once-nameless Magic and Religion that flooded the human world with new meaning. Long after Magic and Religion first came to powerfully impress us, these ancient leaders were joined by a third force, a new leader, one who initially served as servant to the other two. That third leader we now call Art.

Prior to examining Art's role as a leader on the Life side, we should ask what art is, or at least, what is art about? To that end, let me first disagree with Mithen about the nature of visual art symbols. In *The Prehistory of the Mind*, first on his list of properties critical to the development of those symbols is this: "The form of the symbol may be arbitrary to its referent. This is one of the fundamental features of language, but [it] also applies to visual symbols."[9] I must disagree. Consider a bisonlike drawing on an ancient cave wall. Would an arbitrary drawing—say, a stick-figure—serve as well for someone who made such a picture? Hardly. Mithen's second property for a visual art symbol is that it "is created with the intent of communication." This too is specious. The making of a cave-wall image of a bison resulted from an irresistible urge to express something significant. If it was shown to others later, that would have been little part of the initiating impulse.

HOW WE GOT TO BE HUMAN

I'm further constrained to disagree with E. O. Wilson, who writes, "The common property of science and art is the transmission of information, and in one sense the respective modes of science and art can be made logically equivalent."[10] If art transmits information, isn't it odd that college students can't buy "Cliff Notes" digests of Bach and Beethoven? Nor does Wilson explore the sense in which science and art *cannot* be made logically equivalent. Thus, he may not concur with me that the core of art, (subjective) experience, differs profoundly and unexplainably in kind from the center of science, (objective) information.

My basic comprehension of art comes from the reflections of Susanne Langer in *Feeling and Form: A Theory of Art*.[11] Works of art are *nondiscursive* symbols that are quite unlike ordinary discursive language or mathematics. Art symbols are each all of a piece, shaped to be images of high-import aspects of experienced life (for people living in a particular place at a certain time). For most of the last few tens of our many millennia, Art—although it was not known to us as a separate concept—Art showed us images of the glory that was Magic, the grandeur that was Religion, and the terror that was both.

Long before humans came to the scene of Life, expressive activity served part of the function of Art. By projecting ourselves into wolves, we hear it still in their echoing howls. Likewise, the pellmell rushing of apes downhill in response to thunder-and-lightning storms serves as such expressive impractical activity. But wolves have nothing in the way of concepts, and apes don't have a lot, so all of Art is beyond what they can make sense of.

What was the earliest form of Art? Perhaps it was the human activity we now call dance. In *Problems of Art*, Langer tells us that dance, like other forms of Art, is "a perceptible form that expresses the nature of human . . . 'inner life,' the stream of direct experience."[12] She goes on to say, "Subjective existence has a structure; it . . . can be conceptually known. . . . [What] language does not readily do . . . is done by works of art. Such works are expressive form. . . . [They express] the nature of human [experience]."

Cave paintings are among the few things with our imprint from long ago that we now see as Art. They would not have been thought of as Art when they were made. In those days, indentured-servant Art was low-keyed, with no life of its own as it worked to exhibit Magic and Religion. It didn't even have a lowercased name, let alone the luxury (given it here) of the upper-case class. Even though those ancient cave drawings still stir us today, we can't begin to imagine what their import once was, what they might have meant to those who made and saw them then. Much more recent ziggurats and pyramids, whatever their practical purpose, may once have been grand displays of the living power of ancient Ozymandiases who still held their banners (and their heads) high. In the lands of ancient Greece, Art (although still unknown to humans as a general concept) learned to stand by itself, quite apart from Magic and Religion. As human attention shifted from supernatural agents and from human agents with supernatural power, Art started to show people just how lovely at times was human life itself.

Art made high-import things that were beautiful. Think of cathedrals like Chartres. Also, and at the same time, some of those things were frightening, some were grotesque (such as the gargoyles on those cathedrals). However, Art made no things—until recently—that seemed meaningless, empty, insignificant, boring, dull to those living with them. Art made no things that did not seem emblematic of the marvel of supernatural or natural Life.

Let me now ask a critical question: How did Art react to the beginnings of modern Science? For an answer, listen to a couple of wordsmiths who could paint word-pictures.

Here's John Donne, in "An Anatomie of the World," bitterly complaining four centuries ago because Galileo used the newly invented telescope to expand the heavens and gain objective knowledge, thereby destroying subjective knowledge:

> The Sun is lost, and th' earth, and no man's wit
> Can well direct him where to looke for it.

Likewise, here's John Keats in *Lamia* a couple hundred years ago, protesting that Newton "reduced" the rainbow to its prismatic colors. Keats is fearful the price-tag for that knowledge is unbearably high. He suggests that such objective knowledge

> . . . will clip an Angel's wings,
> Conquer all mysteries by rule and line,
> Empty the haunted air, and gnomed mine—
> Unweave a rainbow.

A century ago, pioneer social scientist Max Weber articulated more discursively the concerns of Donne and Keats, concerns that science tends to strip the natural world of its magic and its capacity for meaning, turning life into a tale signifying nothing.

Currently defending Newton and science in *Unweaving the Rainbow* is Richard Dawkins, which is hardly surprising for a scientist and science writer.[13] Nor is it astounding that—much as I'd like help from him in understanding the foibles of the masses of heathen—his preaching is largely to the choir and the converted.

In *Human Nature Mythology*, Kenneth Bock presently says something that, although it doesn't surprise us, is precisely pertinent. With the advance of Science, there was less and less room for "an anthropomorphic God who created and moved all things. . . . [Increasingly, people] found themselves . . . [in a world with] no purpose, no meaning."[14]

The last couple of centuries saw the one-of-a-kind expansion of science. The resulting Western scientific objective world is (dismaying for some but no surprise) orderly, all of a piece, and mechanistic. The space of that world's heavens has swollen to a once-beyond-inference huge structured universe. Its duration has expanded billions of years into the lifeless past, and further billions into a dead—either heat-leveled or space-collapsed—future.

With the same indifferent hand, objective knowledge has shriveled the rich diversity in our understanding of the many who make things happen in the world. From the viewpoint of Science, the shrinking number of living agents in the world is on its way to—I wish I could write it in microprint—to none, to no one at all. For those of us who accept an objective mindset's validity but have no great interest in science as such, physical events in the world around us are no longer the acts of willful agents; rather, they are the consequences of "oh, yeah, blah-blah" cause and effect.

In our—to say "holiest" hardly fits this context, so I'll say—our most "scientific" science, physics, indifferent causality is replaced at the quantum level by something even less like a proper person or living agent: arid and bloodless coin-flipping probability.

Our life sciences—indeed, the very name says it—aspire to explain living creatures and make their bodies' activities and their acts, some in practice and some in principle, predicable consequences of mere causes. The world of the living that we study objectively now definitely includes our very-own human selves. Even consciousness (in my view the main mystery that underlies the widespread belief in a supernatural something—a spirit—within each of us) is under study by some who would reduce it to cause and effect.

"Will we someday build devices that understand symbols, that have symbolic consciousness?" That question is posed by Terrence Deacon in *The Symbolic Species*. His answer is "Yes . . . in the not too distant future. These devices will be made of silicon," they will have sentience in the form of "spontaneously adaptive self-versus-nonself organization."[15] Although he concedes that this is not the same thing as consciousness, it is clear that if it comes, this first rough mechanical beast will be no slouch, moving in our direction, but not yet all the way to us.

Workers in the field of artificial life would like nothing better than the creation of artificial beings that are, in fact, conscious. Here's the telltale title of an example: *The Age of Spiritual Machines*, a book by software scientist Ray Kurzweil.[16] He thinks—hold onto your hat—that we will soon have the ability to live forever! How? By becoming cloned as the software part of a future computer. (Were that to become an option for me, the first thing I'd do is back me up with copies of me, like a billion of 'em.) Living forever, he insists, will be not just objective life like a plant, but also subjective life. Kurzweil thinks this development that he thinks is impending will be as important as the emergence of human intelligence long ago.

If that really represents our future, won't everything, not just everything around us but everything within us, become explainable—reducible to mere predicable mechanism and algorithm—and thus devoid of the often enchanting freedom and the everlasting significance of motivated life? In *A History of the Mind*, Nicholas Humphrey suggests that even if the solution to life, the universe and everything else emerged as some simple fact, by itself that fact might be merely boring.[17] I would add: Unless we get really involved with it subjectively, everything can be a bore.

So is all soon to be lost by the Living spirits—our projections and our subjective images of ourselves—that once ruled in every nook and cranny of a small world? Shall there be victory—totally meaningless to them of course—by the Nonliving, whose legions from the objective viewpoint include the dead? Dylan Thomas once titled a wartime elegy "And Death Shall Have No Dominion." Was that idea no more than edifying wishful thinking? Far worse, are we now to say that soon Life shall have no dominion? Isn't that the vacuous hell where, downhill in a shrinking handbasket, we're dumbfoundedly heading?

To be sure, it is not a pretty picture. Before we even reluctantly accept that answer, let's see where recent Art—which presumably still portrays Life—weighs in on the destruction of meaning by this vast expansion of science. We saw some of Art's early fears in word-pictures from John Donne and Keats. What's happened since?

HOW WE GOT TO BE HUMAN

It wasn't until the middle of the nineteenth century that we see art's reaction to the eighteenth century (the Age of Reason or the Enlightenment) that had followed and expanded the scientific revolution of the seventeenth century. "The moment in which the new painting came into being is the same in which the scientific spirit began to gain the upper hand over the spirit of religion."[18] The new art no longer showed the biblical and mythological stories of prior painting. The new art began with Manet, who painted Olympia (a working woman) and not Venus (a goddess). It continued with the Impressionists and then with the painters whose colors were so vivid that they were called the Wild Beasts (Fauvists). Note, however, that near the end of the nineteenth century, Gauguin dissented from the upbeat view of science and progress. He went to the South Seas to get away from what he saw as the aberration caused by science and to repudiate those who represented the decadent century.

Cubism (1907 to 1914) was the beginning of a further change in what art portrayed, a change that followed the accelerated nineteenth century growth of science. Cubism claimed to be an art of realism. To the extent that "realism" means that Cubist pictures emphasize objective knowledge, that claim is true. When Picasso was an early Cubist, he said, "If a painter asked me what was the first step necessary for painting a table, I should say, measure it." But Cubism is about more than realism and objectivity. The smooth curves and straight edges used in Cubist paintings to represent volumes may also model both the artifacts of technology and the abstract equations that are basic to science.

Cubist pictures also neglect subjective knowledge. According to John Golding's *Cubism*, "it was the . . . disregard of the more sensuous . . . aspects of painting, and the . . . dismissal of all human and associational values . . . that allowed the Cubists to distort and dislocate figures and objects."[19] Consider the paintings of individuals in Cubist works, such as Picasso's *Portrait of Kahnweiler*: It shows little more of Kahnweiler than his eyes, his body is almost indistinguishable from the vague nonhuman objects around him. To me, this suggests the increasing objectlike status of the human body for science. Consider the late phase of Cubism, where ordinary external objects such as bits of paper and poured sand are made part of a picture. They exhibit Cubism's insistence that it not only paints objects but the resulting painting itself is an object, one that includes lessor objects, much like the reductionism of science in finding the components of natural things. Lastly, consider Marcel Duchamp's *Nude*

Living, Lifeless, Dead: Always Trying Concepts

Descending a Staircase. Doesn't it look like a set of body-shaped, connected tin cans in time-lapse photography?

We have with us today some of Cubism's more popular children. We
have, in the land of Oz, the Tin Man, who lacks a heart until a sentimental
wizard with an ear for Life helps him find one. In the land of movies, we
have the perhaps-persons within the bodies of Star Wars' mechanical heroes,
the robotic fire-hydrant-on-wheels, R2D2 and the Tin Man's heir, the android
C3PO. We also have there, and in television, the *Star Trek* "Borg," dronelike
zombies who were once real people, drones who offer homage to Cubism
by inhabiting a huge hive in the shape of a cubic ship. The broad culture
does its best to follow the leaders.

What happened after further expansion of science, and further extension into human lives—and into human deaths—of its robot arm, technology? What happened in the middle of the twentieth century—after two
advanced-weapon world wars and the start of the cold war with its ghastly
nuclear technology—was Abstract Expressionism. Its most famous painter,
Jackson Pollock, gave us paint in deliberately accidentally applied dribbles.

Between the lines of his dripping paintstick, he can be read to say: "We can
no longer make sense of the objective world with its three-dimensional objects
and its lack of human values. Here, instead, is a flat subjective world of spirited
shapes and colors. Let it sink into you. It has meaning inside, meaning that may
be different for each of us." Pollock soon ran dry of drip paintings. He was followed by painters that took a different approach to an objective world, a world
with little in it of much import.

Andy Warhol is viewed by some as our most important artist since Picasso.
Just look at his work. He gave us an art of uncommonly common objects:
painted, air-brushed and silk-screened coke bottles, Brillo Boxes, diagrams of
dance steps, newspaper headlines. He gave us a Marilyn Monroe stripped—not
stripped to display her supreme sensuality, but stripped of her voluptuousness
by reducing her to what seem like a photo-lab-machine-made multitude of tiny,
almost-empty, all-alike images. His multiple Marilyns are as little images of vital
Life as are his painting of tin cans of soup made from dead tomatoes.

Arthur Danto says that Warhol's craftmanship "may reveal us to ourselves as well as anything might: as a mirror held up to nature, they might
serve to catch the conscience of our kings."[20] I think that Warhol's repetitious epiphany, his "revelation of the whatness of a thing" (as James Joyce
used the word), the result of his search for what is significant about life is (in
one or two words that I'll pop into his mouth): "What what?"

HOW WE GOT TO BE HUMAN

In the same period, we had Roy Lichtenstein's gift of ultralight people, mindless and dimensionless cartoon folk, as flat and free of meaningful depth as the paper they occupy. The work of these and other Pop painters of that period doesn't point a finger at the world and say, "Bad, materialistic, not spiritual enough." It says, "This is it, that's all there is."

Don't misunderstand me. I'm no more setting up art as the bad guy than I was putting down science. I'm not questioning that these few and others like them are skilled artists. (Nor, for that matter, does it make sense to ask them what their art is about. The work of artists is to create art, not to make conceptual sense of it.)

About such work, however, certain questions beg for answers. What on earth has art been up to? Has art, in the middle of this century, besmirched itself with all this seeming schmutz? Has it turned tail—to be sure, not as science did, but in its own way? Has it surrendered to the enemy, the indifferent lifeless troops now led by objective science?

No, no! In no way has art turned tail or coat. Art has remained faithful to its age-old intent: To show us how we really feel about what is vital in our time. Unfortunately for us—and perhaps for art—much of the glory and the grandeur that gave us the likes of exorcisms and cathedrals has been leached from our Western life.

Why, then, did art make all these pictures of empty junk, of crud floating on the cup of Life? Why did it make all these images of Nonlife?

With a hand as sure as ever, art followed its mission. It made its empty images precisely because what has been concerning us increasingly are the crushing blows science-led Nonlife has landed on the once overwhelming vitality of Life. We saw that long ago a massive blow was delivered when objective knowledge pushed us into giving up our innocent belief that Life might be endless. We're still punch-drunk with that knowledge of necessary death, of the mortality which—whether or not we have spirits or souls that survive it—takes the life of our bodies without telling us what it all means. Then, only a generation ago, it was frightening to hear repeated Nietzsche's century-old words, "God has died." Increasingly since then, scienceled Nonlife has suggested no less than Life itself is dead. We are dust that, before it returns to dust, is force-marched through time by cause and effect for a few meaningless moments. Art has shown us life's heart-stopping fear that the numbing numb enemy—scienceled Nonlife—has engulfed it.

But is it true? That's what matters. Is art right? Maybe life is not exactly dead, but is it true that we're increasingly forced to believe life has no meaning?

Living, Lifeless, Dead: Always Trying Concepts

Before we even think of accepting that kind of answer, we must look closely at what people now believe. Look first at the vast sections of humanity that do not share the belief system of the Western scientific community. Why don't they accept those beliefs? In *Leaps of Faith*, Nicholas Humphrey answers, "The reason is transparent. Scientific materialism is regarded by many, even by some of its own prophets, as deeply unsatisfying: scary, bewildering, insulting, demeaning, dispiriting, confining."[21] Most non-Western people simply do not accept the idea that the world is run by blind billiard-ball causality, let alone run by probabilistic quantum semireal curiosities. In their view, the cause-and-effect physical world is no more than a set of petty push-buttons slaved to the fingers of Living agents, indeed ever-living, everlasting, ever-present singular or plural supernatural beings.

There happens to be a problem of exactly the same kind right here in the West. As Humphrey also writes, "Few individuals in the modern world can actually be unaware that there are now physicalist explanation for most if not all natural phenomena, not excluding the workings of the human mind. And yet many people—including some of those who are themselves the messengers—continue to ignore the message."[22]

If we look at those who inhabit the Western scientific community that underwrites the forces of the Nonliving cause-and-effect worldview, we can't help wondering: Just how solid is their army? Some of those objective-minded Westerners would have us believe its soldiers consist of those whose clocks are still ticking—the said-to-be-living—and those whose clocks have run down—the formerly said-to-be-living. Perhaps so, perhaps no, perhaps many of those soldiers are less volunteers than they are reluctant draftees, draftees who are ready to desert to irrational Life whenever opportunity presents itself. That objective Atlantic community may be living in a badly weathered Potemkin's Village, a set of fake, dreary, and orderly, empty-in-back, store and house fronts, living in a rickety Hollywood Western set whose parts keep collapsing because no more than a few care.

After all, even in Western society, belief in supernatural Life is widespread. Here are just a few confirming numbers for that last claim, also supplied by Humphrey.[23] United States census data for the 1980s show 95 percent of the people indicated belief in an active God, and 71 percent believed in a soul that survived after death. In the developed world, between one-third and two-thirds of the population believed in such as telepathy, precognition, active-here spirits of the dead, reincarnation, physically effective prayer, remembered past lives, messages from the dead, and ghost-haunted houses.

HOW WE GOT TO BE HUMAN

I offer here a list, a little list to remind us how broad a variety of concepts—from the serious to the silly—we use to support a world that remains full of all kinds of meaningful projected life. We know that widespread here in America is belief in divine creation despite Darwin; as is belief that it is profane to put asunder the sacred union of human egg and seed that bringeth life to us; as is belief in astrological guidance emanating from planets and stars, each with an attitude, like living agents; as is belief in motivated agents who are just Angels; as is belief in Tai Chi meridians on the body, where the insertion of pins and needles will not only tingle and stimulate the production of home-made drugs, but improve the life flow; as is belief in lively extraterrestrial agents hovering right here above River City in order to exchange bits of our life with theirs by sucking up, seducing, and spitting back human lifeforms; as is belief in comatose comets that give us a rare opportunity to sacrifice ourselves in order to gain a new and higher life; as is the little world of odd personal beliefs about interfering agents that, for example, keep some jinxed jocks in unwashed socks for days on end. This list may be limitless, but enough is enough.

So some of us are fearful about science's destruction of meaning, and some of us are not. So where are we, where do we stand? What's with this war between those who believe in the miracle of subjective motivated Life that somehow surrounds us, and those who belief in what sometimes seems a meaningless mishmash of objective causal-statistical Nonlife?

Here's what's with this war. As Homo sapiens with large brains and multiple mindsets that seek and use complex concepts like subjective knowledge and objective knowledge, for us there has always been such a war in our minds, and it continues to this day. As to which side is winning, if either, that's coming right up.

Certainly, vast territories have been taken from the subjective Living by the objective Lifeless. The world external to ourselves, once the domain of magical spirits, now has been lost to the enemy. Born-again Christians, orthodox Jews, fundamentalist Moslems, born-again-and-again Hindus, even heavenless Buddhists, all have long resided in the ranks of the army of the Living. Nonetheless, we don't hear much about their plans to place picket lines around our colleges of physics or chemistry, where objective information is now taught. The subject matters within those institutions are long lost, lost by the Living, lost to the Lifeless.

Living tissue, where Life advocates once could proudly proclaim "This is my own, my native land," came next. The science rider leading the deadly troops of an un-living meaningless universe waved a banner. On it was emblazoned: DNA! Fewer and fewer could say them nay. Every day science makes increasing cause-and-effect sense of nucleotides, cells, tissues, organs, and even of bodies with brains—although not yet of (conscious) minds and spirits. Our healers were once feared and revered medicine men and shamans. They could be seen to plunge a hand into the flesh of a living human body and pull it out bloody with an offending icon or intruding spirit. Those healers are largely gone now, their community inhabited by white-coated often-indifferent-to-their-patient's-humanity wielders of objective magnetic resonance imaging pictures, and of pads for prescribing pills. Both old-time medicine men and modern ones sought to heal. For the old-timers, illness was the result of the work of human or other agents. New-fangled doctors see illness as the result of objective cause-and-effect processes.

What should we make of all the anxiety about science and objective knowledge, all the gloom-and-doom portraits by art of a meaningless Life? We certainly hear a lot of genuine crying. What we hear amounts to crying "Wolf." We hear crying about spilt milk and about falling sky-bits from meteors, repetitious crying from the likes of Andy Warhol: "It's not that the wolf's huffing and puffing, it's not that it's broken the door, it's not even that the wolf is blowing down one house after another. It's that in my house and yours, in all the houses there's no one left alive, there's no one at home except zombies like you and me!"

The fear is real enough, but it's only one side of the story. For the other side, listen to a verbal image from an artist who was not at all intimidated by the objective skepticism of his day a couple hundred years ago. Here is ego-strong William Blake:

> Mock on, mock on, Voltaire, Rousseau;
> Mock on, mock on, 'tis all in vain!
> You throw the sand against the wind
> And the wind blows it back again.

Blake thinks all that distressing objective knowledge is a structure no more substantial than sand, the wind will blow it back. But where does the wind come from, and what is the wind that will overwhelm objective knowledge?

I'll offer a pictorial conceit of my own: The hot wind comes for each of us from a tiny volcano within. With the frequent gas and wind comes occasional boiling lava. The volcanoes are always ready to erupt with the hot magma of concepts based on experienced life, magma that we spew on each other before it becomes solid new land for the Quick, new land to replace that lost to the objective Dead.

At the core of each volcano is, not supernatural Spirit, but consciousness, the mystery that—though we can now think about it better than when first we came to Life near the African Great Rift Valley—still defies explanation. Even in the event that our new Life-defying would-be necromancers—our ladies and gents who create artificial intelligence—even if they could make a new Golem, a conscious robot, the volcanos within us will continue. They will not cease their belching of buoyant gases and burning liquids that still for many seem Divine or Devilish. The lame god, Vulcan, is embodied within each of us humans, ready to lave us with new fountains of archaic lava at any moment, any moment save those when consciousness disappears in deep dreamless sleep.

Talking about sleep, sleep wherein we dream is one of the strong sources of hot subjective lava about supernatural agents. As we saw, in *The Terror That Comes in the Night*, David Hufford examines a terrifying form of paranormal experience that is sometimes called "sleep paralysis with hypnagogic hallucinations," and also called by the name given it in Newfoundland, "the Old Hag."[24] When this experience comes, you feel yourself to be wide awake but terror-stricken and unable to move as a weird creature comes to you in your bed and presses on your chest so that you cannot breath. Hufford's research suggests that fifteen percent or more of people in America experience the Old Hag one or more times. He also mentions somewhat similar forms of paranormal experience, such as night visits from an incubus or a succubus that seeks and usually gets sex with you, out-of-body experiences, and near-death experiences.

To return to my conclusions and put them plainly, our anxieties about science (although accurately portrayed by art) were overstated. Max Weber was right enough about the destruction of much of the frightening, glorious, grand, and terrorizing old growth of religion and magic, but he did not anticipate the new growth that would replace it. Despite the losses of old meaning due to our ever-increasing objective information, new meaning continues to come—to come now, as it always did—from subjective experience, experience of encounters with motivated agents, be they natural or supernatural, real or imagined.

"Big Talk," may at this point come to the minds of more than one reader of what may seem far-fetched prose, "Isn't all this Big Talk nothing more than a war of words?" I must plead, "Guilty as hell, your honor." It is a war of concepts, concepts within each of us, concepts that are denoted with words. Even though no two of us define it the same way, most of us are on both sides of this mother of a war. When we have multiple mindsets, doing that comes naturally, no sweat at all except for the cold flop-sweat of anxiety about a meaningless world.

If you still think this talk of life and nonlife is just talk—talk and no more—it might be worthwhile to think once more. Think of our current struggles with issues of life, and with those of death. Think of our difficulties with stopping the continuation of what starts when human sperm and egg unite. For some, the instant of that union brings into being sacred subjective life, for others (or for the same "some," at other times) it is astonishingly complex nonlife with the potential for future life.

Think of our newly erupting anxiety about a concept we thought dormant, anxiety that cuts to the bone, anxiety about the idea of a human clone. That scientists just might be able to make genetic duplicates of us, machine-engendered duplicates that are not flat like photocopies but as fully dimensioned as we are, that idea of duplication drives a dagger into the heart of our felt uniqueness as living individuals. For some, it even says that monstrous machinations will bring *us* back to another life. Yes, we say, clones may seem like identical twins, but no, it's not the same. Twins are a natural part of artful enigmatic life. Clones are artificial, science-driven; they are dangerous, artifactual, made by nonlife notions.

Think also of our problems with halting the continuation of lives that are empty of meaning and filled with endless pain, and even of ending lives that we call braindead. Note that we have been making a slow and bitterly contested shift, making a shift toward regarding the overdue but not yet here endings of life as not real vital life, although not as non-life. Health care is a life versus nonlife issue that will plague us for decades to come. We cannot bear to treat ourselves as mere nonliving machinery by rationally rationing health care. Nor, on the other hand, can we bear the burden of giving ourselves the limitless care befitting creatures who are alive but briefly and who may be nonliving evermore. We can't cope with the conflict. That is, we can't cope when it suggests robbing young Pete of health care that might have lengthened his life by years or decades, robbing him in order to pay old Paul's medical bills for treatment that may add a few more enfeebled months to his life.

Of course, as some may here protest, science has helped to extend the duration of life. True enough, science has not severed its ties to life; it still goes both ways. Science has vastly increased the number of the living. Although that can scarcely be denied, for countless hundreds of millions of that increased number, science has not provided a good life, let alone a better one.

Whether or not to educate our children in the life sciences, particularly those that teach the views of Darwin and his successors on evolution, is another current life versus nonlife struggle. Many on the side of teaching life-science view those on the other side as being merely misunderstanders of the objective nature and validity of science. That's my view, too, except for the word "merely." Those who would offer a religion-based pseudoscience of how we came to be—instead of accepting Darwin's evolution-without-guiding-agent—find the latter a terrifying threat to "the meaning of life" as they know it. (That some of their children, educated in the life sciences and abandoning traditional beliefs, may in the future adopt one or another of many new belief systems about life is hardly apt to be even cold comfort for them. Rather, it will be hot discomfort to find their young ones adopting spacey new-age beliefs or old live-forever-as-part-of-a-living-universe panaceas.)

As long as we are looking at religion, look also, if you will, at one item of centuries-long dispute between Protestant and Catholic Christianity. This item is the issue of what's required to gain salvation after death. Is it "faith" or is it "works?" Faith, unquestioning belief, is certainly a subjective criterion. Works, what we do, are somewhat objective. Prior to the last one or two millennia, there was no conflict at all here. People did not belief that deity had access to their subjective minds (where faith resided). Also, the early meaning for works was more objective then: Works were not semisubjective moral good deeds, they were acts that benefited the deity or the deity's human agents. Currently, the issue is being resolved in favor of faith, with Catholic doctrine joining that of Protestants. Here we see some of the life forces resolving internal conflict and circling the wagons to defend against the ever encroaching nonlife enemy.

Let's stick with religion a bit longer and take a look at a pair of now-moribund faiths that were around long before the split within Christianity. Middle-Eastern Gnosticism and Persian Manichaeism were not monotheistic. Rather they were based on a dualism, one that set a good spiritual world against a material world of evil forces. The Manichaeans picked up on the Gnostic belief that everything was once divided between the Spiritual kingdom of Light and the Material kingdom of Darkness. The powers of Darkness somehow captured some elements of the Light. We humans and our world were made of these bits of good Spirit that were imprisoned in the evil Matter of the Dark kingdom. I judge these dualistic religions as having developed, in part, as response to the widespread perception that we humans seem to be made of immaterial stuff (that I call consciousness, mind, and spirit) as well as made of real solid matter (that makes up our bodies with brains). The Gnostics and the Manacheains reified this conceptual duality by projecting it outside of themselves. Where I see a subjective Life mindset, they saw a good kingdom of Light and Spirit. Where I conceive an objective Nonlife frame of mind, they perceived an evil Dark kingdom of Matter. They saw unending conflict between the two components of the great divide, as do I, although I don't share their Good Guy versus Bad Guy characterization.

For a look through the eyes of some non-Westerners at the objective and subjective mindsets of humans, look into Evens-Pritchard's study of the Neur people in Africa. From him, we learn that "the Neur believe that there are three parts to the human being: the flesh, the life (or breath), and the soul (or intellect)."[25] In terms of my division between subjective Life and objective Nonlife, the Neur "flesh" can be seen as the objective body and brain of a human. Likewise, the Neur "soul or intellect" is the subjective spirit and mind of a person. What about the remaining part for the Neur, the "life or breath?" It is a reification, a thing-making, of the concept of being alive. I'm sure we in the West once did the same with the idea of life, sometimes believing that life is the *blood*, sometimes that it is the *breath*.

A life versus nonlife issue that preys on us now is that of people versus machines. Our new silicon computers are sapping our faith in living minds. The algorithmic programs that run on computers are the apotheosis of nonliving cause-and-effect procedures. They show that semiconductor machines can do much of our work, thus hinting that we ourselves may be only semiliving.

The logic of computer efficiency (itself a nonlife concept) still mesmerizes us with its tick-talk pendulum beneath its time-clock face. But if the fraction of our populous that does not profit from science-originated growth and efficiency increases significantly, we may hear from Luddite throwers of number-crunchers into trash-compressors.

Okay, so there's this subjective-objective struggle in our minds, shouldn't we be doing something about it? I think so, sure, but do what? If we belief something, say, that white is White, and black Black, shouldn't we assert it forcefully? Remember the words we once heard from presidential candidate

HOW WE GOT TO BE HUMAN

Barry Goldwater, "Extremism in the defense of liberty is no vice. . . . Moderation in the pursuit of justice is no virtue." Recall as well Jack Kennedy's inaugural address, where he said the United States (in it's cold war with the Soviet Union) "would pay any price, bear any burden." Well, no, that's not the kind of thing I have in mind. (A subjective mindset may not see the same liberty and justice, or price and burden, as does an objective frame of mind.) I prefer Richard Nixon's inaugural words, "We cannot expect to make everyone our friend, but we can try to make no one our enemy."

We should strive for more communication *between* these two subminds, we should lighten up on the black-and-white impatience and moralism that is often present on both sides. Does that seem too little, too modest a goal? I think it's a grand goal, one that's almost—considering how frequently and easily we think out of both sides of our minds without realizing it—almost grandiose. All of us—including true believers—have some ability to tone down the over-the-top rhetoric and allow more open-minded dialogue between our subjective and objective frames of mind.

With regard to these Life-Nonlife battles, let me ask a question: What might make us pry apart the clamshell compartments in our minds so that the contestants might makes less war, and if not make love, then at least might enjoy more conceptual intercourse? To do that, we need what we already have, another agent within our minds. We are all the owners of another submind, a social one, whereby outsiders can strongly influence us, and we likewise sway them.

All we humans live with each other in a *social world* that is real in its own way and corresponds to our social mindset. What kind of reality am I suggesting here? If you believe, as I do, that there is a natural world within which are interdependant objective and subjective realms, what can social reality be? Can there be an additional real realm, such as the Platonic one many mathematicians believe exists for numbers and other mathematic objects? No, the social world is no new realm. It is an area common to the objective and subjective worlds, a place partly objective and partly subjective. In *The Construction of Social Reality,* John Searle provides us with a detailed guide to this large territory where (subjective) experience and (objective) information interact strongly.[26] (Of course, the two intereact to some extent everywhere—neither the subjective realm nor the objective one are ever pure and unadulterated.)

This social world is the subject matter for social scientists. Their work is so difficult because that realm is only somewhat objective. In *Culture and*

Enchantment, Mark Schneider looks at the territory between objective scientific knowledge and sometimes-edifying social knowledge.[27] As we saw above, where the material is too tough for objective explanation, he finds what he calls *enchanted* phenomena. Clifford Geertz, in *The Interpretation of Cultures,* advocates cultural anthropology as a field that *interprets* various cultural structures as suppliers of meaning, rather than as an *explanatory* science seeking evidence of causes.[28] It is with edifying and meaningful social help, help from our societies and help from us for our societies, that we all might learn not merely to tolerate divergences but to deal with them.

Somewhat related views may be seen in Steven Gould's *Rocks of Ages: Science and Religion in the Fullness of Life.* I take his thesis here as *edifying,* as providing the rest of us with social help by suggesting there *should* be dialogue and there *need* not be conflict between science and religion; those two conceptual structures *should* occupy "nonoverlapping magisteria."[29]

It's worth asking: Are the two (which are part of our objective and subjective mindsets) *really* not overlapping? If science and religion are to have a dialogue, it has to be about something they each have opinions about, so mustn't they overlap?

I think our social mindsets generate answers to nonscientific social conceptual questions like these. As with *social* concepts in general (be they edifying or terrifying), we are only partially in the objective realm here. We are also in the subjective realm. In this social domain, we try to influence each other, but seldom can we agree completely. In my own opinion, the conflict between science and religion is quite real. I also think we can and should be civil about that conflict. With social help, we are each capable of allowing more civil discourse between not only our subjective and objective viewpoints but between all of our confederate states of minds. That may be worthwhile for all of us. In fact, it may be a large part of civilizing ourselves.

Where does all this leave us? Where will any of it take us? Where will our worlds go when we who now occupy them are gone? Is there anywhere we will go when we're gone? In *The Symbolic Species,* Terrence Deacon comments, "As a species, we seem to be preoccupied with ends. . . . We struggle in vain to comprehend the implications of our own impending cessation of life."[30] To which, I can imagine Amen—the ancient Egyptian god of life and reproduction—responding: "That's life, amen."

NOTES

1. Howard W. Hintz, "Whitehead's Concept of Organism and the Mind-Body Problem," in *Dimensions of Mind*, ed. Sidney Hook (New York: Collier Books, Division of MacMillan Publishers, 1973), p. 103.

2. Steven Mithen, *The Prehistory of the Mind* (London: Thames and Hudson Ltd, 1996), p. 151.

3. Leda Cosmides and John Tooby, "Origins of Domain Specificity: The Evolution of Functional Organization," in *Mapping the Mind*, ed. Lawrence A. Herschfeld and Susan A. Gelman (New York: Cambridge University Press, 1994), p. 102.

4. Susanne K. Langer, *Mind: An Essay on Human Feeling*, vol. 3 (Baltimore, Md.: The Johns Hopkins University Press, 1982), pp. 4–5.

5. Mark A. Schneider, *Culture and Enchantment* (Chicago, Ill.: The University of Chicago Press, 1993).

6. Carl Sagan, *The Demon-haunted World: Science as a Candle in the Dark* (New York: Random House, 1996), p. 9.

7. Mithen, *The Prehistory of the Mind*, p. 213.

8. Peter Baumgartner and Sabine Payr, eds. *Speaking Minds: Interviews with Twenty Eminent Cognitive Scientists* (Princeton: Princeton University Press, 1995), p. 259.

9. Mithen, *The Prehistory of the Mind*, p. 157.

10. Edward O. Wilson, *Consilience* (New York: Vintage Books, 1999), p. 128.

11. Susanne K. Langer, *Feeling and Form: A Theory of Art* (New York: Charles Scribner's Sons, 1953).

12. Susanne K. Langer, *Problems of Art* (New York: Charles Scribner's Sons, 1957), p. 7.

13. Richard Dawkins, *Unweaving the Rainbow* (Boston: Houghton Mifflin, 1998).

14. Kenneth Bock, *Human Nature Mythology* (Chicago: University of Illinois Press, 1994), p. 80.

15. Terrence W. Deacon, *The Symbolic Species* (New York: Norton, 1997), p. 455.

16. Ray Kurzweil, *The Age of Spiritual Machines* (New York: Viking, 1999).

17. Nicholas Humphrey, *A History of the Mind* (New York: Simon & Schuster, 1992), p. 19.

18. Gaetan Picon, *The Birth of Modern Painting* (New York: Rizzoli International Publications, 1972), p. 37.

19. John Golding, *Cubism* (Cambridge, Mass.: Harvard University Press, 1988), p. 70.

20. Arthur Danto, "The Artworld," in *Pop Art: A Critical History*, ed. Steven Henry Madoff (Berkeley, Calif.: University of California Press, 1997), p. 278.

---

21. Nicholas Humphrey, *Leaps of Faith* (New York: Basic Books, 1996), p. 6.

22. Ibid., p. 6.

23. Ibid., p. 3.

24. David J. Hufford, *The Terror That Comes in the Night* (Pennsylvania: University of Pennsylvania Press, 1982), p. 245.

25. Pals, *Seven Theories of Religion*.

26. John Searle, *The Construction of Social Reality* (New York: Free Press, 1997).

27. Schneider, *Culture and Enchantment*.

28. Clifford Geertz, *The Interpretation of Cultures* (New York: BasicBooks, 1973), p. 434.

29. Steven Jay Gould, *Rocks of Ages: Science and Religion in the Fullness of Life* (New York: Ballantine, 1999).

30. Deacon, *The Symbolic Species*, p. 433.

BIBLIOGRAPHY

Aitchison, Jean. *The Seeds of Speech*. New York: Cambridge University Press, 1996.

Altmann, Gerry T. M. *The Ascent of Babel: An Exploration of Language, Mind, and Understanding*. New York: Oxford University Press, 1997.

Armstrong, David F., William C. Stokoe, and Sherman E. Wilcox. *Gesture and the Nature of Language*. New York: Cambridge University Press, 1995.

Bard, Kim A. "Intentional Behavior and Intentional Communication in Young Free-Ranging Orangutans." *Child Development* 63 (1992): 1186–97.

Barkow, Leda Cosmides, and John Tooby, eds. *The Adapted Mind*. New York: Oxford University Press, 1992.

Bateson, Gregory. *Mind and Nature: A Necessary Unity*. New York: E. P. Dutton, 1979.

Baumgartner, Peter, and Sabine Payr, eds. *Speaking Minds: Interviews with Twenty Eminent Cognitive Scientists*. Princeton, N.J.: Princeton University Press, 1995.

Beyerstein, Barry L. "Why Bogus Therapies Seem To Work." *Skeptical Inquirer* (September/October 1997).

Bickerton, Derek. *Language and Species*. Chicago: University of Chicago Press, 1990.

Blackmore, Susan. *The Meme Machine*. New York: Oxford University Press, 1999.

Block, Ned. "On a Confusion about a Function of Consciousness." *Behavioral and Brain Sciences* 18 (1995): 227–87.

Bock, Kenneth. *Human Nature Mythology*. Chicago, Ill: University of Illinois Press, 1994.

Bonner, John T. *The Evolution of Culture in Animals*. Princeton, N.J.: Princeton University Press, 1980.

Brothers, Leslie. *Friday's Footprint: How Society Shapes the Human Mind*. New York: Oxford University Press, 1997.

Brown, Michael H. *The Search for Eve*. New York: Harper & Row, 1990.

Budiansky, Stephen. *The Covenant of the Wild*. New York: William Morrow, 1992.

———. *If A Lion Could Talk*. New York: The Free Press, 1998.

Bullough, Vern L, and Bonnie Bullough. *Sexual Attitudes*. Amherst, N.Y.: Prometheus Books, 1995.

Buranhult, Goran, ed. *The First Humans: Human Origins and History to 10,000 BC*. New York: HarperCollins Publishers, 1993.

Burkert, Walter. *Creation of the Sacred*. Cambridge, Mass.: Harvard University Press, 1996.

Cairns-Smith, A. G. *Evolving the Mind: On the Nature of Matter and the Origin of Consciousness*. New York: Cambridge University Press, 1996.

Calvin, William H. *The Ascent of Mind*. New York: Bantam Books, 1991.

———. *How Brains Think*. New York: BasicBooks, 1996.

Campbell, Keith. *Body and Mind*. Notre Dame, Ind.: University of Notre Dame Press, 1970.

Candland, Douglas Keith. *Feral Children and Clever Animals: Reflections on Human Nature*. New York: Oxford University Press, 1993.

Casti, John L. *Paradigms Lost: Tackling the Unanswered Mysteries of Modern Science*. New York: William Morrow and Company, 1989.

Chalmers, David J. *The Conscious Mind: In Search of a Fundamental Theory*. New York: Oxford University Press, 1996.

Cheney, Dorothy L., and Robert M. Seyfarth. *How Monkeys See The World*. Chicago, Ill.: University of Chicago Press, 1990.

Churchland, Paul M. *Matter and Consciousness*. Cambridge, Mass.: The MIT Press, 1988.

Coppens, Yves. "East Side Story: The Origin of Humankind." *Scientific American* (May 1994).

Cosmides, Leda, and John Tooby. "Origins of domain specificity: The evolution of functional organization." In *Mapping the Mind*. Edited by Lawrence A. Herschfeld and Susan A. Gelman. New York: Cambridge University Press, 1994.

Damasio, Antonio R. *Descartes' Error: Emotion, Reason, and the Human Brain*. New York: G. P. Putnam's Sons, 1994.

Davis, Murray S. *Smut: Erotic Reality/Obscene Ideology*. Chicago: University of Chicago Press, 1983.

Dawkins, Marian Stamp. *Through Our Eyes Only? The Search for Animal Consciousness*. New York: W. H. Freeman and Company, 1993.

Dawkins, Richard. *The Selfish Gene*. New York: Oxford University Press, 1979.

———. *The Blind Watchmaker*. New York: W. W. Norton, 1987.

———. *River out of Eden*. New York: BasicBooks, 1995.

———. *Climbing Mount Improbable*. New York: W. W. Norton, 1996.

———. *Unweaving the Rainbow*. Boston: Houghton Mifflin, 1998.

Deacon, Terrence W. *The Symbolic Species*. New York: W. W. Norton, 1997.

Dennett, Daniel C. "Precis of The Intentional Stance." *Behavioral and Brain Sciences* 11 (1988): 495–546.

———. *Consciousness Explained*. New York: Little Brown and Company, 1991.

———. *Darwin's Dangerous Idea*. New York: Simon & Schuster, 1996.

———. *Kinds of Minds*. New York: Basic Books, 1996.

de Waal, Frans B. M. *Chimpanzee Politics*. New York: Harper & Row, 1982.

———. *Peacemaking Among Primates*. Cambridge, Mass.: Harvard University Press, 1989.

———. *Good Natured*. Cambridge, Mass.: Harvard University Press, 1996.

———. *Bonobo: The Forgotten Ape*. Berkeley, Calif.: University of California press, 1997.

Diamond, Jared. *The Third Chimpanzee*. New York: HarperPerennial, 1992.

———. *Why is Sex Fun?: The Evolution of Human Sexuality*. New York: Basic Books, 1997.

———. *Guns, Germs, and Steel*. New York: W.W. Norton, 1997.

Donald, Merlin. *Origins of the Modern Mind*. Cambridge, Mass.: Harvard University Press, 1991.

Dreyfus, Hubert J. *What Computers Can't Do: The Limits of Artificial Intelligence*. New York: Harper & Row, 1972.

Dreyfus, Hubert J., and Stuart E. Dreyfus. *Mind over Machine*. New York: The Free Press, 1986.

Edelman, Gerald M. *Neural Darwinism*. New York: Basic Books, 1987.

———. *The Remembered Present: A Biological Theory of Consciousness*. New York: Basic Books, 1989.

———. *Bright Air, Brilliant Fire: On the Matter of the Mind*. New York: Basic Books, 1992.

Edey, Maitland A., and Donald C. Johanson. *Blueprints, Solving the Mystery of Evolution*. New York: Little, Brown and Company, 1989.

Eldredge, Niles. *Dominion: Can Nature and Culture Co-exist*. New York: Henry Holt & Company, 1995.

———. *Re-inventing Darwin*. New York: John Wiley and Sons, 1995.

———. *Life in the Balance*. Princeton, N.J.: Princeton University Press, 1998.

Eliade, Mircea. *Shamanism*. Princeton, N.J.: Princeton University Press, 1964.

Ferrari, M., and R. J. Sternberg, eds. *Self-Awareness: Its Nature and Development*. Guilford Press, 1998.

Ferry, Georgina, ed. *The Understanding of Animals*. Oxford: Basil Blackwell Limited, 1984.

Fisher, Helen E. *The Sex Contract*. New York: William Morrow and Company, 1982.

———. *Anatomy of Love*. New York: W. W. Norton and Company, 1992.

Flanagan, Owen. *Consciousness Reconsidered*. Cambridge, Mass.: The MIT Press, 1992.

Fodor, Jerry A. *The Modularity of Mind*. Cambridge, Mass.: The MIT Press, 1983.

Fossey, Dian. *Gorillas in the Mist*. Boston, Mass.: Houghton Mifflin, 1983.

Friedrich, Heinz, ed. *Man and Animal: Studies in Behavior*. New York: St. Martin's Press, 1972.

Gallup, Gordon G., Jr. "Self-recognition: Research strategies and experimental design." *Self-awareness in Animals and Humans*. Edited by Sue Taylor Parker, Robert W. Mitchell, and Maria L. Boccia. New York: Cambridge University Press, 1994.

Gardner, Howard. *The Mind's New Science*. New York: Basic Books, Inc., 1985.

Gazzaniga, Michael S. *Nature's Mind*. New York: BasicBooks, 1992.

Geertz, Clifford. *The Interpretation of Cultures*. New York: BasicBooks, 1973.

———. *Local Knowledge*. New York: Basic Books, 1983.

Gellner, Ernest. *Plough, Sword and Book*. London: Collins Harvill, 1988.

Gibbs, Jack P. *Control: Sociology's Central Notion*. Chicago, Ill.: University of Illinois Press, 1989.

Goodall, Jane. *In The Shadow Of Man*. Boston, Mass.: Houghton Mifflin, 1971.

———. *The Chimpanzees of Gombe*. Cambridge, Mass.: Harvard University Press, 1986.

———. *Through A Window*. Boston, Mass.: Houghton Mifflin, 1990.

Gould, James L., and Carol Grant Gould. *The Animal Mind*. New York: Scientific American Library, 1984.

Gould, Stephen Jay. *The Mismeasure of Man*. New York: W. W. Norton and Company, 1981.

———. *Bully for Brontosaurus*. New York: W. W. Norton, 1991.

———. *Full House*. New York: Harmony Books div. Crown Publishers, 1996.

———. "Evolution: The Pleasures of Pluralism." *The New York Review of Books* (June 26, 1997).

———. *Rock of Ages*. New York: Ballantine, 1999.

Gribbon, John. *In Search of Schrodinger's Cat*. New York: Basic Books, 1984.

Griffin, Donald R. *The Question of Animal Awareness*. New York: The Rockefeller University Press, 1981.

———. *Animal Thinking*. Cambridge, Mass.: Harvard University Press, 1984.

———. *Animal Minds*. Chicago: The University of Chicago Press, 1992.

Harris, Marvin. *Our Kind*. New York: Harper and Row, 1989.

Harris, Randy Allen. *The Linguistics Wars*. New York: Oxford University Press, 1993.

Harrison, Peter. "Do Animals Feel Pain?" *Philosophy* 66 (1991): 25–40.

Haugeland, John, ed. *Mind Design: Philosophy, Psychology, Artificial Intelligence*. Cambridge, Mass.: The MIT Press, 1981.

Bibliography

Herbert, Nick. *Quantum Reality*. Garden City, N.Y.: Anchor Press, 1985.

———. *Elemental Mind*. New York: Penguin Books, 1993.

Hirschfeld, Lawrence A., and Susan A. Gelman, eds. *Mapping the Mind*. New York: Cambridge University Press, 1994.

Hook, Sidney, editor. *Dimensions of Mind*. New York: Collier Books, Division of MacMillan Publishing, 1973.

Hrdy, Sarah Blaffer, and Patrica L. Whitten. "Patterning of Sexual Activity." in *Primate Societies*. Edited Barbara B. Smuts, Dorothy L. Cheney, Robert M. Seyfarth, Richard W. Wrangham, and Thomas T. Struhsaker. Chicago: University of Chicago Press, 1986.

Hufford, David J. *The Terror That Comes in the Night*. Pennsylvania: University of Pennsylvania Press, 1982.

Humphrey, Nicholas. *Consciousness Regained*. New York: Oxford University Press, 1984.

———. *A History of the Mind*. New York: HarperCollins Publishers, 1992.

———. *Leaps of Faith*. New York: Basic Books, 1996.

Jerrison, Harry J. "Evolutionary Biology of Intelligence: The Nature of the Problem." *Intelligence and Evolutionary Biology*. Edited by Harry J. Jerison and Irene Jerison. Springer-Verlag, 1988.

———. *Brain Size and The Evolution of Mind*. New York: American Museum of Natural History, 1991.

Johanson, Donald. *Blueprints: Solving the Mystery of Evolution*. New York: Little Brown, 1989.

———. *Ancestors: In Search of Human Origins*. New York: Villard Books, 1994.

———, and Maitland Edey. *Lucy: The Beginnings of Humankind*. New York: Simon and Schuster, 1981.

Johanson, Donald, and James Shreeve. *Lucy's Child*. New York: William Morrow & Company, 1989.

Johanson, Donald, and Blake Edgar. *From Lucy to Language*. New York: Simon & Schuster, 1996.

Johnson, Mark. *The Body in the Mind*. Chicago, Ill.: University of Chicago Press, 1987.

Jonas, Hans. *The Phenomenon of Life: Toward a Philosophical Biology*. Chicago, Ill.: The University of Chicago Press, 1966.

Kakar, Sudhir. *Intimate Relations: Exploring Indian Sexuality*. Chicago, Ill.: University of Chicago Press, 1989.

Kenny, Anthony. *The Metaphysics of Mind*. New York: Oxford University Press, 1992.

Kummer, Hans. *Primate Societies*. Chicago, Ill.: Aldine Atherton, Inc., 1971.

Kurtz, Paul. *The Transcendental Temptation*. Amherst, N.Y.: Prometheus Books, 1991.

Kurzweil, Ray. *The Age of Spiritual Machines*. New York: Viking, 1999.

La Barre, Weston. *The Human Animal*. Chicago, Ill.: The University of Chicago Press, 1954.

Lakoff, George. *Women, Fire, and Dangerous Things*. Chicago, Ill.: University of Chicago Press, 1987.

Lakoff, George, and Mark. Johnson. *Metaphors We Live By*. Chicago, Ill.: University of Chicago Press, 1980.

Lambert, David. *The Field Guide to Early Man*. New York: Facts On File, Inc., 1987.

Langer, Susanne K. *Philosophy in a New Key: A Study in the Symbolism of Reason, Rite, and Art*. Cambridge, Mass.: 1942.

———. *Feeling and Form: A Theory of Art*. New York: Charles Scribner's Sons, 1953.

———. *Problems of Art*. New York: Charles Scribner's Sons, 1957.

———. *Mind: An Essay on Human Feeling*. 3 vols. Baltimore, Md.: The Johns Hopkins University Press, 1967, 1972, 1982.

Langton, Christopher G., ed. *Artificial Life*. New York: Addison Wesley, 1989.

Larick, Roy, and Russell L. Ciochon. "The African Emergence and Early Asian Dispersals of the Genus Homo." *American Scientist* (November/December 1996).

Leakey, Richard, and Roger Lewin. *Origins Reconsidered*. New York: Doubleday, 1992.

Leavens, David A., William D. Hopkins, and Kim A. Bard. "Indexical and Referential Pointing in Chimpanzees." *Journal of Comparative Psychology* 110, no. 4 (1996): 346–53.

Leeming, David Adams. *Encyclopedia of Creation Myths*. New York: ABC-CLIO, Inc., 1994.

Libet, Benjamin. "Subjective Referral of the Timing for a Conscious Sensory Experience." *Brain* 102 (1979): 193.

Lieberman, Philip. *Uniquely Human: The Evolution of Speech, Thought, and SElf-less Behavior*. Cambridge, Mass.: Harvard University Press, 1991.

Lumsden, Charles J., and Edward O. Wilson. *Promethean Fire*. Cambridge, Mass.: Harvard University Press, 1983.

Mandelbaum, David G. *Society in India*. Berkeley, Calif.: University of California Press, 1970.

Marks, Charles E. *Commissurotomy Consciousness and Unity of Mind*. Cambridge, Mass.: The MIT Press, 1981.

McCrone, John. *The Ape That Spoke: Language and the Evolution of the Human Mind*. New York: William Morrow and Company, 1991.

———. *The Myth of Irrationality*. New York: Carroll & Graf Publishers, 1993.

McGrew, William C., Linda F. Marchant, and Toshisada Nishida, eds. *Great Ape Societies*. Cambridge: Cambridge University Press, 1996.

Masson, Jeffrey Moussaieff. *Dogs Never Lie About Love*. New York: Crown, 1997.

Masson, Jeffrey Moussaieff, and Susan McCarthy. *When Elephants Weep: The Emotional Lives of Animals*. New York: Delacorte Press, 1995.

Milani, Myrna M. *The Body Language and Emotions of Cats*. New York: William Morrow and Company, 1987.

Mills, Cynthia. "Unusual Suspects." *The Sciences*. New York Academy of Sciences, July/August 1997.

Minsky, Marvin. *The Society of Mind*. New York: Simon and Schuster, 1986.

Mithen, Steven. *The Prehistory of the Mind*. London: Thames and Hudson, 1996.

Morris, Desmond. *The Naked Ape*. New York: McGraw-Hill Book Company, 1967.

———. *Catlore*. New York: Crown Publishers, Inc., 1987.

Mortenson, Joseph. *Whale Songs and Wasp Maps*. New York: E. P. Dutton, 1987.

Moses, Louis J. Foreword to *Self-Awareness in Animals and Humans*. Edited by Sue Taylor Parker, Robert W. Mitchell, and Maria L. Boccia. New York: Cambridge University Press, 1984.

Ng, Yew-Kwang. "Towards Welfare Biology: Evolutionary Economics of Animal Consciousness and Suffering." *Biology and Philosophy* 10 (1995): 255–85.

Nishida, Toshisida, and Mariko Hiraiwa-Hasegawa. "Chimpanzees and Bonobos: Cooperative Relationships among Males." *Primate Societies*. Chicago, Ill.: University of Chicago Press, 1986.

Noble, William, and Iain Davidson. *Human Evolution, Language, and Mind*. New York: Cambridge University Press, 1996.

Norretranders, Tor: *The User Illusion: Cutting Consciousness Down to Size*. New York: Viking, 1998.

O'Keefe, Daniel Lawrence. *Stolen Lightening: The Social Theory of Magic*. New York: Vintage Books, 1983.

Ornstein, Robert E. *The Psychology of Consciousness*. New York: Harcourt Brace Jovanovich, Inc., 1977.

———. *The Evolution of Consciousness*. New York: Prentice Hall, 1991.

Osgood, Charles E., William H. May, and Murray S. Miron. *Cross-Cultural Universals of Affective Meaning*. Chicago, Ill.: University of Illinois Press, 1975.

Pals, Daniel L. *Seven Theories of Religion*. New York: Oxford University Press, 1996.

Parker, Sue Taylor, and Robert W. Mitchell. "Evolving Self-awareness." *Self Awareness in Animals and Humans*. Edited by Sue Taylor Parker, Robert W. Mitchell, and Maria L. Boccia. New York: Cambridge University Press, 1994.

Penrose, Roger. *The Emperor's New Mind*. New York: Oxford University Press, 1989.

———. *Shadows of the Mind: A Search for the Missing Science of Consciousness*. New York: Oxford University Press, 1994.

Pinker, Steven. *The Language Instinct*. New York: William Morrow and Company, 1994.

———. *How The Mind Works*. New York: W. W. Norton, 1997.

Plotkin, Henry. *Darwin Machines and the Nature of Knowledge*. Cambridge, Mass.: Cambridge University Press, 1994.

———. *Evolution in Mind*. Cambridge, Mass.: Harvard University Press, 1998.

Quiatt, Duane, and Vernon Reynolds. *Primate Behavior*. New York: Cambridge University Press, 1993.

Radner, Daisie, and Michael Radner.*Animal Consciousness*. Amherst, N.Y.: Prometheus Books, 1989.

Rancour-Laferriere, Daniel. *Signs of the Flesh*. Bloomingtion, Ind. : Indiana University Press, 1992.

Reinisch, June M., with Ruth Beasley. *The Kinsey Institute New Report on Sex*. New York: St. Martin's Press, 1990.

Ridley, Matt. *The Red Queen: Sex and the Evolution of Human Nature*. New York: Macmillan Publishing, 1993.

———. *The Origins of Virtue*. New York: Viking, 1997.

Rochberg-Halton, Eugene. *Meaning and Modernity: Social Theory in the Pragmatic Attitude*. Chicago, Ill.: University of Chicago Press, 1986.

Rose, Anthony L. "Orangutan, Science, and Collective Reality." In *The Neglected Ape*. Edited by Ronald D. Nadler, et al. New York: Plenum Press, 1996.

Rose, Steven. *Lifelines: Biology Beyond Determinism*. New York: Oxford University Press, 1998.

Rose, Steven, ed. *From Brains to Consciousness?* Princeton, N.J.: Princeton University Press, 1998.

Rosenfeld, Robert P. "Parsimony, Evolution, and Animal Pain." *Between Species* 9, no. 3 (summer 1993): 133–37.

Sagan, Carl. *The Dragons of Eden*. New York: Random House, 1977.

———. *The Demon-haunted World: Science as a Candle in the Dark*. New York: Random House, 1996.

Savage-Rumbaugh, E. Sue, Shelly L. Williams, Takeshi Furuichi, and Takayoshi Kano. "Language perceived: Paniscus branches out." In *Great Ape Societies*. Edited William C. McGrew, Linda F. Marchant, and Toshisada Nishida. New York: Cambridge University Press, 1996.

Savage-Rumbaugh, Sue, Stuart G. Shanker, and Talbot J. Taylor. *Apes, Language, and the Human Mind*. Hew York: Oxford University Press, 1998.

Schneider, Mark A. *Culture and Enchantment*. Chicago: University of Chicago Press, 1993.

Schwartz, Jeffrey H. *The Red Ape: Orang-utans and Human Origins*. Boston, Mass.: Houghton Mifflin Company, 1987.

Searle, John R. *Minds, Brains and Science*. Cambridge, Mass.: Harvard University Press, 1984.

———. *The Rediscovery of the Mind*. Cambridge, Mass.: The MIT Press, 1994.

———. "The Mystery of Consciousness." *The New York Review of Books*, 2–16 November 1995.

———. *The Construction of Social Reality*. New York: Free Press, 1997.

Sheets-Johnstone, Maxine. *The Roots of Thinking*. Philadelphia: Temple University Press, 1990.

————. *The Roots of Power*. Chicago, Ill.: Open Court Publishing Company, 1994.

Shreeve, James. *The Neandertal Enigma*. New York: William Morrow & Company, 1995.

Silverman, David, and Brian Torode. *The Material Word: Some theories of language and its limits*. London: Routledge & Kegan Paul, 1980.

Small, Meredith F. *What's Love Got to Do with It?: The Evolution of Human Mating*. New York: Doubleday, 1995.

Smith, Curtis G. *Ancestral Voices: Language and the Evolution of Human Consciousness*. Englewood Cliffs, N.J.: Prentice-Hall, 1985.

Smith, W. John. *The Behavior of Communicating: An Ethological Approach*. Cambridge, Mass.: Harvard University Press, 1977.

Snow, C. P. *The Two Cultures and the Scientific Revolution*. New York: Cambridge University Press, 1959.

Sober, Elliott, and David Sloan Wilson. *Unto Others: The Evolutionary Psychology of Unselfish Behavior*. Cambridge, Mass.: Harvard University Press, 1998.

Stringer, Christopher, and Clive Gamble. *In Search of the Neanderthals*. New York: Thames and Hudson, 1993.

Stringer, Christopher, & Robin McKie. *African Exodus: The Origins of Modern Humanity*. New York: Henry Holt, 1997.

Strum, Shirley C. *Almost Human: A Journey into the World of Baboons*. London: Elm Tree Books, 1987.

Sulloway, Frank. *Born To Rebel: Birth Order, Family Dynamics, and Creative Life*. New York: Pantheon, 1996.

Symons, Donald. *The Evolution of Human Sexuality*. New York: Oxford University Press, 1979.

Takahata, Yukio, Hiroshi Ihobe, and Gen'ishi Idani. "Comparing Copulations of Chimpanzees and Bonobos: Do Females Exhibit Proceptivity or Receptivity?" In *Great Ape Societies*. Edited by William C. McGrew, Linda F. Marchant, and Toshisada Nishida. New York: Cambridge University Press, 1996.

Tattersall, Ian. *The Fossil Trail*. New York: Oxford University Press, 1995.

————. *Becoming Human: Evolution and Human Uniqueness*. New York: Harcourt Brace, 1998.

Tavris, *The Mismeasure of Woman*. New York: Simon & Schuster, 1992.

Taylow, Gordon Rattray. *The Natural History of the Mind*. New York: E. P. Dutton, 1979.

Taylor, Timothy. *The Prehistory of Sex*. New York: Bantam Books, 1998.

Thomas, Elizabeth Marshall. *The Hidden Life of Dogs*. Boston, Mass.: Houghton Mifflin Co., 1993.

Tomasello, Michael, & Luigia Camaioni. "A Comparison of the Gestural Communication of Apes and Human Infants." *Human Development* 40 (1997):7–24.

Townsend, John. *What Women Want—What Men Want*. New York: Oxford University Press, 1998.

Trinkaus, Erik, and Pat Shipman. *The Neandertals*. New York: Alfred A. Knopf, 1993.

Van Shaik, Carel P., and Jan A. R. A. M. Van Hooff. "Toward an Understanding of the Orangutan's Social System." In *Great Ape Societies*. Edited by William C. McGrew, Linda F. Marchant, and Toshisada Nishida. New York: Cambridge University Press, 1996.

Vyse, Stuart A. *Believing in Magic: The Psychology of Superstition*. New York: Oxford University Press, 1997.

Walker, Stephen. *Animal Thought*. London: Routledge & Kegan Paul, 1983.

Watts, David P. "Comparative Socio-ecology of Gorillas." In *Great Ape Societies*. New York: Cambridge University Press, 1996.

White, Frances J. "Comparative Socio-ecology of Pan Paniscus." In *Great Ape Societies*. Edited by William C. McGrew, Linda F. Marchant, and Toshisada Nishisada. New York: Cambridge University Press, 1996.

Wilson, Edward O. *Sociobiology: The New Synthesis*. Cambridge, Mass.: Harvard University Press, 1975.

———. *The Diversity of Life*. Cambridge, Mass.: Harvard University Press, 1992.

———. *Consilience*. New York: Knopf, 1998.

Wilson, Peter J. *Man: The Promising Primate*. New Haven: Yale University Press, 1980.

Wolfe, Alan. *The Human Difference: Animals, Computers, and the necessity of Social Science*. Berkeley, Calif.: University of California Press, 1993.

Wrangham, Richard, and Dale Peterson. *Demonic Males*. Boston, Mass.: Houghton Mifflin, 1996.

Wright, Robert. *The Moral Animal*. New York: Pantheon Books, 1994.

Zeller, Ann C. "Communication by Sight and Smell." In *Primate Societies*. Chicago: University of Chicago Press, 1987.

INDEX